Mark Georg Schiebl

Thermodynamically consistent space-time discretization of non-isothermal mechanical systems in the framework of GENERIC

Schriftenreihe des Instituts für Mechanik
Karlsruher Institut für Technologie (KIT)

Band 7

Herausgeber:
Prof. Dr.-Ing. habil. Peter Betsch
Prof. Dr.-Ing. habil. Thomas Seelig

Eine Übersicht aller bisher in dieser Schriftenreihe erschienenen Bände finden Sie am Ende des Buches.

Thermodynamically consistent space-time discretization of non-isothermal mechanical systems in the framework of GENERIC

by
Mark Georg Schiebl

KIT Scientific Publishing

Karlsruher Institut für Technologie
Institut für Mechanik

Thermodynamically consistent space-time
discretization of non-isothermal mechanical
systems in the framework of GENERIC

Zur Erlangung des akademischen Grades eines Doktor-Ingenieurs
von der KIT-Fakultät für Bauingenieur-, Geo- und Umweltwissenschaften
des Karlsruher Instituts für Technologie (KIT) genehmigte Dissertation

von Mark Georg Schiebl, M.Sc.

Tag der mündlichen Prüfung: 16. April 2021
Referent: Prof. Dr.-Ing. habil. Peter Betsch
Korreferent: Prof. Dr.-Ing. habil. Michael Groß

Impressum

Karlsruher Institut für Technologie (KIT)
KIT Scientific Publishing
Straße am Forum 2
D-76131 Karlsruhe

KIT Scientific Publishing is a registered trademark
of Karlsruhe Institute of Technology.
Reprint using the book cover is not allowed.

www.ksp.kit.edu

Print on Demand 2021 – Gedruckt auf FSC-zertifiziertem Papier

ISSN 2363-4936
ISBN 978-3-7315-1117-5
DOI 10.5445/KSP/1000136011

This thesis is dedicated to my grandfather Georg Schiebl.

Abstract

The present thesis addresses the design of structure-preserving numerical methods that emanate from the general equation for non-equilibrium reversible-irreversible coupling (GENERIC) formalism.

For conservative mechanical systems, the GENERIC reduces to the purely Hamiltonian case. Considering this special case first, novel energy-momentum (EM) consistent time-stepping schemes in the realm of molecular dynamics are introduced that take periodic boundary conditions, three-body potentials and interatomic functional potentials into account. This method is thermodynamically consistent because it preserves the energy and the symmetries of the system, independent of the time-step size.

Afterwards, thermodynamical systems are considered in the framework of the GENERIC, and a new variational formulation for large-strain thermoelasticity is proposed. This variational formulation emanates from the GENERIC framework and allows for the free choice of the thermodynamic state variable among three options: (i) the absolute temperature, (ii) the internal energy density, or (iii) the entropy density. In this context the notion "GENERIC-consistent space discretization" is introduced, facilitating the design of novel energy-momentum-entropy (EME) consistent schemes. This method is thermodynamically consistent in the sense that the solution strictly obeys the first and second law of thermodynamics and the symmetries of such a system, independent of the size of the time-step.

Finally, large-strain thermo-viscoelasticity is considered in the context of the GENERIC framework. A mixed finite element approach for the discretization in space is proposed that incorporates a GENERIC-consistent space discretization. Depending on the choice of the thermodynamic state variable, the plain mid-point rule already yields partially structure-preserving schemes on its own. Thus, this newly developed GENERIC-based weak form is particularly well suited for the design of structure-preserving methods. In all cases, numerical investigations are presented that confirm the theoretical findings and shed light on the numerical stability of the newly developed schemes.

Keywords: structure-preserving space-time integration, GENERIC, molecular dynamics, periodic boundary conditions, interatomic potentials, thermoelasticity, thermo-viscoelasticity, finite element method, finite deformation

Zusammenfassung

Die vorliegende Dissertation behandelt die Entwicklung von strukturerhaltenden numerischen Methoden, welche aus dem GENERIC (Akkronym für general equation for non-equilibrium reversible-irreversible coupling) Formalismus hervorgehen.

Für konservative mechanische Systeme reduziert sich das GENERIC auf die hamiltonische Beschreibung. Zunächst werden für diesen Spezialfall neuartige energy-momentum (EM) konsistente Zeitschrittverfahren für den Bereich molekulardynamischer Systeme eingeführt, welche periodische Randbedingungen, Dreikörperpotentiale und interatomare Funktionalpotentiale berücksichtigen. Eine derartige Methode ist thermodynamisch konsistent, da sie die Energie und die Symmetrien einer solchen Struktur unabhängig von der Zeitschrittgröße bewahrt.

Anschließend wird der Fall von thermodynamischen Systemen im GENERIC Formalismus betrachtet und eine neue Variationsformulierung für die Thermoelastizität unter Berücksichtigung großer Deformationen vorgeschlagen. Diese Variationsformulierung entstammt dem GENERIC-Gerüst und ermöglicht die freie Wahl der thermodynamischen Zustandsvariable unter folgenden drei Optionen: (i) der absoluten Temperatur, (ii) der inneren Energiedichte, oder (iii) der Entropiedichte. In diesem Zusammenhang wird der Begriff der "GENERIC-konsistenten räumlichen Diskretisierung" eingeführt, welcher den Entwurf neuartiger energy-momentum-entropy (EME) konsistenter Schemata ermöglicht. Eine solche vorgeschlagene Methode ist thermodynamisch konsistent, da die Lösung streng dem ersten und zweiten Hauptsatz der Thermodynamik und den Symmetrien eines solchen Systems folgt, unabhängig von der Größe des Zeitschritts.

Schließlich wird die Thermo-Viskoelastizität unter Berücksichtigung großer Deformationen im Kontext des GENERIC-Gerüsts betrachtet. Für die Diskretisierung im Raum wird ein Ansatz mit gemischten finiten Elementen vorgeschlagen, welcher sich einer GENERIC-konsistenten räumlichen Diskretisierung bedient. Abhängig von der Wahl der thermodynamischen Variable liefert die einfache Mittelpunktregel bereits teilweise strukturerhaltende Schemata. Damit eignet sich die neuartige GENERIC-basierte schwache Form ideal für die Entwicklung strukturerhaltender Verfahren. In allen Fällen werden numerische Untersuchungen vorgestellt, die die theoretischen Erkenntnisse bestätigen und Aufschluss über die numerische Stabilität der neu entwickelten Verfahren geben.

Schlüsselwörter: Strukturerhaltende Raum-Zeit Integration, GENERIC, Molekulardynamik, periodische Randbedingungen, interatomare Potentiale, Thermoelastizität, Thermo-Viskoelastizität, Finite Elemente Methode, große Deformationen

Acknowledgements

Over the past six years, I have had the privilege of working as a research associate and doctoral candidate at the Institute for Mechanics (IFM) of the Karlsruhe Institute of Technology (KIT). Financial support for this research was provided by the Deutsche Forschungsgemeinschaft (DFG, German Research Foundation) within the Project BE 2285/13-1 entitled "Thermodynamisch konsistente Diskretisierungsverfahren für thermomechanisch gekoppelte Probleme" (Projectnumber 388118188) which is gratefully acknowledged.

First and foremost, I want to express my deepest gratitude to my supervisor Prof. Dr.-Ing. habil. Peter Betsch for offering me this great opportunity and for his generous support, which he has given me to the greatest extent. Since my time as a master's degree candidate he has been an exceptional mentor in all respects and his passion for mechanics and numerical methods has inspired and motivated me ever since. Without his support throughout numerous discussions and his invaluable guidance this work could not have been possible and I am very grateful for that.

My special thank also goes to Prof. Dr.-Ing. habil. Michael Groß for acting as the second examiner of this thesis.

Further, I wish to express my gratitude to all my colleagues and friends at the institute, who supported me in both research and teaching over the last years. In particular, I would like to thank Dr.-Ing. Marlon Franke for our successful collaboration and especially for sharing his knowledge in Linux with me. A very special thank you goes to Dr.-Ing. Alexander Janz for his support in research and teaching, the countless tips and advice I received from him, and the many after-work dart sessions we shared. Further, I would like to thank the colleagues and members of the "IFM running club": Jonas Hund, Tobias Laschütza, Robin Pfefferkorn, Vanessa Susana Valdes y Beck and Paul Wasmer not only for their irreplaceable support in research and teaching but also for pushing my limits on the tracks near the University. Furthermore, I would like to thank Timo Ströhle, who accompanied me on almost every walk to the bakery "La Minzbrueck" and for the numerous meaningful conversations we shared. Next, I would like to thank Simeon Schneider for all the intense and insightful discussions, which we could build up starting almost always from simple mechanical and mathematical considerations.

Additionally, I want to thank my other colleagues Julian Bauer and Moritz Hille and former colleagues Michael Strobl and Friedemann Streich for their support and contribution to the pleasant working atmosphere at the institute. Moreover, I would like to thank Gabriele Herrmann and Rosemarie Krikis for their administrative assistance and guidance and

Klaus Neidhardt for his technical support. A special thanks goes to Florian Schrammer for his unconditional support and his good-hearted spirit.

Acknowledgment is also given to Prof. Dr.-Ing. habil. Christian Hesch for his helpful advice and technical support on the ESRA-project. In this regard, I would like to thank Dr.-Ing. Maik Dittmann, Dr.-Ing. Melanie Krüger, Jonathan Schulte, Stefan Schuß and Felix Schmidt for the fruitful collaboration on this project.

Further, I would like to pay my special regards to Prof. Dr. Ignacio Romero. I am very grateful to have had the opportunity to work with him and his group for six months during my research stay in Madrid. From the beginning I was welcomed with open arms and all the discussions and his invaluable insights on mechanics and beyond made my stay a memorable experience. The support of the Karlsruhe House of Young Scientists (KHYS) for the stay abroad is gratefully acknowledged.

Of course, I cannot forget friends who went through hard times but also celebrated each accomplishment together: Marc Galley, Fabian Haas, Geza Hildenbrand, Dr.-Ing. Sven Nagel, Gustav Strauß, Dr.-Ing. David Walter, to name a few of them.

Finally, and most importantly, I would like to express my deepest gratitude towards my entire family. I must thank my parents, Erma Schiebl and Dr.-Ing. Hans Schiebl, and my sister Amira Schiebl for their continuous support, their unconditional love and their endless patience they have given me throughout my life. I am forever thankful to them for giving me the opportunities and experiences that have made me who I am.

Karlsruhe, 2021 Mark Schiebl

Contents

List of Figures

List of Tables

1. Introduction[1]

Mathematical models are extensively used by engineers to gain insight into the quantitative and structural behavior of a system. Although simple systems can be understood and analyzed more easily, complexity arises when a more accurate representation of reality is required, necessitating the replacement of pencil and paper by large computer simulations in favor of more powerful design tools.

At the heart of every dynamical computer simulation lies the numerical method, also known as the "integrator", which constructs an approximation of the exact solution by providing discrete snapshots of the system over time for a given problem. Today, driven by the exponential growth of computational power, the qualitative properties of the integrator itself become vital to the success of the simulation, as longer simulation durations require more stable numerical methods. While the equations of motion for a dynamic system may "hide" important properties of the physical system, such as conservation laws, these properties are not inherited by the numerical method in general, which can cause unphysical time evolution and thus introduces a source for numerical instabilities. One feasible approach that counters this physical inconsistency is the use of subtle numerical methods that are numerically accurate and capable of reliably capturing the main characteristics of the underlying physical system. This motivated the development of structure-preserving numerical methods, also known as "geometric integrators", for a wide array of fields in applied and theoretical science. Geometric integrators are designed to respect the fundamental physics of the underlying mathematical model, ranging from time reversibility to first integrals, by preserving the geometric properties in a discrete setting. Most attention has been focused hereby on Hamiltonian or Lagrangian systems due to their well-understood mathematical (geometric) structure. Based on the common underlying structure, structure-preserving methods can be easily generalized from simple systems such as spring-mass arrays to complex nonlinear elastodynamical structures. Therefore, reports of excellent long-term behavior and numerical stability for a huge variety of systems have laid the foundation for the success of this method.

[1] This chapter is based on the introductions given in [1–4]

1.1. Structure-preserving integration

Structure-preserving numerical methods have played a major role in the development of numerical time integration methods for decades, and monographs such as [5–7] beautifully demonstrate the advancement of this field. While numerical integrators have traditionally been viewed solely as an approximation to a time-continuous problem, geometric integrators consider the integration scheme itself as a discrete dynamical system with its own characteristics, including balance laws, symmetries, symplecticity, reversibility, and many more. Even though it was developed within a different context, the well-known mid-point rule can be viewed as a member of this general class because it preserves the symmetry and the symplecticity.

Different classes of structure-preserving numerical methods can be distinguished by the features inheritated from the continuous problem and the applied numerical techniques. The earliest developments of structure-preserving numerical methods can be traced back to symplectic methods, first introduced in [8]. Early schemes of this kind were explicit symplectic integrators [9] and have been applied to various problems of celestial mechanics (see, e.g., [10, 11]), followed by implicit schemes [12, 13] based on [14]. Later, a symplectic family of Runge-Kutta schemes were constructed independently in [15, 16].

For mechanical systems, most researchers have focused their attention on momenta, energy, and symplectic structures. Unfortunately, simultaneously preserving the momenta, energy and symplecticity for a fixed time-step method is not possible (for more details, see [17]). Therefore, many geometric integrators in the realm of Lagrangian and Hamiltonian mechanics can be gathered around two classes: variational integrators and energy-momentum (EM) integrators.

1.1.1. Variational integrators

A more contemporary approach towards symplectic methods is based on a variational nature and is thus termed "variational integrator". Variational integrators preserve the symplecticity and the momenta of the underlying system and, due to its variational nature, allows the extension to non-conservative systems by utilizing the discrete counterpart of the Lagrange-d'Alembert principle. The origin of this method can be traced back to [18] and [19, 20]. Based on those concepts, a theory of discrete Lagrangian and Hamiltonian mechanics was provided in [21], which has been applied to mechanical problems with multi-symplectic geometry [22, 23], dynamical systems evolving on nonlinear manifolds [24, 25], structural elements [26–28], contact and impact problems [29], multibody dynamics and control [30, 31], stochastic differential equations [32], constrained and forced problems [33] and dynamic of fluids [34] (see, e.g., [35] for more examples).

Sharp phase portraits and long numerical simulations demonstrate the strong performance of this method, especially in explicit calculations. Attempts have been made to express thermoelasticity in the framework of Lagrangian/Hamiltonian mechanics by introducing the concept of thermal displacement [36]. Unfortunately, expressing Fourier's law of heat conduction within this framework remains an unsolved issue [35][2]. In addition, the loss of symplecticity and thus the loss of one key feature of variational integrators is a consequence of dissipation. For that matter, it seems that structure-preserving numerical methods for thermodynamical systems fit better within the energy-momentum integrator approach, which has been preferentially chosen for this work. Nevertheless, besides the preservation of the symplectic structure and the momenta, variational integrators are known for their improved long-term energy behavior[3] compared to classical methods and are excellent numerical time-stepping schemes.

1.1.2. Energy-momentum integrators

EM algorithms are an additonal, frequently used class of structure-preserving integration schemes which, in contrast to variational integrators (symplectic-momentum integrators), preserve the momenta and the energy of the system in the discrete setting. This method has its roots in [40–44], whereas the first energy-momentum scheme for nonlinear elastodynamics was proposed in [45]. Based on these previous reports, a systematic approach using the *discrete gradient* operator was developed in [46]. This second-order accurate operator ensures the preservation of the energy and the momenta by design and, due to its systemized construction, its remarkable robustness, and its good qualitative accuracy, has made the energy-momentum integrator a popular choice for simulating the governing equations of particle dynamics [40, 41], rigid bodies [43], nonlinear solid mechanics [45, 47, 48], nonlinear shells [49–51] and rods [52–54] multi-body dynamics [55, 56], gradient systems [57], and general PDEs [58]. For a comprehensive overview of energy-momentum integrators in the context of Hamiltonian dynamics, see [59].

1.1.3. Dissipation

The conservation of energy is not always a desired feature of Hamiltonian or Lagrangian systems because occasionally there is a need for dissipation in resolving high-frequency modes, which is often related to the spatial discretization of the problem. These "spurious" high frequencies are not present in real systems due to physical damping. However, physical damping is seldom addressed in the mathematical modeling and is usually compensated for by introducing controllable numerical dissipation schemes. The first well-known representatives of such schemes can be found in the family of the Newmark method [60], the

[2] Efforts have been made in [37] to extend the Hamiltonian to account for the second sound phenomenon by adding a dissipative entropy-flux perturbation term and utilizing the Lagrange-D'Alembert principle. A similar approach has been followed for Fourier-type heat conduction in [38, 39]. It is not clear to the author if and how this adiabatic method can be generalized for arbitrary thermodynamical processes.

[3] For conservative systems, variational integrators capture the evolution of a "perturbated" Hamiltonian exactly. This modified energy-level set is, most likely, closely related to the true energy-level set and explains the characteristic energy oscillation of this method.

HHT method developed in [61], or the Wilson-θ method [62] among other methods that have been successfully implemented in common software packages. Unfortunately, these classical dissipation schemes, being developed in the linear regime, lose their dissipative character in the nonlinear regime [63]. In that sense, several energy-decaying numerical methods have been proposed [64–67].

Nevertheless, the philosophy of the current work is to incorporate physical dissipative processes into the mathematical model instead of considering numerical dissipation schemes. Most real structures can only be idealized as Hamiltonian or Lagrangian systems, yet are of relevance in many areas of science and engineering.

Polymeric solids and rubber-like materials are a perfect example of structures that exhibit viscoelastic material features and are highly temperature sensitive, thus being subjected to viscous dissipation and dissipation due to the conduction of heat. Consequently, a large class of problems involving dissipative mechanisms can not be expressed within the framework of Hamiltonian or Lagrangian mechanics, including, for example, the aforementioned Fourier-type heat conduction, which arises in a wide range of industrial applications, including the following:

- **Civil engineering:** Tuned mass damper (bridges, high-rise buildings), dissipative connections in high-rise buildings for seismic protection, and welding
- **Mechanical engineering:** Shock absorbers (automotive and railway), brake discs, welding, and energy dissipation in vehicle crashes

Unfortunately, structure-preserving integrators for thermodynamical systems have received little attention, one reason being the lack[4] of a well-understood geometric structure guiding the construction of such numerical method. Numerous attempts have been made to extend structure-preserving time-stepping schemes to the domain of non-conservative mechanical systems, such as port-Hamiltonian systems [68], viscoelasticity [69, 70], elasto-plasticity [71], thermoelasticity [72, 73], thermo-viscoelasticity [74–76], isothermal and non-isothermal fiber-reinforced continuas [77–80]. Unlike the Lagrangian/Hamiltonian case, these structure-preserving schemes do not emerge from a unifying theory.

However, the metriplectic structure of the general equation for non-equilibrium reversible-irreversible coupling (GENERIC) formalism provides a double-generator framework that expresses thermomechanical models of dissipative materials (or generalized standard materials) in a unifying formalism. Its formulation is based on an additive decomposition of the time-evolution equations into a reversible part and a dissipative part. It can be seen as a natural extension of Hamiltonian mechanics to the dissipative regime (see, e.g., [81] for a relation to port-Hamiltonian systems).

Originally developed in the context of complex fluids [82, 83], GENERIC has been applied to a vast range of problems (predominantly in the domain of complex fluids), such as the reptation model for entangled linear polymers [84], polymer blends [85], colloidal

[4] At least to the computational mechanics' community

suspensions [86], two-phase systems [87], relativistic hydromechanics [88], discrete formulations of hydromechanics for simulations [89], thermodynamically guided simulations [89], diffusion through polymeric membranes [90], rheological model of suspension of red blood cells [91], structure-preserving neural networks [92], and the theory of quantum systems [93], to mention only a few examples. We refer to the book by Öttinger [94] for a comprehensive account of the GENERIC formalism up to the year 2005 and to [95] for a huge list of advanced developments up to the year 2017.

Focusing on thermodynamic models for non-isothermal solids, as one of the main themes of the present work, an early application of the GENERIC formalism to finite-strain thermoelasticity is credited to [96], albeit in an Eulerian setting which is quite uncommon in solid mechanics, and to [97] for continuum damage mechanics. Later, the Lagrangian setting was used in [98, 99] and [100] to develop the GENERIC framework for non-isothermal solid mechanics. Moreover, [98, 99] preferred to use the absolute temperature as the thermodynamic state variable, while in [100] a special form of GENERIC is devised that makes the free choice of the thermodynamic state variable possible.

In the field of computational solid mechanics, Romero [101, 102] recognized at an early stage the great potential of the GENERIC framework for the design of structure-preserving time-stepping schemes. Since the GENERIC framework automatically ensures the thermodynamic admissibility of the time-evolution equations, it provides an ideal starting point for the development of thermodynamically consistent (TC) integrators. The solutions of this integrator could be classified as thermodynamically guided simulations, in analogy to the thermodynamically guided simulation of consistent coarse-graining schemes (e.g., [103, 104]), as TC integrators comply with the first and the second law of thermodynamics, independent of the size of the time-step. Therefore, TC integrators may also be termed "energy-entropy" integrators.

If TC integrators also respect the symmetries of the underlying mechanical system, they can be viewed as an extension to the dissipative regime of Energy-Momentum (EM) integrators previously developed for Hamiltonian systems with symmetry. The GENERIC framework also facilitates a concise characterization of momentum maps and associated conservation properties (see [105]).

Hence, the GENERIC formalism provides a common structure for non-conservative systems, similarly to Hamiltonian dynamics for conservative problems, and can therefore be used in a similar manner to guide the design of structure-preserving numerical schemes for thermodynamical problems. Consequently, GENERIC provides a solid theoretical foundation for the design of energy-momentum-entropy (EME) methods, as has been shown in [102] for finite-strain thermo-elasticity and in [106] for finite-strain thermo-viscoelasticity.[5]

[5] Of course, the GENERIC framework is not a prerequisite for the development of structure-preserving numerical methods for non-isothermal solid mechanics. In the context of coupled thermomechanical problems, alternative procedures have been proposed, such as in [72], [75], [76] and [73].

1.2. Objectives

In this work, we aim to formulate, analyze, implement, and verify second-order structure-preserving numerical methods by employing the GENERIC structure. The present work can be divided into two parts:

- The first part (Chapter 3) deals with structure-preserving numerical methods for molecular dynamical systems. It extends the scope of application for Energy-Momentum integrators to problems with periodic domains by making use of the discrete gradient operator [46], targeting the specific geometric structure of molecular dynamical systems.
- The second part (Chapters 4 to 6) of this thesis adresses structure-preserving numerical methods for thermo-viscoelastic continuum dynamics. The focus hereby is on smooth irreversible effects, and therefore effects like plastic or damage transformations are not considered.

1.2.1. Energy-momentum conserving integration schemes for molecular dynamics

Given the favorable properties of EM methods, it is remarkable that they have not received more attention in the field of molecular dynamics, at least when investigating the use of implicit time integration schemes for obtaining numerical solutions. The governing equations of the latter are essentially Hamiltonian and fit seamlessly in the framework developed since the 1970's for integrating these types of problems, while preserving the energy and the momenta. It would seem natural that integration schemes designed to preserve the main invariants of the motion would give accurate predictions of the thermodynamic averages, which are of interest in many practical and theoretical situations [107–111], but have rarely been studied [112]. Instead, molecular dynamics codes seem to favor the use of the explicit Verlet method or symplectic methods, mainly due to their smaller computational cost as compared with implicit schemes. Although popular explicit methods have desirable properties in terms of computational cost, accuracy, and geometry preservation, they lack energy conservation, a key invariant that is most important in the simulation of microcanonical ensembles. A thorough investigation of the accuracy of energy and momentum conserving schemes in capturing the statistical behavior of atomistic systems for long periods of simulation is lacking. Preliminary results [112] are promising, but much testing and validation is still required.

The development and implementation of energy and momentum conserving algorithms in the context of molecular dynamics has three specific issues that affect the discretization of the equations and their analysis—issues that do not appear in their application to nonlinear solids, shells, rods, etc., or any of the other systems for which the use of these methods is widespread. The first critical issue is the treatment of periodic boundary conditions. These are almost invariably required for the study of average properties in particle systems [110,

111, 113], and demand a careful analysis, especially to ascertain whether they spoil the conserving properties of the method or not. Taking this into account requires to consider the geometry and topology of the periodic configuration space, and might also affect the accuracy of the integration scheme.

The second issue that needs to be carefully dealt with is the use of conserving schemes in the context of three-body potentials. These functions are employed in modeling angle interactions in atomic bonding [114, 115], and their impact on the global behavior of some systems is so critical that it must to be accounted for. In fact, atomic systems with potentials of this type allow for large relative motions among the particles, and this is precisely the area where conserving schemes have shown their superiority with regard to other implicit integrators.

The third aspect that requires a detailed analysis is the application of conserving schemes to mechanical systems in which the potential is based on cluster functionals [116–120]. These effective potentials are often required for the correct modeling of complex binding among metallic atoms and again require a careful study when used in combination with conserving schemes. While pair potentials of the Lennard-Jones type [121] have been employed together with EM conserving schemes [41], their formulation for cluster potentials needs to be specifically addressed.

Here, we formulate energy and momentum conserving schemes for simulating the dynamics of atomic systems. Some of the methods discussed have already been employed in the literature, and we identify new ones. In all cases, we explain how the three critical issues identified before (i.e., periodic boundary conditions, three-body potentials and functional potentials) affect their formulation. To the author's knownledge, none of these have been previously studied.

1.2.2. Structure-preserving integration of large-strain thermo-viscoelasticity in the framework of GENERIC

The aforementioned GENERIC-based numerical methods are subject to two serious limitations: the first limitation is related to the use of the entropy density as the thermodynamic state variable, and the second limitation is due to the fact that the GENERIC formulation focuses on closed systems and thus does not account for boundary conditions. Of course, both points limit the applicability of numerical methods for the solution of initial boundary value problems in thermo-mechanics.

The first limitation has been addressed in [122, 123] by using the absolute temperature as the thermodynamic state variable in the underlying GENERIC formulation. In the context of discrete systems, the free choice of the thermodynamic state variable has been recently addressed in [124]. The second limitation has been circumvented in [106] by applying the Lagrange multiplier method to enforce temperature boundary conditions.

We completely resolve the aforementioned limitations of GENERIC-based numerical methods for finite-strain thermo-elasticity by proposing a novel GENERIC-based variational formulation that makes the free choice of the thermodynamic state variable possible. In particular, one may choose (i) the internal energy density, (ii) the entropy density, or (iii) the absolute temperature as the thermodynamic state variable. Moreover, boundary conditions are taken into account by applying a generalized GENERIC formulation for open systems.

This novel, GENERIC-based variational formulation is discretized in time using the well-known mid-point rule. Depending on the choice of the thermodynamic state variable, partially structure-preserving schemes are obtained. For example, choosing the internal energy density as the thermodynamic state variable leads to an EM scheme. On the other hand, choosing the entropy density as the thermodynamic state variable yields a momentum-entropy (ME) scheme. However, despite their partially structure-preserving properties, all of the mid-point type schemes turn out to be prone to numerical instabilities. These observations led to the conjecture that only fully structure-preserving schemes guarantee numerical stability for dissipative systems in the same way that EM schemes do for Hamiltonian systems.

In this context, the GENERIC-based weak form provides an ideal starting point for the development of EME schemes. First, the GENERIC-based weak form is discretized in space, resulting in a *GENERIC-consistent space discretization.* Then, the semi-discrete system is discretized in time by applying the partitioned discrete gradient operator in the sense of [125] leading to three different EME schemes. These EME schemes satisfy a specific Lyapunov-type stability estimate and thus do not exhibit any numerical instabilities. Hence, we develop novel EME schemes for large-strain thermoelasticity.

Finally, we extend the applicability of this method to the realm of thermo-viscoelasticity. In particular, we aim at a material formulation of isotropic large-strain thermo-viscoelasticity which makes the free choice of the thermodynamic state variable possible. To this end, we build on previous work by [99], who laid the theoretical basis for the GENERIC description of thermo-viscoelasticity.

1.3. Overview

The thesis is organized into seven chapters.
Each chapter can be read independently—therefore, repetitions are inevitable.[6]

Chapter 1 covers the motivation, the contemporary state of structure-preserving methods, and the objectives of this work.

[6] A consistent notation throughout the present work would result in a tiresome reading experience, but notational shifts between the chapters are kept to a minimum depending on the focus therein.

Chapter 2 introduces the GENERIC framework briefly. In Section 2.1 the framework is presented for finite dimensional systems. This includes its relation to Hamilton's equations of motion. Furthermore, we introduce the notation of *GENERIC-consistent space discretization*. Then, in Section 2.2, the structure of the building blocks of the GENERIC formalism is discussed when transformed to a different set of variables.

Chapter 3 addresses the formulation and analysis of EM conserving time integration schemes in the context of particle dynamics, and in particular atomic systems. In Section 3.1, we review the basic topology of periodic systems in order to clearly define the distance function. Section 3.2 introduces particle dynamics, with special attention to its Hamiltonian structure. Next, in Section 3.3, conserving time integration schemes are presented for particle systems in periodic domains, restricted to those with pair potentials. These are extended to systems with angle potentials in Section 3.4 and to atomic systems with functional potentials in Section 3.5.

Chapter 4 proposes a new GENERIC-based variational formulation for finite-strain thermo-elastodynamics. First, the GENERIC framework is dealt within bracket form for closed systems. Therefore, in Section 4.1, we start with finite-strain elastodynamics and subsequently present in Section 4.2 its extension to thermo-elastodynamics. Then, in Section 4.3, the transition to open systems is performed. In the following Section 4.4, the resulting GENERIC-based weak form of the initial boundary value problem (IBVP) at hand is then discretized in time and space. Representative numerical examples are presented in Section 4.5.

Chapter 5 designs novel EME methods for finite-strain thermo-elastodynamics. Section 5.1 briefly recapitulates the variational formulation of large-strain thermo-elasticity. Then, in Section 5.2, the GENERIC-based weak form is discretized in space, resulting in a GENERIC-consistent space discretization. In addition to that, the main balance laws to be preserved under discretization are outlined. In Section 5.3, the semi-discrete system is further discretized in time, leading to three alternative EME schemes. Section 5.4 contains representative numerical examples that confirm both the structure-preserving features and the enhanced numerical stability of the newly developed EME schemes when compared to the standard time integration schemes developed in Chapter 4.

Chapter 6 introduces in Section 6.1 the bracket form of the GENERIC formalism for thermo-viscoelasticity with heat conduction. In particular, in Section 6.2, a new material version of the inelastic dissipative bracket is proposed which allows for the free choice of the thermodynamic state variable. In Section 6.3, a mixed finite element approach is proposed that yields a GENERIC-consistent semi-discrete form of the evolution equations.

The mid-point type discretization in time is dealt with in Section 6.4, leading to alternative partially structure-preserving schemes. Numerical investigations are presented in Section 6.5.

Chapter 7 summarizes the findings. The thesis closes by drawing conclusions and giving an outlook on interesting research topics that follow the direction of this work.

2. A brief introduction to the GENERIC framework

The general equation for non-equilibrium reversible-irreversible coupling framework (GENERIC) is a double-generator formalism that expresses thermomechanical models of dissipative materials (or generalized standard materials) in a unifying framework. Its formulation is based on an additive decomposition of the time-evolution equations into a reversible part and a dissipative part. While the reversible part is generated by the total energy of the system, the irreversible part is generated by the total entropy.

The GENERIC framework is typically presented in two alternative forms: an operator version (i) in which the entries of the operator matrices are either generalized functions or differential operators, and (ii) a bracket version. Moreover, the GENERIC formulation typically focuses on *closed* systems. That is, neither thermal nor mechanical interactions with the surrounding of the system are considered. This Chapter summarizes the GENERIC framework for finite dimensional systems, which will be repeatedly referred to throughout the following chapters. For a comprehensive account of the GENERIC framework see, e.g., [94].

The fundamental concept of the GENERIC framework for finite dimensional systems also applies (with corresponding modifications) to infinitesimal dimensional systems, which will be considered in Chapters 4 to 6. In this connection the notion "GENERIC consistent space discretization" is introduced, which facilitates the design of energy-momentum-entropy (EME) consistent schemes. For now we will consider the case of isolated systems, for which the GENERIC framework was originally developed for. In the case of isolated (or closed) systems, boundary effects are considered to be irrelevant.

2.1. GENERIC framework for finite dimensional systems

The GENERIC framework hinges on an additive decomposition of the evolution equations into reversible and irreversible parts. Assuming that the state of a finite-dimensional system is characterized by the state vector $\mathbf{z}(t) \in \mathbb{R}^n$, the time-evolution of the discrete system is described by

$$\frac{\mathrm{d}\mathbf{z}}{\mathrm{d}t} = \mathbf{L}\partial_{\mathbf{z}}\mathcal{E} + \mathbf{M}\partial_{\mathbf{z}}\mathcal{S} , \tag{2.1}$$

which corresponds to the operator representation (i). In eq. (2.1) reversible processes are generated by the total energy $\mathcal{E}(\mathbf{z})$, while dissipative processes are generated by the total entropy $\mathcal{S}(\mathbf{z})$. In this connection, $\mathbf{L}(\mathbf{z})$ is the Poisson matrix, which must be skew-symmetric, whereas $\mathbf{M}(\mathbf{z})$ is the dissipative matrix, which must be symmetric and positive semi-definite. Two degeneration (or non-interaction) conditions have to hold. Namely,

$$\mathbf{L}\partial_{\mathbf{z}}\mathcal{S} = \mathbf{0} , \tag{2.2}$$

and

$$\mathbf{M}\partial_{\mathbf{z}}\mathcal{E} = \mathbf{0} . \tag{2.3}$$

Using the GENERIC formulation (2.1) guarantees that the following two fundamental properties of a closed thermomechanical system are satisfied. Firstly, in accordance with the first law of thermodynamics, the total energy is conserved. That is,

$$\frac{\mathrm{d}\mathcal{E}}{\mathrm{d}t} = \partial_{\mathbf{z}}\mathcal{E} \cdot \frac{\mathrm{d}\mathbf{z}}{\mathrm{d}t} = \partial_{\mathbf{z}}\mathcal{E} \cdot \mathbf{L}\partial_{\mathbf{z}}\mathcal{E} + \partial_{\mathbf{z}}\mathcal{E} \cdot \mathbf{M}\partial_{\mathbf{z}}\mathcal{S} = 0 . \tag{2.4}$$

Note that on the right-hand side of the last equation, the skew-symmetry of the Poisson matrix has been used along with degeneration condition (2.3). Secondly, in compliance with the second law of thermodynamics, the total entropy should be a non-decreasing function of time. In fact, using (2.1) leads to

$$\frac{\mathrm{d}\mathcal{S}}{\mathrm{d}t} = \partial_{\mathbf{z}}\mathcal{S} \cdot \frac{\mathrm{d}\mathbf{z}}{\mathrm{d}t} = \partial_{\mathbf{z}}\mathcal{S} \cdot \mathbf{L}\partial_{\mathbf{z}}\mathcal{E} + \partial_{\mathbf{z}}\mathcal{S} \cdot \mathbf{M}\partial_{\mathbf{z}}\mathcal{S} \geq 0 . \tag{2.5}$$

Here, degeneration condition (2.2) has been used, along with the positive semi-definiteness of the dissipative matrix. The properties of $\mathbf{L}$ and $\mathbf{M}$ can also be conveniently discussed in terms of two brackets

$$\begin{aligned} \{\mathcal{A}, \mathcal{B}\} &= \partial_{\mathbf{z}}\mathcal{A} \cdot \mathbf{L}\partial_{\mathbf{z}}\mathcal{B} , \\ [\mathcal{A}, \mathcal{B}] &= \partial_{\mathbf{z}}\mathcal{A} \cdot \mathbf{M}\partial_{\mathbf{z}}\mathcal{B} . \end{aligned} \tag{2.6}$$

Hereby are $\mathcal{A}(\mathbf{z})$ and $\mathcal{B}(\mathbf{z})$ arbitrary functions of the state variables. The brackets are termed Poisson and dissipative bracket, respectively. Using eq. (2.1) these brackets lead to the following evolution equation for an arbitrary function $\mathcal{A}$

$$\frac{\mathrm{d}\mathcal{A}}{\mathrm{d}t} = \{\mathcal{A}, \mathcal{E}\} + [\mathcal{A}, \mathcal{S}] , \tag{2.7}$$

which corresponds to bracket version (ii) of the GENERIC framework. The antisymmetry property of the Poisson matrix can now be stated as

$$\{\mathcal{A}, \mathcal{B}\} = -\{\mathcal{B}, \mathcal{A}\}\,. \tag{2.8}$$

Further, the Poisson bracket has to satisfy the product (or Leibniz) rule and the Jacobi identity. Those properties are well-known from the Poisson bracket of classical mechanics and express the essence of reversible dynamics, see e.g., [126].

The symmetry of the dissipative matrix $\mathbf{M}$ can be formulated in terms of the dissipative bracket as

$$[\mathcal{A}, \mathcal{B}] = [\mathcal{B}, \mathcal{A}]\,. \tag{2.9}$$

The positive semi-definiteness can be expressed as

$$[\mathcal{A}, \mathcal{A}] \geq 0\,. \tag{2.10}$$

Finally, the two degeneration conditions can be restated in terms of the bracket representation

$$\begin{aligned} \{\mathcal{A}, \mathcal{S}\} &= 0\,, \\ [\mathcal{A}, \mathcal{E}] &= 0\,. \end{aligned} \tag{2.11}$$

2.1.1. Connection to Hamilton's equations of motion

If the system under consideration is purely mechanical and conservative, the choice

$$\mathbf{z} = \left(\mathbf{x}^1, \quad \ldots, \quad \mathbf{x}^N, \mathbf{p}^1, \quad \ldots, \quad \mathbf{p}^N\right),$$

where $\mathbf{x}^a$ is the a-th position, $\mathbf{p}^a$ the a-th momentum and $N \in \mathbb{Z}_+$, results in a zero dissipative matrix $\mathbf{M}$ and a Poisson matrix $\mathbf{L}$ of the form

$$\mathbf{L}(\mathbf{z}) = \left[\begin{array}{c|c} 0 & \mathrm{I} \\ \hline -\mathrm{I} & 0 \end{array}\right], \tag{2.12}$$

where 0 is the zero and I is the identity matrix with appropriate dimension depending on the system. By this means the state vector $\mathbf{z}$ has to be viewed as a column vector. The previous matrix is the well-known canonical symplectic matrix. In this case the evolution equation of the GENERIC coincide with Hamilton's equation of motion, which will be considered in Chapter 3.

2.1.2. GENERIC-consistent space discretization

If an infinite dimensional system at hand is discretized in space in such a way that the GENERIC structure for finite dimensional systems is preserved, then we speak of a *GENERIC-consistent space discretization.* Such a method is devised in Chapters 5 and 6 of the present work.

2.2. Transformation of variables

By reasons of taste or convenience for the choice of the specific variables, it is important to address the relation between the building blocks of the GENERIC for different sets of state variables. We therefore consider a one-to-one transformation $\mathbf{z} \mapsto \mathbf{z}'$. As for the scalar energy and entropy generators $\mathcal{E}$ and $\mathcal{S}$ we assume a simple transformation behavior

$$\begin{aligned} \mathcal{E}'(\mathbf{z}') &= \mathcal{E}(\mathbf{z}(\mathbf{z}')) \,, \\ \mathcal{S}'(\mathbf{z}') &= \mathcal{S}(\mathbf{z}(\mathbf{z}')) \,. \end{aligned} \tag{2.13}$$

Applying a transformation of variables in eq. (2.1) using eq. (2.13) results in the following transformation laws for the Poisson and the dissipative matrix

$$\begin{aligned} \mathbf{L}'(\mathbf{z}') &= \partial_{\mathbf{z}}\mathbf{z}' \cdot \mathbf{L}(\mathbf{z}(\mathbf{z}')) \cdot \partial_{\mathbf{z}}\mathbf{z}'^{T} \,, \\ \mathbf{M}'(\mathbf{z}') &= \partial_{\mathbf{z}}\mathbf{z}' \cdot \mathbf{M}(\mathbf{z}(\mathbf{z}')) \cdot \partial_{\mathbf{z}}\mathbf{z}'^{T} \,. \end{aligned} \tag{2.14}$$

Considering that

$$\begin{aligned} \mathcal{A}'(\mathbf{z}') &= \mathcal{A}(\mathbf{z}(\mathbf{z}')) \,, \\ \mathcal{B}'(\mathbf{z}') &= \mathcal{B}(\mathbf{z}(\mathbf{z}')) \,, \end{aligned} \tag{2.15}$$

holds, then the transformation laws eq. (2.14) lead to

$$\{\mathcal{A}', \mathcal{B}'\} = \partial_{\mathbf{z}'}\mathcal{A}' \cdot \mathbf{L}'\partial_{\mathbf{z}'}\mathcal{B}' = (\partial_{\mathbf{z}}\mathcal{A} \cdot \mathbf{L}\partial_{\mathbf{z}}\mathcal{B})\,|_{\mathbf{z}=\mathbf{z}(\mathbf{z}')} = \{\mathcal{A}, \mathcal{B}\}|_{\mathbf{z}=\mathbf{z}(\mathbf{z}')} \,, \tag{2.16}$$

and

$$[\mathcal{A}', \mathcal{B}'] = \partial_{\mathbf{z}'}\mathcal{A}' \cdot \mathbf{M}'\partial_{\mathbf{z}'}\mathcal{B}' = (\partial_{\mathbf{z}}\mathcal{A} \cdot \mathbf{M}\partial_{\mathbf{z}}\mathcal{B})\,|_{\mathbf{z}=\mathbf{z}(\mathbf{z}')} = [\mathcal{A}, \mathcal{B}]|_{\mathbf{z}=\mathbf{z}(\mathbf{z}')} \,. \tag{2.17}$$

In other terms: the Poisson bracket and the dissipative bracket are independent of the particular choice of state variables.

The time evolution of the system is invariant with respect to the choice of a different set of variables. However, the building blocks of the GENERIC formalism, the matrices $\mathbf{L}$, $\mathbf{M}$ and the gradients $\partial_{\mathbf{z}}\mathcal{E}$, $\partial_{\mathbf{z}}\mathcal{S}$ may have a simpler form, which can be of advantage for the formulation of a time-stepping integration scheme. Further, as an one-to-one transformation preserves the brackets, the transformations can be very useful when considering the verification of the Jacobi identity (see, e.g., [94]).

3. Energy-momentum conserving integration schemes for molecular dynamics[1]

In this chapter we study the dynamic motion of systems of particles. This type of problems is of great importance for the simulation of matter at the atomic scale and a very large body of references study details pertaining to their numerical solution and the information that can be extracted from these simulations.

Such systems can be considered as a prototype of a classical mechanical system: N particles in a periodic domain. The state vector of such a system assumes the form

$$\mathbf{z} = \left(\mathbf{x}^1, \quad \dots, \quad \mathbf{x}^N, \mathbf{p}^1, \quad \dots, \quad \mathbf{p}^N\right),$$

to be introduced in the subsequent. Hereby the state vector has to be viewed as a column vector. As only reversible processes are present, the GENERIC evolution eq. (2.1) boils down to Hamilton's equations of motion

$$\frac{\mathrm{d}\mathbf{z}}{\mathrm{d}t} = \mathbf{L}\partial_{\mathbf{z}}\mathcal{E}\ , \tag{3.1}$$

where $\mathbf{L}$ is the symplectic matrix eq. (2.12), or equivalently the equations of motion in Poisson bracket form

$$\frac{\mathrm{d}\mathcal{A}}{\mathrm{d}t} = \{\mathcal{A}, \mathcal{E}\}\ . \tag{3.2}$$

Therefore, solely the total energy $\mathcal{E}$ (which is the Hamiltonian $\mathcal{H}$ in this case) in connection with the Poisson bracket $\{\cdot, \cdot\}$ describes the evolution of such systems, where the Poisson bracket is given by

$$\{\mathcal{A}, \mathcal{B}\} = \sum_{a=1}^{N} \left(\partial_{\mathbf{x}^a}\mathcal{A} \cdot \partial_{\mathbf{p}^a}\mathcal{B} - \partial_{\mathbf{p}^a}\mathcal{A} \cdot \partial_{\mathbf{x}^a}\mathcal{B}\right)\ , \tag{3.3}$$

which is the well known Poisson bracket of classical mechanics. Depending on the specific structure of the total energy, different interatomic interactions can be modelled. Further,

[1] This chapter is based on [3]

the GENERIC/Hamiltonian framework guides the construction of structure-preserving numerical methods, as we will show in the following sections.

We address the formulation and analysis of energy and momentum conserving time integration schemes and identify three critical aspects of these models that demand a careful analysis when discretized: (i) the treatment of periodic boundary conditions, (ii) the formulation of approximations of systems with three-body interaction forces, and (iii) their extension to atomic systems with functional potentials. These issues, and in particular their interplay with energy-momentum (EM) integrators, are studied in detail. Novel expressions for these time integration schemes are proposed and numerical examples are given to illustrate their performance.

3.1. The topology of periodic domains

Many particle systems of interest are formulated in periodic domains. These allow to study large systems by only discretizing representative volumes, much smaller in size, while hopefully not losing too much information. In this section we gather some topological and geometrical facts of periodic domains that will be necessary to analyze numerical methods.

We start by considering a periodic three-dimensional box $\mathcal{B}$ of side length L, noting that all the results are applicable to systems in one and two dimensions, with the corresponding modifications. This box is isomorphic to the torus $\mathbb{T}^3$ (see e.g., [108, 110]), which itself can be identified with the product manifold $S^1 \times S^1 \times S^1$. Hence, each point $\boldsymbol{\xi} \in \mathbb{T}^3$ can be uniquely characterised by three angles (α, β, γ) and the complete manifold is covered by a single chart.

Numerical methods defined on $\mathbb{T}^3$ pose difficulties that can be alleviated by mapping this set into a more convenient one. For that, let us first define the following equivalence relation on $\mathbb{R}^3$: two points $\mathbf{x}, \mathbf{y} \in \mathbb{R}^3$ are defined to be equivalent, and indicated as $\mathbf{x} \sim \mathbf{y}$, if there exists a triplet of integers $\mathbf{z} \in \mathbb{Z}^3$ such that $\mathbf{x} = \mathbf{y} + L\mathbf{z}$. Using this equivalence relation, we can define the quotient space $\mathbb{P} := \mathbb{R}^3/\mathbb{Z}^3$ that is homeomorphic to the torus. In what follows the equivalent class of a point $\mathbf{x} \in \mathbb{R}^3$ will be denoted as $[\mathbf{x}] \in \mathbb{P}$.

When dealing with systems of particles in periodic domains one has to choose one of the two homeomorphic descriptions described above, namely, the torus and $\mathbb{P}$. From the computational point of view, employing the latter has many advantages. The first one is that given a distance on $\mathbb{R}^3$, this quotient space naturally inherits a distance, and thus a topology. In terms of the standard Euclidean distance $d(\cdot,\cdot) : \mathbb{R}^3 \times \mathbb{R}^3 \mapsto \mathbb{R}^+ \cup \{0\}$ we can define $d_T(\cdot,\cdot) : \mathbb{P} \times \mathbb{P} \mapsto \mathbb{R}^+ \cup \{0\}$ by the relation

$$d_T([\mathbf{x}], [\mathbf{y}]) = \inf_{\mathbf{x}\in[\mathbf{x}],\mathbf{y}\in[\mathbf{y}]} d(\mathbf{x}, \mathbf{y}) \,. \tag{3.4}$$

Abusing slightly the notation, from this point on we will write $d_T(\mathbf{x}, \mathbf{y})$ instead of $d_T([\mathbf{x}], [\mathbf{y}])$. The second advantage of such a choice is that, for each equivalent class $[\mathbf{x}]$, there exists a unique point $\bar{\mathbf{x}} \in [\mathbf{x}] \cap [-L/2, L/2)^3$ that serves as identifier of the whole class which, in practical terms, implies that all operations need to be performed as with standard points in a cubic box. This identifier can be found using a projection operator

$$\boldsymbol{\pi} : \mathbb{R}^3 \mapsto \mathcal{B} , \tag{3.5}$$

defined as

$$\bar{x}_j = \pi(\mathbf{x})_j = x_j - \mathtt{floor}\left(\frac{x_j + L/2}{L}\right) L , \tag{3.6}$$

where $x_j, j = 1, 2, 3$ denote the Cartesian coordinates of the point $\mathbf{x}$ and $\mathtt{floor} : \mathbb{R} \mapsto \mathbb{Z}$ is the function that gives the largest integer smaller than or equal to a given real number. See Fig. 3.1 for an illustration of the projection operator.

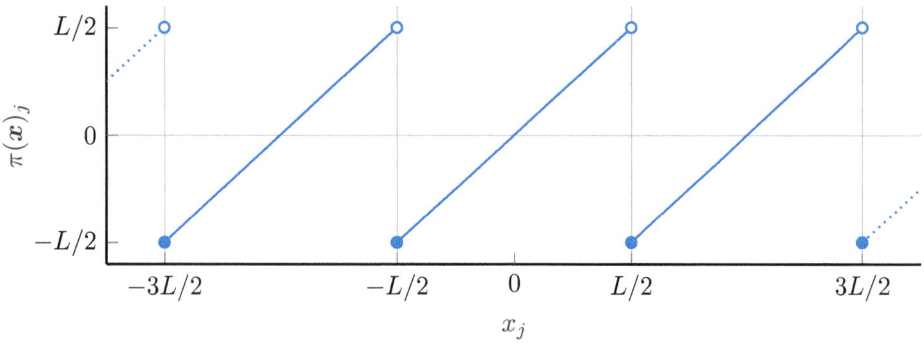

Figure 3.1.: Graph of the projection operator $\boldsymbol{\pi}$ restricted to one of the three coordinates of Euclidean space.

The projection map $\boldsymbol{\pi}$ determines some of the properties and limitations of the numerical methods employed on systems with periodic boundary conditions, and it is helpful to summarize some of its properties:

1. $\boldsymbol{\pi}$ is a nonlinear, surjective, projection, that is $\boldsymbol{\pi} \circ \boldsymbol{\pi} = \boldsymbol{\pi}$.
2. The point $\boldsymbol{\pi}(\mathbf{x})$ is the closest one to the origin among all points in $[\mathbf{x}]$, that is,

$$\boldsymbol{\pi}(\mathbf{x}) = \arg \inf_{\mathbf{x} \in [\mathbf{x}]} d(\mathbf{x}, \mathbf{0}) . \tag{3.7}$$

3. In general, for arbitrary $\mathbf{x}, \mathbf{y} \in \mathbb{R}^3$,

$$d_T(\mathbf{x}, \mathbf{y}) = |\boldsymbol{\pi}(\mathbf{x} - \mathbf{y})| \neq |\boldsymbol{\pi}(\mathbf{x}) - \boldsymbol{\pi}(\mathbf{y})| . \tag{3.8}$$

4. The map $\boldsymbol{\pi}$ is C^∞ except on the planes $x_i = L/2 + k_i L$, with $k_i \in \mathbb{Z}$ and $i = 1, 2, 3$, where it is discontinuous. Away from these planes, the gradient $\partial_{\mathbf{x}} \boldsymbol{\pi}$ is the identity $\mathbf{I} : \mathbb{R}^3 \mapsto \mathbb{R}^3$. A one-dimensional illustration of the gradient $\partial_{\mathbf{x}} \boldsymbol{\pi}$ is given in Fig. 3.2.

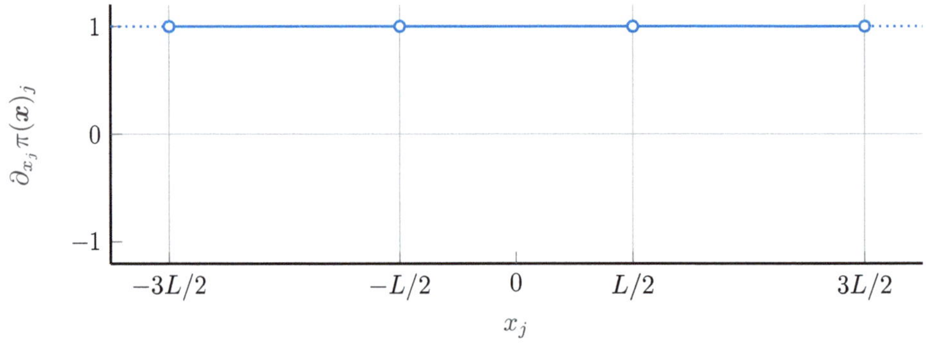

Figure 3.2.: Graph of the gradient of the projection operator $\partial_{\mathbf{x}} \boldsymbol{\pi}$ restricted to one of the three coordinates of Euclidean space.

5. For any $\mathbf{x} \in \mathbb{R}^3$

$$\boldsymbol{\pi}(\mathbf{x}) = -\boldsymbol{\pi}(-\mathbf{x}) . \tag{3.9}$$

6. For any $\mathbf{x}, \mathbf{y} \in \mathbb{R}^3$ and $\mathbf{a} \in \mathbb{R}^3$

$$d_T(\mathbf{x} + \mathbf{a}, \mathbf{y} + \mathbf{a}) = d_T(\mathbf{x}, \mathbf{y}) . \tag{3.10}$$

However, in general, if $\mathbf{Q} \in SO(3)$,

$$d_T(\mathbf{x}, \mathbf{y}) \neq d_T(\mathbf{Q}\mathbf{x}, \mathbf{Q}\mathbf{y}) . \tag{3.11}$$

Hence, in contrast with the Euclidean distance, the function $d_T(\cdot, \cdot)$ is not invariant under the action of the special orthogonal group.

In this section we provide the basic ingredients that describe the dynamics of particulate systems. This dynamical system is governed by Hamilton's equations of motion, and most of its complexity comes from the particle interactions, as given by the potential energy. Here we present a fairly general class of potentials that will be examined more carefully in Sections 3.3, 3.4, and 3.5.

3.2. Particle dynamics: basic description

3.2.1. System description

Let us consider a system of N particles labeled $a = 1, 2, \ldots, N$ moving inside a periodic box $\mathcal{B}$. Let the mass of the a-th particle be denoted as m^a, its position as $\mathbf{x}^a$, and its velocity as $\mathbf{v}^a = \dot{\mathbf{x}}^a$, where the dot indicates the derivative with respect to time.

A system of particles such as the one introduced possesses a kinetic energy defined by

$$T := \sum_{a=1}^{N} \frac{1}{2} m^a |\mathbf{v}^a|^2 . \tag{3.12}$$

Introducing the momentum $\mathbf{p}^a = m^a \mathbf{v}^a$ of particle a the kinetic energy reads

$$T = \sum_{a=1}^{N} \frac{1}{2} \frac{1}{m^a} |\mathbf{p}^a|^2 , \tag{3.13}$$

and the potential energy is given by

$$V = \hat{V}(\{\mathbf{x}^a\}_{a=1}^{N}) , \tag{3.14}$$

modeling the energetic interactions among all the particles. This potential energy is always a model of the true interatomic interaction and as such there exists a large number of simple *effective* potentials that have proven their value in different contexts (gases, fluids, organic molecules, metals, etc.).

3.2.2. Equations of motion

The dynamics of systems of particles may be expressed in the GENERIC format which is governed by

$$\frac{\mathrm{d}\mathbf{z}}{\mathrm{d}t} = \mathbf{L} \partial_{\mathbf{z}} \mathcal{E} , \tag{3.15}$$

where the total energy $\mathcal{E} = T + V$ is given as the sum of kinetic and potential energy. As no irreversible effects are present the GENERIC framework coincide with Hamilton's equation of motion (see Section 2.1.1). Eq. (3.15) gives rise to the following evolution equation for the state variables

$$\begin{aligned} \dot{\mathbf{x}}^a &= \frac{1}{m^a} \mathbf{p}^a , \\ \dot{\mathbf{p}}^a &= -\partial_{\mathbf{x}^a} \hat{V}(\{\mathbf{x}^b\}_{b=1}^{N}) . \end{aligned} \tag{3.16}$$

The gradient of the potential energy is related to the forces acting on the particles, and we define

$$\mathbf{f}^a := -\partial_{\mathbf{x}^a} \hat{V}(\{\mathbf{x}^b\}_{b=1}^N) , \tag{3.17}$$

to be the resultant of all forces applied on particle a.

These standard equations need to be carefully studied since the topology of $\mathcal{B}$ is not identical to that of Euclidean space and the notion of derivative has to be re-examined. For the moment being, let us assume that this object is well-defined, deferring until Section 3.3 a more detailed inspection.

3.2.3. Conserved quantities

Eqs. (3.16) describe the motion of systems of particles that often possess first integrals, that is, conserved quantities along their trajectories. These quantities are of great relevance to understand the qualitative dynamics of the system, to develop controls, etc. These *momentum maps* are related to the symmetries of the equations, according to Noether's theorem (see, e.g., [126]). We review them very briefly, since they have a direct impact in the formulation of conserving schemes.

We consider only potential energies with translational invariant, that is, functions $\hat{V}$ such that for every vector $\mathbf{c} \in \mathbb{R}^3$ satisfy

$$\hat{V}(\{\mathbf{x}^a\}_{a=1}^N) = \hat{V}(\{\mathbf{x}^a + \mathbf{c}\}_{a=1}^N) . \tag{3.18}$$

Differentiating both sides of this equation with respect to $\mathbf{c}$ and setting later $\mathbf{c} = \mathbf{0}$ we obtain the relation

$$\mathbf{0} = \sum_{a=1}^N \partial_{\mathbf{x}^a} \hat{V}(\{\mathbf{x}^b\}_{b=1}^N) = -\sum_{a=1}^N \mathbf{f}^a . \tag{3.19}$$

This invariance condition is related to the conservation of the linear momentum of the system, defined as

$$\mathbf{L} = \sum_{a=1}^N \mathbf{p}^a . \tag{3.20}$$

The time derivative of this quantity follows from the definition of particle momentum and eq. (3.19):

$$\dot{\mathbf{L}} = \sum_{a=1}^N \dot{\mathbf{p}}^a = \sum_{a=1}^N \mathbf{f}^a = \mathbf{0} . \tag{3.21}$$

Systems of particles moving in the Euclidean space $\mathbb{R}^n$, with $n = 2$ or 3, often conserve angular momentum. This is a consequence of the rotational invariance of the potential energy. However, systems defined on a periodic domain do not preserve it, in general (see, e.g., [127, 128]). One way to explain this loss of symmetry is by noting that the projection (3.6) is not rotationally invariant (see eq. (3.11)) and thus when used in the definition of the potential energy, it spoils the invariance of the whole system.

The total energy of the system is given as the sum of kinetic energy T and potential energy V. The time derivative of the energy can be evaluated using the equations of motion (3.16), giving

$$\dot{\mathcal{E}} = \sum_{a=1}^{N} m^a \mathbf{v}^a \cdot \dot{\mathbf{v}}^a - \sum_{a=1}^{N} \mathbf{f}^a \cdot \dot{\mathbf{x}}^a = \sum_{a=1}^{N} \mathbf{f}^a \cdot \mathbf{v}^a - \sum_{a=1}^{N} \mathbf{f}^a \cdot \dot{\mathbf{x}}^a = 0\,, \tag{3.22}$$

proving that the total energy must be a first integral of the motion. Given the relevance of the aforementioned conservation laws, numerical schemes have been proposed that attempt to preserve them. In particular, energy-momentum schemes, the ones under study in this Chapter, have been designed to integrate the equations of Hamiltonian systems while preserving both linear and angular momentum, in addition to the total energy. From the previous discussion, however, it follows that when dealing with periodic systems, one might focus on the preservation of linear momentum and energy, only.

3.2.4. The hierarchical definition of the potential energy

The potential energy of a system of particles is a function with the general form given in eq. (3.14) satisfying the invariance condition (3.18). A hierarchy of functions of growing complexity can be defined considering interactions involving an increasing number of particles. This is abstractly expressed as

$$V = V_0 + \sum_{\substack{a,b=1 \\ a\neq b}}^{N} V_2(\mathbf{x}_a, \mathbf{x}_b) + \sum_{\substack{a,b,c=1 \\ a\neq b\neq c}}^{N} V_3(\mathbf{x}_a, \mathbf{x}_b, \mathbf{x}_c) + \dots\,, \tag{3.23}$$

where V_k is a function involving k–tuples of atoms and satisfying eq. (3.18). The formulation of accurate potentials is an active field of research and we limit our exposition to the most common types. The reader may consult standard references for a detailed motivation and derivation of other types (e.g., [120]).

A convenient way of formulating potential functions that are translationally invariant is to include atomic interactions only via the distance between pairs of particles. In general, this would mean that the functions V_k employed in eq. (3.23) must be of the form

$$\begin{aligned} V_2(\mathbf{x}_a, \mathbf{x}_b) &= \tilde{V}_2(d(\mathbf{x}_a, \mathbf{x}_b))\,, \\ V_3(\mathbf{x}_a, \mathbf{x}_b, \mathbf{x}_c) &= \tilde{V}_3(d(\mathbf{x}_a, \mathbf{x}_b), d(\mathbf{x}_b, \mathbf{x}_c), d(\mathbf{x}_c, \mathbf{x}_a))\,, \end{aligned} \tag{3.24}$$

and similarly for higher order terms.

3.2.5. Dynamics in periodic domains

Eqs. (3.16) define the motion of particles in periodic domains, but special care has to be taken with the definition of the potential energy and its derivative.

With respect to the potential, we note that, when formulating the dynamics of particles in periodic domains, the distance function $d(\cdot, \cdot)$ in eq. (3.24) should be replaced with the distance $d_T(\cdot, \cdot)$ defined in eq. (3.4).

An aspect with important practical implications is that hierarchical potential functions of the form (3.23) are invariably defined employing a *cut-off* radius that effectively limits the number of particles that interact with those within that distance, in the sense of $d_T(\cdot, \cdot)$. Moreover, in order to avoid the singularities in the definition of the gradient of this distance, the cut-off radius is always chosen to be strictly smaller than $L/2$ (see Fig. 3.2). Equivalently, the dimension L of the periodic box must be selected larger than twice the cut-off radius. Under this condition, we observe that a collection of N particles in a periodic box $\mathcal{B}$ is a mathematical representation of an infinite domain consisting of boxes of dimension $L \times L \times L$ that repeat themselves in the three directions of space.

The formulation of energy and momentum conserving schemes in this kind of domains must take these two remarks into consideration, and we explore them in the following sections, starting from the simplest potential function possible.

In this section we study the formulation of energy and momentum conserving algorithms for systems of particles in periodic domains where the potential energy includes only pairwise interactions. In terms of practical applications, only the simplest potentials belong to this class (for example, Lennard-Jones'). They are only accurate for modeling noble gases, but are very often employed for benchmarking and the study of numerical methods.

3.3. Energy-momentum methods for periodic systems with pairwise interactions

3.3.1. Equations of motion

We consider again a system of N particles in a periodic box $\mathcal{B}$ of side L with equations of motion (3.16) and an effective potential

$$V = \frac{1}{2} \sum_{\substack{a,b=1 \\ b \neq a}}^{N} \tilde{V} \left(d_T(\mathbf{x}^a, \mathbf{x}^b) \right) , \tag{3.25}$$

where $\tilde{V} : \mathbb{R}^+ \cup \{0\} \mapsto \mathbb{R}$. Using the definitions

$$\bar{\mathbf{r}}^{ab} := \boldsymbol{\pi}(\mathbf{x}^b - \mathbf{x}^a) \qquad \text{and} \qquad \bar{r}^{ab} = d_T(\mathbf{x}^a, \mathbf{x}^b) := |\bar{\mathbf{r}}^{ab}| , \tag{3.26}$$

the forces (3.17) deriving from a pairwise potential can be written as

$$\mathbf{f}^a = \sum_{\substack{b=1 \\ a \neq b}}^{N} \mathbf{f}^{ab} , \qquad \text{with} \qquad \mathbf{f}^{ab} = \tilde{V}'(\bar{r}^{ab}) \frac{\bar{\mathbf{r}}^{ab}}{\bar{r}^{ab}} . \tag{3.27}$$

For the following sections we further define

$$\mathbf{r}^{ab} := \mathbf{x}^b - \mathbf{x}^a \qquad \text{and} \qquad r^{ab} = d(\mathbf{x}^a, \mathbf{x}^b) = |\mathbf{r}^{ab}| . \tag{3.28}$$

3.3.2. Time discretization

We consider now the integration in time of the equations of motion (3.16) of a system in the periodic box $\mathcal{B}$ and effective potential (3.25). To approximate their solution we will employ implicit time stepping schemes that partition the integration interval $[0, T]$ into disjoint subintervals $[t_n, t_{n+1}]$ with $t_n = n\Delta t$, and Δt being the time step size, assumed to be constant to simplify the notation. In the algorithms defined below, we will use the notation $\mathbf{x}_n$ to denote the approximation to $\mathbf{x}(t_n)$, and similarly for the velocity. Moreover, the symbol $f_{n+\alpha}$ will denote the convex combination $(1-\alpha)f_n + \alpha f_{n+1}$ for any variable f and $0 \leq \alpha \leq 1$.

3.3.2.1. Mid-point scheme

The canonical mid-point rule approximates the equations of motion (3.16) by the implicit formula

$$\begin{aligned} \frac{\mathbf{x}^a_{n+1} - \mathbf{x}^a_n}{\Delta t} &= \mathbf{v}^a_{n+1/2} \,, \\ m^a \frac{\mathbf{v}^a_{n+1} - \mathbf{v}^a_n}{\Delta t} &= \sum_{\substack{b=1 \\ a \neq b}}^{N} \mathbf{f}^{ab}_{\text{MP}} \,. \end{aligned} \tag{3.29}$$

Here we introduced $\mathbf{f}^{ab}_{\text{MP}}$, the mid-point approximation of the force acting on particle a due to the presence of particle b, that is,

$$\mathbf{f}^{ab}_{\text{MP}} := \tilde{V}'(\bar{r}^{ab}_{n+1/2}) \frac{\bar{\mathbf{r}}^{ab}_{n+1/2}}{|\bar{\mathbf{r}}^{ab}_{n+1/2}|} \,, \tag{3.30}$$

with

$$\begin{aligned} \bar{\mathbf{r}}^{ab}_{n+1/2} &= \boldsymbol{\pi}(\mathbf{x}^b_{n+1/2} - \mathbf{x}^a_{n+1/2}) \,, \\ \bar{r}^{ab}_{n+1/2} &= d_T(\mathbf{x}^a_{n+1/2}, \mathbf{x}^b_{n+1/2}) \,. \end{aligned} \tag{3.31}$$

As in the continuous case, the condition

$$\mathbf{f}^{ab}_{\text{MP}} = -\mathbf{f}^{ba}_{\text{MP}} \,, \tag{3.32}$$

holds due to eq. (3.9). The properties of the mid-point rule are well known. For example, this method preserves the total liner momentum of the system, defined at an instant t_n as

$$\mathbf{L}_n = \sum_{a=1}^{N} m^a \mathbf{v}^a_n \,. \tag{3.33}$$

To prove this property it suffices to verify

$$\frac{\mathbf{L}_{n+1} - \mathbf{L}_n}{\Delta t} = \sum_{a=1}^{N} m^a \frac{\mathbf{v}^a_{n+1} - \mathbf{v}^a_n}{\Delta t} = \sum_{\substack{a,b=1 \\ a \neq b}}^{N} \mathbf{f}^{ab}_{\text{MP}} = \mathbf{0} \,, \tag{3.34}$$

where we have employed eq. (3.32).

3.3.2.2. Energy and momentum conserving discretization

It is possible to construct a perturbation of the mid-point rule that, in addition to preserving the linear momentum of the system, preserves its total energy. The key ingredient of such methods is the so called *discrete gradient* operator, an approximation to the gradient that guarantees the strict conservation of energy and momentum along the discrete trajectories generated by the integrator.

Conserving integrators for problems in molecular dynamics have been studied since the 1970's [40, 41, 46], although never for systems with periodic boundary conditions, with the exception of [112]. In all of these works, the conserving schemes are variations of the mid-point rule (3.29) of the form

$$\begin{aligned} \frac{\mathbf{x}^a_{n+1} - \mathbf{x}^a_n}{\Delta t} &= \mathbf{v}^a_{n+1/2} \,, \\ m^a \frac{\mathbf{v}^a_{n+1} - \mathbf{v}^a_n}{\Delta t} &= -\mathsf{D}_{\mathbf{x}^a} V, \end{aligned} \tag{3.35}$$

where $\mathsf{D}_{\mathbf{x}^a}$ is precisely the discrete gradient operator, an algorithmic approximation to the derivative $\partial_{\mathbf{x}^a}$. In analogy to expression (3.30), the discrete gradient defines a force contribution, to be specified later, such that

$$\mathsf{D}_{\mathbf{x}^a} V = -\sum_{\substack{b=1 \\ a \neq b}}^{N} \mathbf{f}^{ab}_{\text{algo}} \,. \tag{3.36}$$

If the following condition holds

$$\mathbf{f}^{ab}_{\text{algo}} = \mathbf{f}^{ab}_{\text{MP}} + \mathcal{O}(\Delta t^2) \,, \tag{3.37}$$

the second order accuracy of the mid-point rule will be preserved. If we want the new method to preserve linear momentum we note from the previous section that it suffices that the pairwise forces $\mathbf{f}^{ab}_{\text{algo}}$ mimic the symmetry condition (3.32), that is,

$$\mathbf{f}^{ab}_{\text{algo}} = -\mathbf{f}^{ba}_{\text{algo}} \,. \tag{3.38}$$

Any second order perturbation of $\mathbf{f}^{ab}_{\text{MP}}$ with this property will result in a second order accurate integrator that preserves linear momentum. The "classical" EM method is constructed in such a way, and preserves, in addition to the total energy, the linear and angular momenta of the system, the latter being important in domains without periodic boundary conditions [46]. For problems in molecular dynamics interacting through pair potentials and posed on periodic domains, the "classical" EM method is based on the discrete gradient (3.36) with

$$\mathbf{f}^{ab}_{\text{algo}} = \mathbf{f}^{ab}_{\text{algo}}(\mathbf{r}^{ab}_n, \mathbf{r}^{ab}_{n+1}) := \frac{\tilde{V}^{ab}_{n+1} - \tilde{V}^{ab}_n}{\bar{r}^{ab}_{n+1} - \bar{r}^{ab}_n} \frac{\bar{\mathbf{r}}^{ab}_{n+1} + \bar{\mathbf{r}}^{ab}_n}{\bar{r}^{ab}_{n+1} + \bar{r}^{ab}_n} \,, \tag{3.39}$$

where we have introduced the notation

$$\tilde{V}^{ab}_n = \tilde{V}(\bar{r}^{ab}_n) \,, \tag{3.40}$$

and where the appropriate limit must be taken in eq. (3.39) when $|\bar{r}^{ab}_{n+1} - \bar{r}^{ab}_{n}| \mapsto 0$. This form of the discrete gradient is frequently cited in the literature of integration algorithms (see, e.g., [112]) and is responsible for the conservation properties of the method, also in periodic domains. However, it is not the only possible form and, in fact, it can be shown that there are an infinite number of discrete gradients [129], some of which can be more easily extended to the periodic case. More precisely, a second class of EM schemes follows from a new definition of the algorithmic approximation of the pairwise forces:

$$\mathbf{f}^{ab}_{\text{algo}}(\mathbf{r}^{ab}_{n}, \mathbf{r}^{ab}_{n+1}) := \mathbf{f}^{ab}_{\text{MP}} + \frac{\tilde{V}^{ab}_{n+1} - \tilde{V}^{ab}_{n} + \mathbf{f}^{ab}_{\text{MP}} \cdot (\mathbf{r}^{ab}_{n+1} - \mathbf{r}^{ab}_{n})}{|\mathbf{r}^{ab}_{n+1} - \mathbf{r}^{ab}_{n}|}\, \mathbf{n} \,, \tag{3.41}$$

where $\mathbf{n}$ is the normalized direction given by

$$\mathbf{n} = \frac{\mathbf{r}^{ab}_{n+1} - \mathbf{r}^{ab}_{n}}{|\mathbf{r}^{ab}_{n+1} - \mathbf{r}^{ab}_{n}|} \,. \tag{3.42}$$

The expression (3.41) results from perturbing the force $\boldsymbol{f}^{ab}_{\text{MP}}$ in the direction that produces work and then imposing precisely that this work should coincide with the difference $\tilde{V}^{ab}_{n+1} - \tilde{V}^{ab}_{n}$, an idea that has been previously employed in the literature in order to formulate EM methods [129]. The proposed definition corrects the pairwise force between the particles a and b with a "small" term in the direction $\mathbf{n}$ depending on *unprojected* relative positions. This is the result of the fact that the first equation in (3.35) is not posed in the quotient space but rather in the full $\mathbb{R}^3$. It might be argued that such a correction is nonphysical because it is not defined on the quotient space, where the problem is posed. While this is true, the velocity equation is not posed on this space from the outset, and the proposed correction results from this mismatch.

The EM force given in eq. (3.41) is symmetric in a and b. Hence, the method defined by (3.35) and (3.41) preserves linear momentum. To show that the method indeed preserves exactly the total energy exactly, it suffices to take the dot product of the $(3.35)_1$ with the left hand side of $(3.35)_2$, and vice versa, and then add the result over all particles, that is,

$$\begin{aligned}
\sum_{a=1}^{N} m^a \frac{\mathbf{x}^a_{n+1} - \mathbf{x}^a_n}{\Delta t} \cdot \frac{\mathbf{v}^a_{n+1} - \mathbf{v}^a_n}{\Delta t} &= \sum_{a=1}^{N} m^a \mathbf{v}^a_{n+1/2} \cdot \frac{\mathbf{v}^a_{n+1} - \mathbf{v}^a_n}{\Delta t} \\
\sum_{a=1}^{N} m^a \frac{\mathbf{v}^a_{n+1} - \mathbf{v}^a_n}{\Delta t} \cdot \frac{\mathbf{x}^a_{n+1} - \mathbf{x}^a_n}{\Delta t} &= \sum_{\substack{a,b=1 \\ a\neq b}}^{N} \mathbf{f}^{ab}_{\text{algo}} \cdot \frac{\mathbf{x}^a_{n+1} - \mathbf{x}^a_n}{\Delta t} \,.
\end{aligned} \tag{3.43}$$

Subtracting $(3.43)_2$ from $(3.43)_1$ gives

$$\frac{T_{n+1} - T_n}{\Delta t} - \sum_{\substack{a,b=1 \\ a\neq b}}^{N} \mathbf{f}^{ab}_{\text{algo}} \cdot \frac{\mathbf{x}^a_{n+1} - \mathbf{x}^a_n}{\Delta t} = 0 \,, \tag{3.44}$$

where the total discrete kinetic energy at time t_n is given by

$$T_n = \sum_{a=1}^{N} \frac{1}{2} m^a \mathbf{v}_n^a \cdot \mathbf{v}_n^a \,. \tag{3.45}$$

In the spirit of eq. (3.41), further rewriting results in

$$\begin{aligned} \sum_{\substack{a,b=1\\a\neq b}}^{N} \mathbf{f}_{\text{algo}}^{ab} \cdot \frac{\mathbf{x}_{n+1}^a - \mathbf{x}_n^a}{\Delta t} &= \frac{1}{\Delta t} \sum_{\substack{a,b=1\\a<b}}^{N} \mathbf{f}_{\text{algo}}^{ab} \cdot (\mathbf{x}_{n+1}^a - \mathbf{x}_n^a - \mathbf{x}_{n+1}^b + \mathbf{x}_n^b) \\ &= -\frac{1}{\Delta t} \sum_{\substack{a,b=1\\a<b}}^{N} \mathbf{f}_{\text{algo}}^{ab} \cdot (\mathbf{r}_{n+1}^{ab} - \mathbf{r}_n^{ab}) \,. \end{aligned} \tag{3.46}$$

Therefore, a necessary and sufficient condition for energy conservation is that the following directionality condition is satisfied

$$\sum_{\substack{a,b=1\\a<b}}^{N} \mathbf{f}_{\text{algo}}^{ab} \cdot (\mathbf{r}_{n+1}^{ab} - \mathbf{r}_n^{ab}) = \frac{1}{2} \sum_{\substack{a,b=1\\a\neq b}}^{N} \left(\tilde{V}(\bar{r}_{n+1}^{ab}) - \tilde{V}(\bar{r}_n^{ab}) \right) \,. \tag{3.47}$$

It is important to emphasize that the proof is based on the inner product between the algorithmic EM force and the *unprojected* relative position vectors. Finally, to show that the EM scheme (3.41) is indeed a second-order accurate method, it suffices to prove that the correction term in the definition (3.41) is of size $\mathcal{O}(\Delta t^2)$. Making use of the relation

$$\mathbf{f}_{\text{MP}}^{ab} = -\partial_{\mathbf{x}^a} \tilde{V}(|\bar{\mathbf{r}}_{n+1/2}^{ab}|) = \partial_{\mathbf{r}^{ab}} \tilde{V}(|\bar{\mathbf{r}}_{n+1/2}^{ab}|) \,, \tag{3.48}$$

a Taylor series expansion around the point $\mathbf{r}_{n+1/2}^{ab}$ gives

$$\tilde{V}_{n+1}^{ab} - \tilde{V}_n^{ab} = \mathbf{f}_{\text{MP}}^{ab} \cdot (\mathbf{r}_{n+1}^{ab} - \mathbf{r}_n^{ab}) + \mathcal{O}(\Delta t^3) \,. \tag{3.49}$$

Then, since the direction vector of the correction has size

$$\frac{\mathbf{r}_{n+1}^{ab} - \mathbf{r}_n^{ab}}{|\mathbf{r}_{n+1}^{ab} - \mathbf{r}_n^{ab}|} = \mathcal{O}(1) \,, \tag{3.50}$$

and $|\mathbf{r}_{n+1}^{ab} - \mathbf{r}_n^{ab}|$ is $\mathcal{O}(\Delta t)$, we conclude that the correction term is indeed $\mathcal{O}(\Delta t^2)$.

3.3.2.3. Time reversibility

Time-reversible (or symmetric) integration schemes are often favored for the approximation of Hamiltonian systems for two main reasons. First, the Hamiltonian flow itself is symmetric, so it is desirable that its numerical approximation also possesses this property. Second, symmetric numerical schemes are known to have several favorable properties [7], especially in long-term simulations. The class of EM integration schemes defined in this section have also this property. This is a direct consequence of the time reversibility of the algorithmic approximation of the pairwise forces, namely,

$$\mathbf{f}^{ab}_{\text{algo}}(\mathbf{r}^{ab}_{n}, \mathbf{r}^{ab}_{n+1}) = \mathbf{f}^{ab}_{\text{algo}}(\mathbf{r}^{ab}_{n+1}, \mathbf{r}^{ab}_{n}) \,, \tag{3.51}$$

that is trivially satisfied by both (3.39) and (3.41).

3.3.3. Interatomic potential

For the following numerical examples we consider the well-known Lennard-Jones potential [121] with $r = \bar{r}^{ab}$, that is,

$$\tilde{V}(r) = 4\varepsilon \left[\left(\frac{\sigma}{r}\right)^{12} - \left(\frac{\sigma}{r}\right)^{6} \right] \,, \tag{3.52}$$

where ε and σ are material constants.

3.3.3.1. Cut-off distance considerations

In the Lennard-Jones potential, atomic interactions between distant particles are negligible. For this reason, a cut-off distance r_c is often introduced beyond which the interaction is completely ignored (see, e.g., [108, 110]). However, simply trimming the Lennard-Jones potential beyond the cut-off distance leads to a discontinuity in this function at $r = r_c$ that might affect the properties of the integration scheme. Since the derivative of the potential enters the equations of motion (3.16), this discontinuity precludes the computation of the interatomic force at $r = r_c$. The discontinuity can be avoided, first, by shifting the potential function by the amount $V(r_c)$, leading to the shifted potential (SP)

$$\tilde{V}_{\text{SP}}(r) = \begin{cases} V(r) - V(r_c) & \text{if } r < r_c \,, \\ 0 & \text{if } r \geq r_c \,. \end{cases} \tag{3.53}$$

The derivative of this function at $r = r_c$ is still not defined and neither is the force. To resolve this physical inconsistency one can introduce a shifted and linearly truncated potential (SF), which is equivalent to a shift in the force (see, e.g., [110, 128]), and given by

$$\tilde{V}_{\text{SF}}(r) = \begin{cases} V(r) - V(r_c) - (r - r_c)V'(r_c) & \text{if } r < r_c \,, \\ 0 & \text{if } r \geq r_c \,. \end{cases} \tag{3.54}$$

One can further introduce a quadratic correction term that yields the shifted and quadratically truncated potential (STF), which is equivalent to a shift and a linear truncation in the force,

$$\tilde{V}_{\text{STF}}(r) = \begin{cases} V(r) - V(r_c) - (r - r_c)V'(r_c) - \frac{1}{2}(r - r_c)^2 V''(r_c) & \text{if } r < r_c\,, \\ 0 & \text{if } r \geq r_c\,. \end{cases} \tag{3.55}$$

This potential is twice differentiable. Due to its higher smoothness, it is better suited for structure-preserving schemes than the standard potential since it eliminates numerical oscillations in the energy evolution that sometimes appear when employing non-smooth potentials. As mentioned in Section 3.2.5, it is important to stress out that the cut-off distance must not be greater than $L/2$ for consistency with the minimum image convention. Due to the quadratic term in the corrected potential, a cut-off radius of $r_c = 5\sigma$ is suggested.

3.3.4. Numerical evaluation

All numerical examples are based on a set of dimensionless units. The implementation of periodic boundary conditions follows the details of Section 3.1 (see also [108, 110, 111, 113]), and the order of magnitude of the time step sizes has been selected using standard criteria from molecular dynamics simulations (see, e.g., [130] and references therein.)

3.3.4.1. Accuracy study

In the first numerical example, we consider a two-dimensional box $[-L/2, L/2]^2$ with two particles. The initial positions and velocities are given, respectively, by

$$\begin{aligned} \mathbf{x}^1 &= (0,0)^{\mathrm{T}}\,, & \mathbf{v}^1 &= (0,0)^{\mathrm{T}}\,, \\ \mathbf{x}^2 &= (1.9,1)^{\mathrm{T}}\,, & \mathbf{v}^2 &= (5,0)^{\mathrm{T}}\,, \end{aligned}$$

where the first particle is constrained to remain on the center of the box by holding fixed its degrees of freedom and only the second particle is allowed to move freely. For this simulation we used the Lennard-Jones potential with a simple spherical cut-off distance of $r_c = 2.5\sigma$. The numerical values of the remaining parameters of the simulation can be found in Table 3.1.

Material parameters	ε	20	Final configuration and trajectory
	σ	1	
Mass	m^a	0.06	
Side length	L	12	
Newton tolerance	-	10^{-6}	
Simulation duration	T	0.8	
Reference time step size	Δt_{ref}	0.0005	
Time step size	Δt	0.004,0.005,0.00625 0.008,0.1,0.125,0.16 0.02,0.025	

Table 3.1.: Accuracy study: Data used in the simulation

Next we compare the "classical" EM (3.39), the newly defined EM (3.41), and the mid-point rule. For that, we compare the errors in the position and the linear momentum made by these three methods when using increasingly smaller time steps. In the two analyses, we use as reference the mid-point rule solution and define the errors, respectively, as

$$\mathrm{e}_x^{(\mathrm{MP})} = \frac{||\mathbf{x}^a - \mathbf{x}_r^a||_2}{||\mathbf{x}_r^a||_2}, \quad \mathrm{e}_{\mathrm{p}}^{(\mathrm{MP})} = \frac{||\mathbf{p}^a - \mathbf{p}_r^a||_2}{||\mathbf{p}_r^a||_2}, \tag{3.56}$$

where $\square_r^a$, with $\square \in \{\mathbf{x}, \mathbf{p}\}$, is the solution at time T calculated with the mid-point rule using the reference time step size Δt_{ref} and where $\square^a$ is the solution of the considered scheme at time T for each time step size Δt. Fig. 3.3 confirm that all the schemes under consideration are second order accurate.

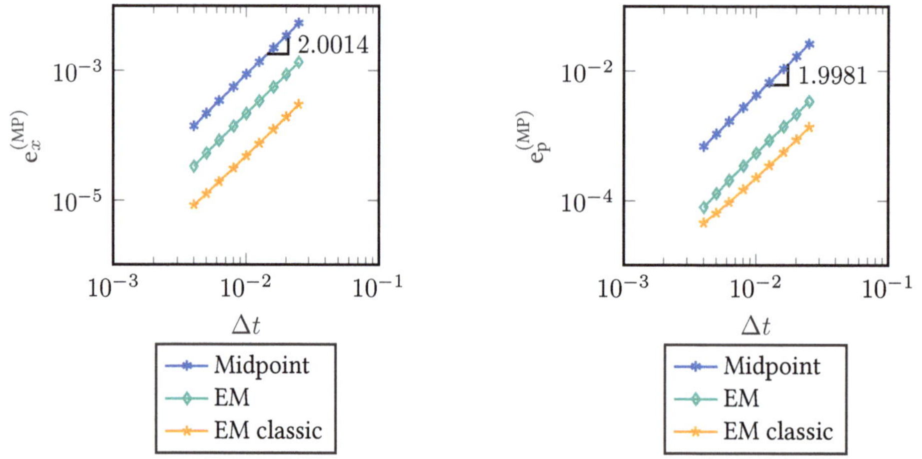

Figure 3.3.: Accuracy study: Relative error in the position w.r.t. mid-point rule (left), Relative error in the linear momentum w.r.t. mid-point rule (right)

3.3.4.2. Energy consistency study

The second numerical example investigates the energy conservation properties of the integrators described in Section 3.3. We consider 150 arbitrarily positioned particles inside a three-dimensional periodic box $[-L/2, L/2]^3$ such that the initial distance between the particles is greater than $2^{1/6}\sigma$. Starting from rest, the kinetic energy of the system will rise until the system is in equilibrium due to the random order of the particles. For this simulation, we consider the shifted and quadratically truncated Lennard-Jones potential (STF) with a spherical cut-off distance of $r_c = 5\sigma$. Numerical values for the parameters of the simulation can be found in Table 3.2.

Material parameters	ε	2	Initial configuration
	σ	1	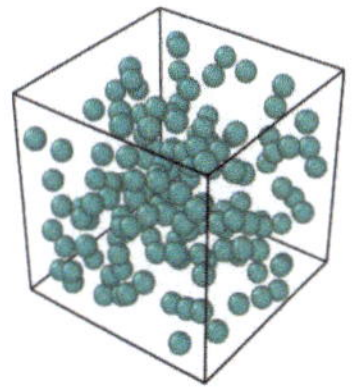
Mass	m^a	1	
Sidelength	L	12	
Newton tolerance	-	10^{-9}	
Simulation duration	T	40	
Time steps	Δt	0.08	

Table 3.2.: Energy consistency study: Data used in the simulation

For this relatively large time step size, the mid-point rule introduces energy into the system leading to an energy blow-up and eventually to a termination of the simulation, indicated with a vertical line in Fig. 3.4 (left).

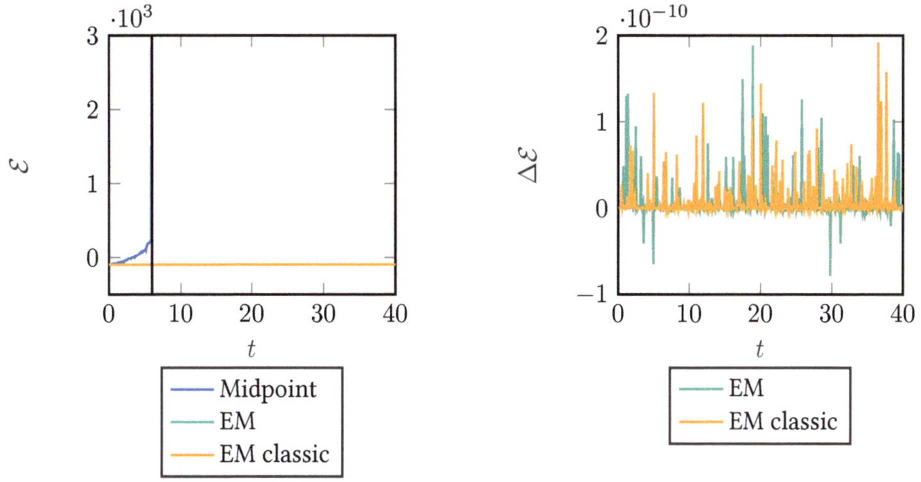

Figure 3.4.: Energy consistency study: Total energy (left), Total energy difference (right)

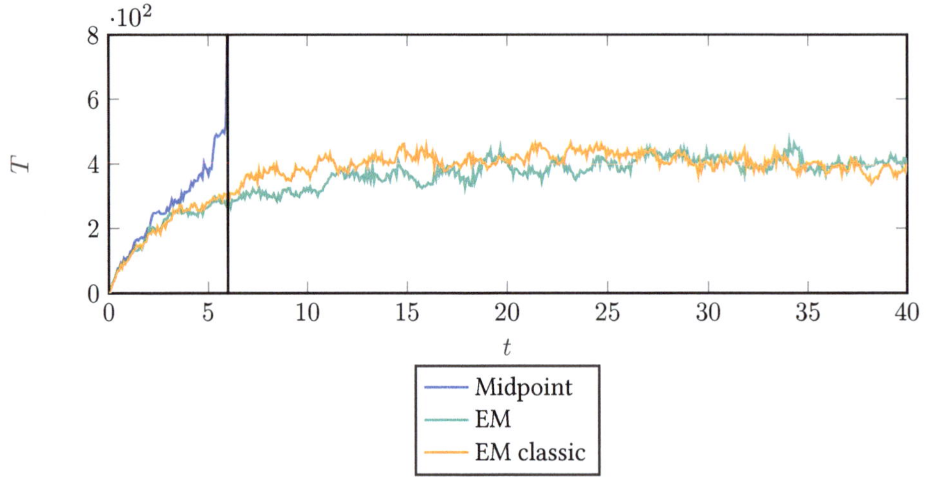

Figure 3.5.: Energy consistency study: Kinetic energy

Both EM methods described, on the other hand, preserve the total energy up to machine precision for the whole duration of the simulation, see Fig. 3.4 (right). Hence, as long as the nonlinear iterative solver finds the step update, the solution of the EM will never blow up. Of course, if the time step size is fixed to value that is too large, the iterative solver will fail to solve the update equations and the method will not be able to proceed. Fig. 3.5 shows the evolution of the kinetic energy throughout the simulation. Since this energy is proportional to the temperature in the system, the figure reveals that only the EM methods can compute the evolution of the system until it reaches equilibrium, for the chosen time step size.

We compare the newly proposed EM method with an explicit integrator, namely, the symplectic velocity-Verlet scheme provided by the molecular dynamics code LAMMPS[2] [131]. For this example we use the shifted and linearly truncated potential (SF) instead of the shifted and quadratically truncated potential (STF), since the former is available in the software package. Moreover, we extended the simulation duration to $T = 80$. Keeping the time step size of the EM method $\Delta t_{\mathrm{EM}} = 0.08$ constant, the time step size of the velocity-Verlet integrator Δt_{vV} is reduced until the energy fluctuations become sufficiently small. As Fig. 3.6 reveals, the time step size employed for the explicit method is $\Delta t_{\mathrm{vV}} = 0.0008$, 100 times smaller than the time step size of the EM method. This is the time step size required to keep the relative energy fluctuations in the explicit solution approximately below $0.1‰$.

[2] https://lammps.sandia.gov

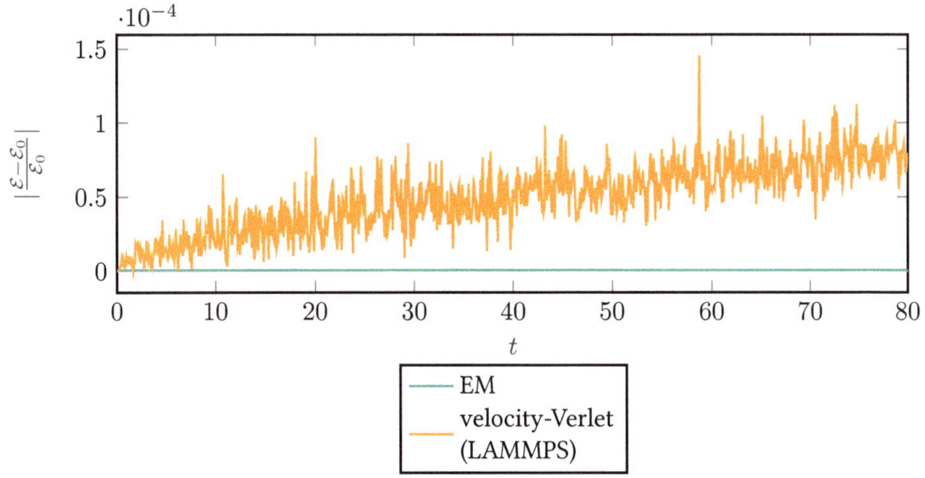

Figure 3.6.: Energy consistency study: Relative energy drift for $\Delta t_{\mathrm{EM}} = 100\Delta t_{\mathrm{vV}}$

Regarding the computational cost per step, the proposed implicit integrator is naturally more expensive than the explicit scheme, but comparable to that of other second-order implicit methods like the mid-point rule.

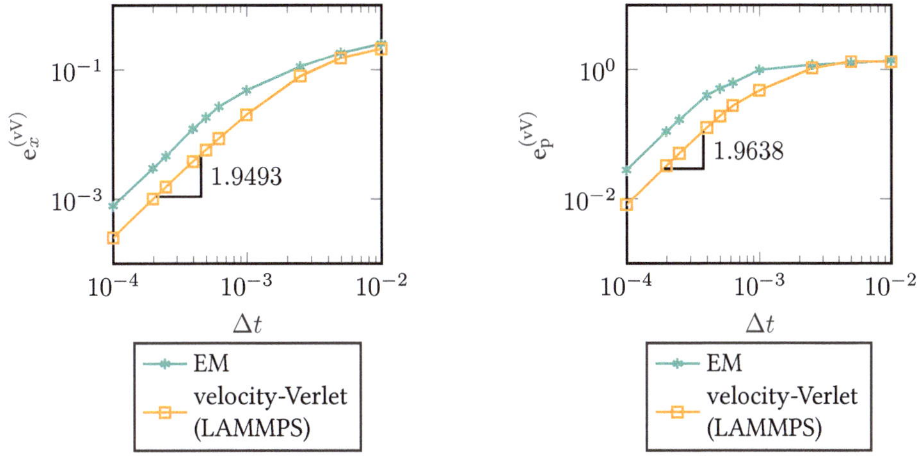

Figure 3.7.: Energy consistency study: Relative error in the position w.r.t. velocity-Verlet scheme (left), Relative error in the momentum w.r.t. velocity-Verlet scheme (right)

In addition to the previous investigation, we perform an accuracy comparison between the EM and the velocity-Verlet methods. We use again the system described in Table 3.2 and obtain a reference solution with the explicit method and a small time step $\Delta t_{\mathrm{ref}} = 0.000001$.

The displacement and momentum error measures of eq. (3.56) will be used to compare the accuracy of the two integrators for the time step sizes $\Delta t = \{0.0001, 0.0002, 0.00025, 0.0004, 0.0005, 0.00625, 0.001, 0.0025, 0.004, 0.005, 0.01\}$ and a final simulation time $T = 4$. As shown in Fig. 3.7, both schemes are second order accurate in the position and the linear momentum.

3.4. Energy-momentum methods for periodic systems with three-body interactions

In Section 3.3 we derived the equation of motion of a system of N particles moving inside a period box $\mathcal{B}$ where the interactions were based on pair potentials. For materials with strong covalent-bonding character, however, we further need to incorporate a bond-angle dependency to the effective potential. This can be archived by including three-body terms in the expression of the potential.

3.4.1. Equation of motion

We consider now a system of N particles in a periodic box $\mathcal{B}$ of side L with equations of motion (3.16) and an effective potential that depends only on the interactions among all triplets of particles. Such a potential must be of the form

$$V = \frac{1}{3!} \sum_{\substack{a,b,c=1 \\ a \neq b \neq c}}^{N} \tilde{V}\left(d_T(\mathbf{x}^a, \mathbf{x}^b), d_T(\mathbf{x}^a, \mathbf{x}^c), d_T(\mathbf{x}^b, \mathbf{x}^c)\right) , \tag{3.57}$$

where $\tilde{V} : \mathbb{R}^+ \cup \{0\} \times \mathbb{R}^+ \cup \{0\} \times \mathbb{R}^+ \cup \{0\} \mapsto \mathbb{R}$ is a three-body potential between the a-th, b-th, and the c-th particle. The total kinetic energy of the system is given by eq. (3.13). The forces acting on the particles defined by eq. (3.17) can be written using the definitions in eq. (3.26) as

$$\mathbf{f}^a = \sum_{\substack{b=1 \\ a \neq b}}^{N} \mathbf{f}^{ab} \qquad \text{with} \qquad \mathbf{f}^{ab} := \varphi^{ab} \frac{\bar{\mathbf{r}}^{ab}}{\bar{r}^{ab}} . \tag{3.58}$$

The strength of the force is now obtained by

$$\varphi^{ab} = \sum_{\substack{c=1 \\ a \neq b \neq c}}^{N} \partial_{\bar{r}^{ab}} \tilde{V}(\bar{r}^{ab}, \bar{r}^{ac}, \bar{r}^{bc}) . \tag{3.59}$$

3.4.2. Time discretization

Next, we consider the integration in time of the equations of motion (3.16) of a system in the periodic box $\mathcal{B}$ and effective potential (3.57) and employ the time integration strategy outlined in Section 3.3.2.

3.4.2.1. Mid-point scheme

The canonical mid-point rule approximates the equation of motion by the implicit formula (3.29), where the mid-point approximation of the force acting on the a-th particle in the direction of the b-th particle is given by

$$\begin{aligned} \mathbf{f}^{ab}_{\text{MP}} &= \varphi^{ab}_{\text{MP}} \frac{\bar{\mathbf{r}}^{ab}_{n+1/2}}{|\bar{\mathbf{r}}^{ab}_{n+1/2}|} \,, \\ \varphi^{ab}_{\text{MP}} &= \sum_{\substack{c=1 \\ a\neq b\neq c}}^{N} \partial_{\bar{r}^{ab}_{n+1/2}} \tilde{V}\left(\bar{r}^{ab}_{n+1/2}, \bar{r}^{ac}_{n+1/2}, \bar{r}^{bc}_{n+1/2}\right) \,. \end{aligned} \tag{3.60}$$

As in the continuous case, the weak law of action and reaction is satisfied and therefore the approximation preserves the total linear momentum of the system, see Section 3.3.2.1.

3.4.2.2. Energy and momentum conserving discretization

As in Sections 3.3.2.2, it is possible to construct a perturbation of the mid-point rule which, in addition to preserving the total linear momentum, preserves the total energy. Instead of the discrete gradient operator the *partitioned discrete gradient* operator [46] will now be employed, which is an approximation similar to the discrete gradient but is applicable to functions with more than one independent variable. To this end, let us rewrite the potential energy function (3.57) in terms of $N(N-1)/2$ independent variables, e.g.,

$$\tilde{V} = \tilde{V}(\bar{r}^{12}, \bar{r}^{13}, \ldots, \bar{r}^{1N}, \bar{r}^{23}, \ldots, \bar{r}^{N-1,N}) = \tilde{V}(\{\bar{r}^{ab}\}) \,, \tag{3.61}$$

where the double indexed set is given by

$$\{\bar{r}^{ab}\} = \{\bar{r}^{ab} \mid a, b \in (1, \ldots, N),\ a < b\} \,. \tag{3.62}$$

To simplify the definition of the partitioned discrete gradient, it proves useful to re-label the interatomic distances $\bar{r}^{ab}$ using only their position in the array $\{\bar{r}^{ab}\}$. Therefore, a single indexed set is defined by

$$\{\bar{r}^{\alpha}\} = \{\bar{r}^{\alpha} \mid \alpha \in (1, \ldots, N(N-1)/2)\} \,. \tag{3.63}$$

Note that here an ordering of the $\binom{N}{2}$ pairs (a, b) has been established. For example, the map $(a, b) \mapsto \alpha$ could be chosen to be lexicographic (e.g., [132, pg. 43]). Then, the potential energy can be expressed, abusing slightly the notation, as

$$V = \tilde{V}(\{\bar{r}^\alpha\}) . \tag{3.64}$$

For a potential function like this one, the discrete gradient operator is defined as

$$\mathsf{D}_{\mathbf{x}^a} V = -\sum_{\substack{b=1 \\ a\neq b}}^{N} \mathbf{f}^{ab}_{\text{algo}} = -\sum_{\substack{b=1 \\ a\neq b}}^{N} \frac{1}{2}\left(\mathbf{f}^{ab}_{n,n+1} + \mathbf{f}^{ab}_{n+1,n}\right) , \tag{3.65}$$

with the contributions

$$\begin{aligned} \mathbf{f}^{ab}_{n,n+1} &= \mathbf{f}^{ab}_{\text{MP}} + \frac{\tilde{V}^\alpha_{n,n+1}(\bar{r}^\alpha_{n+1}) - \tilde{V}^\alpha_{n,n+1}(\bar{r}^\alpha_{n}) - \mathbf{f}^{ab}_{\text{MP}} \cdot (\mathbf{r}^{ab}_{n+1} - \mathbf{r}^{ab}_{n})}{|\mathbf{r}^{ab}_{n+1} - \mathbf{r}^{ab}_{n}|} \mathbf{n} , \\ \mathbf{f}^{ab}_{n+1,n} &= \mathbf{f}^{ab}_{\text{MP}} + \frac{\tilde{V}^\alpha_{n+1,n}(\bar{r}^\alpha_{n+1}) - \tilde{V}^\alpha_{n+1,n}(\bar{r}^\alpha_{n}) - \mathbf{f}^{ab}_{\text{MP}} \cdot (\mathbf{r}^{ab}_{n+1} - \mathbf{r}^{ab}_{n})}{|\mathbf{r}^{ab}_{n+1} - \mathbf{r}^{ab}_{n}|} \mathbf{n} , \end{aligned} \tag{3.66}$$

for which we introduced the compact notation

$$\begin{aligned} \tilde{V}^\alpha_{n,n+1}(\bar{r}^\alpha) &= \tilde{V}(\bar{r}^1_n, \ldots, \bar{r}^{\alpha-1}_n, \bar{r}^\alpha, \bar{r}^{\alpha+1}_{n+1}, \ldots, \bar{r}^{N(N-1)/2}_{n+1}) , \\ \tilde{V}^\alpha_{n+1,n}(\bar{r}^\alpha) &= \tilde{V}(\bar{r}^1_{n+1}, \ldots, \bar{r}^{\alpha-1}_{n+1}, \bar{r}^\alpha, \bar{r}^{\alpha+1}_n, \ldots, \bar{r}^{N(N-1)/2}_n) . \end{aligned} \tag{3.67}$$

To show that the proposed integrator exactly preserves the total linear momentum, it suffices to follow the proof outlined in Section 3.3.2.2 and details are omitted. Similarly, to prove the energy conservation property, it is sufficient to show that

$$\sum_{\substack{a,b=1 \\ a<b}}^{N} \mathbf{f}^{ab}_{\text{algo}} \cdot (\mathbf{r}^{ab}_{n+1} - \mathbf{r}^{ab}_{n}) = \frac{1}{3!} \sum_{\substack{a,b,c=1 \\ a\neq b\neq c}}^{N} \left(\tilde{V}(\bar{r}^{ab}_{n+1}, \bar{r}^{ac}_{n+1}, \bar{r}^{bc}_{n+1}) - \tilde{V}(\bar{r}^{ab}_{n}, \bar{r}^{ac}_{n}, \bar{r}^{bc}_{n})\right) . \tag{3.68}$$

A straightforward manipulation shows that the proposed method with algorithmic forces given by eq. (3.65) satisfies this condition.

Remark 1. Many three-body potentials are expressed in terms of the bond angles $\bar{\theta}^{bac}$ at particle a, between the bonds ab and ac. Since the angle itself can be written in terms of the distances, that is,

$$\bar{\theta}^{bac} = g(\bar{r}^{ab}, \bar{r}^{ac}, \bar{r}^{bc}) = \arccos\left(\frac{(\bar{r}^{ab})^2 + (\bar{r}^{ac})^2 - (\bar{r}^{bc})^2}{2(\bar{r}^{ab})(\bar{r}^{ac})}\right) , \tag{3.69}$$

the composition $\tilde{V} \circ g$ has again the structure of the potential (3.61) and thus the partitioned discrete gradient operator defined before can be employed without modifications.

Remark 2. In problems without periodic boundary conditions, using eq. (3.28), the EM method reads

$$\mathsf{D}_{\mathbf{x}^a} V = -\sum_{\substack{b=1\\a\neq b}}^{N} \mathbf{f}^{ab}_{\text{algo}} = -\sum_{\substack{b=1\\a\neq b}}^{N} \frac{1}{2}\left(\mathbf{f}^{ab}_{n,n+1} + \mathbf{f}^{ab}_{n+1,n}\right) , \tag{3.70}$$

with the contributions

$$\begin{aligned} \mathbf{f}^{ab}_{n,n+1} &= \frac{\tilde{V}^{\alpha}_{n,n+1}(r^{\alpha}_{n+1}) - \tilde{V}^{\alpha}_{n,n+1}(r^{\alpha}_{n})}{r^{\alpha}_{n+1} - r^{\alpha}_{n}} \frac{\mathbf{r}^{ab}_{n+1} + \mathbf{r}^{ab}_{n}}{r^{ab}_{n+1} + r^{ab}_{n}} , \\ \mathbf{f}^{ab}_{n+1,n} &= \frac{\tilde{V}^{\alpha}_{n+1,n}(r^{\alpha}_{n+1}) - \tilde{V}^{\alpha}_{n+1,n}(r^{\alpha}_{n})}{r^{\alpha}_{n+1} - r^{\alpha}_{n}} \frac{\mathbf{r}^{ab}_{n+1} + \mathbf{r}^{ab}_{n}}{r^{ab}_{n+1} + r^{ab}_{n}} , \end{aligned} \tag{3.71}$$

where we used the compact notation

$$\begin{aligned} \tilde{V}^{\alpha}_{n,n+1}(r^{\alpha}) &= \tilde{V}(r^{1}_{n}, \dots, r^{\alpha-1}_{n}, r^{\alpha}, r^{\alpha+1}_{n+1}, \dots, r^{N(N-1)/2}_{n+1}) , \\ \tilde{V}^{\alpha}_{n+1,n}(r^{\alpha}) &= \tilde{V}(r^{1}_{n+1}, \dots, r^{\alpha-1}_{n+1}, r^{\alpha}, r^{\alpha+1}_{n}, \dots, r^{N(N-1)/2}_{n}) . \end{aligned} \tag{3.72}$$

This EM method preserves the total angular momentum in addition to the total energy and the total linear momentum. It can be used, e.g., for the simulation of bonded three-body interactions between macromolecules.

3.4.3. Interatomic potential

For our numerical simulation we consider the Stillinger-Weber potential [115], which includes two- and three-body contributions

$$\tilde{V} = \frac{1}{2!}\sum_{\substack{a,b=1\\a\neq b}}^{N} \varepsilon \tilde{f}_2(\bar{r}^{ab}/\sigma) + \frac{1}{3!}\sum_{\substack{a,b,c=1\\a\neq b\neq c}}^{N} \varepsilon \tilde{f}_3(\bar{r}^{ab}/\sigma, \bar{r}^{ac}/\sigma, \bar{r}^{bc}/\sigma) . \tag{3.73}$$

The pair contribution is given by

$$\tilde{f}_2(\hat{r}) = \begin{cases} A\,(B\hat{r}^{-q} - \hat{r}^{-p})\, g_2(\hat{r}) & \text{if}\, \hat{r} < a , \\ 0 & \text{if}\, r \geq a , \end{cases} \tag{3.74}$$

where the hats indicate the normalization of the distances by σ. After composing with the law of cosines (3.69), the three-body contribution takes the form

$$\tilde{f}_3(\hat{r}^{ab}, \hat{r}^{ac}, \hat{r}^{bc}) = h(\hat{r}^{ab}, \hat{r}^{ac}, \hat{r}^{bc}) + h(\hat{r}^{ab}, \hat{r}^{bc}, \hat{r}^{ac}) + h(\hat{r}^{ac}, \hat{r}^{bc}, \hat{r}^{ab}) , \tag{3.75}$$

with

$$h(\hat{r}^1, \hat{r}^2, \hat{r}^3) = \begin{cases} \lambda \left(\frac{(\hat{r}^1)^2 + (\hat{r}^2)^2 - (\hat{r}^3)^2}{2\hat{r}^1\hat{r}^2} + \frac{1}{3} \right)^2 g_3(\hat{r}^1, \hat{r}^2) & \text{if } \hat{r}^1 < a \text{ and } \hat{r}^2 < a \;, \\ 0 & \text{otherwise} \;. \end{cases} \tag{3.76}$$

Additionally, we introduce the functions

$$\begin{aligned} g_2(\hat{r}) &= \exp\left([\hat{r} - a]^{-1}\right) \;, \\ g_3(\hat{r}^1, \hat{r}^2) &= \exp\left(\gamma[\hat{r}^1 - a]^{-1} + \gamma[\hat{r}^2 - a]^{-1}\right) \;. \end{aligned} \tag{3.77}$$

Here $\tilde{\theta}^{bac} = g(\bar{r}^{ab}, \bar{r}^{ac}, \bar{r}^{bc}) = \arccos(-1/3) \approx 109.47^\circ$ minimizes the function h given in eq. (3.76), that corresponds to the underlying diamond structure of silicon. From this reasoning it follows that the function h penalizes bond-angles which differ from the ones in this crystal structure. The parameters in the potential are A, B, a, ε, σ, q, p, λ, and γ.

3.4.4. Numerical evaluation

We consider now the numerical solution of systems of particles with the Stillinger-Weber effective potential. The pairwise contributions to the potential are discretized according to Section 3.3; the remaining three-body interactions are defined as in Section 3.4.2.2.

3.4.4.1. Accuracy study

We consider in this example a three-dimensional box $[-L/2, L/2]^3$ filled with 5 particles. The initial configuration of the system has a particle at the center of the box and the rest of the particles form bonds with the first one with angle $\arccos(-1/3)$. In addition, these four particles are at distance 1 from the center. Starting from rest, the motion of the system results from the non-vanishing interacting forces among particles away from the center. Further parameters of the simulation can be found in Table 3.3.

Material parameters	ε	1	Initial configuration
	A	0	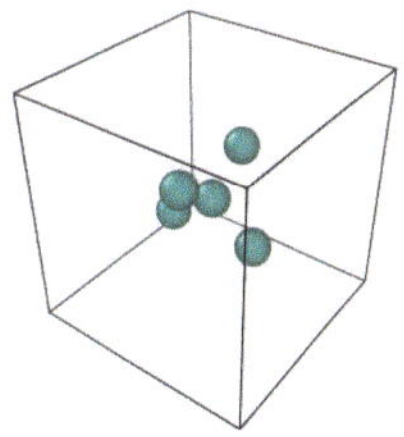
	B	0	
	σ	1	
	λ	21	
	p	0	
	q	0	
	a	1.8	
	γ	1.2	
Mass	m^a	1	
Side length	L	4	
Newton tolerance	-	10^{-8}	
Simulation duration	T	0.7	
Reference time step size	Δt_{ref}	0.001	
Time step size	Δt	0.01, 0.0125,0.016 0.02,0.25,0.04,0.05,0.08	

Table 3.3.: Accuracy study: Data used in the simulation

As in Section 3.3.4.1, we perform first an accuracy analysis of the EM integrator using the mid-point rule as a reference. This study is carried out using ten different time step sizes for both integrators and employing the error measures eq. (3.56). Fig. 3.8 reveals that both schemes are second order accurate.

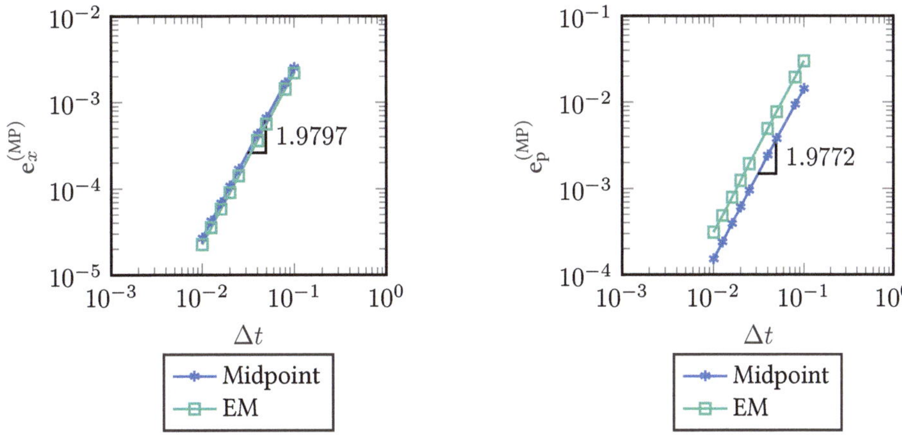

Figure 3.8.: Accuracy study: Relative error in the position w.r.t. mid-point rule (left), Relative error in the linear momentum w.r.t. mid-point rule (right)

3.4.5. Energy consistency study

We consider now 64 atoms inside a three-dimensional box $[-L/2, L/2]^3$, initially arranged in a perfect diamond cubic lattice structure.

Material parameters	ε	1	Initial configuration
	A	7.049556277	
	B	0.6022245584	
	σ	1	
	λ	210	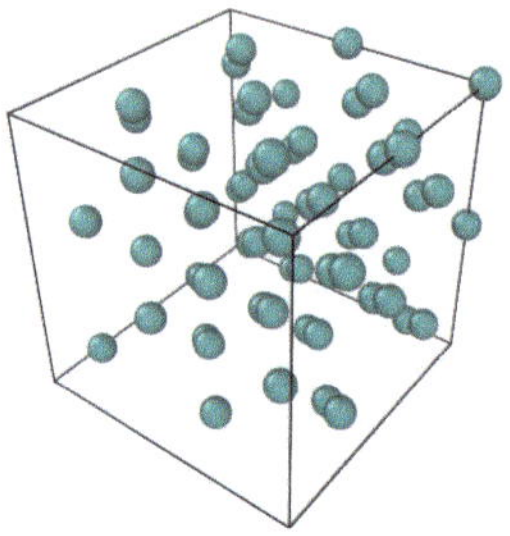
	p	4	
	q	0	
	a	2	
	γ	1.2	
Mass	m^a	1	
Sidelength	L	4	
Newton tolerance	-	10^{-9}	
Simulation duration	T	8	
Time step size	Δt	0.04	

Table 3.4.: Energy consistency study: Data used in the simulation

This lattice will be disrupted during the simulation as we consider an initial velocity associated to each atom such that the total initial kinetic energy is approximately 768.22. For the chosen time step size, the mid-point rule clearly violates the conservation of the total energy, see Fig. 3.9, leading to an energy blow-up and finally to a termination of the simulation, indicated by the black line on the same Figure.

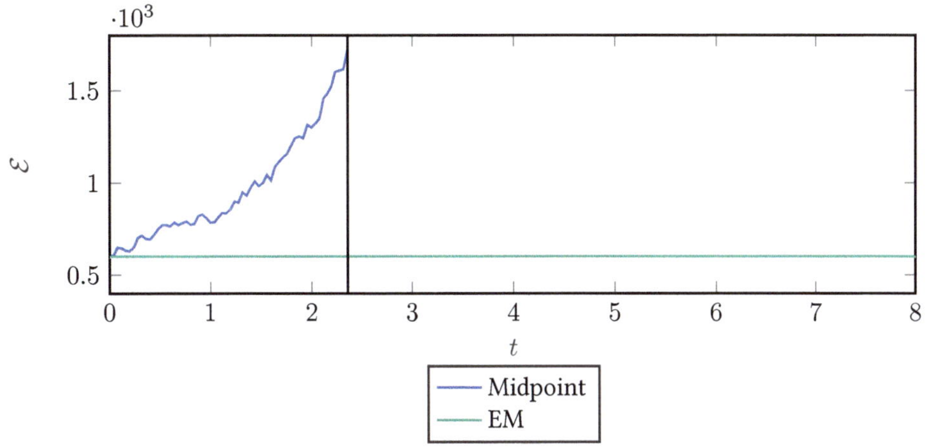

Figure 3.9.: Energy consistency study: Total energy

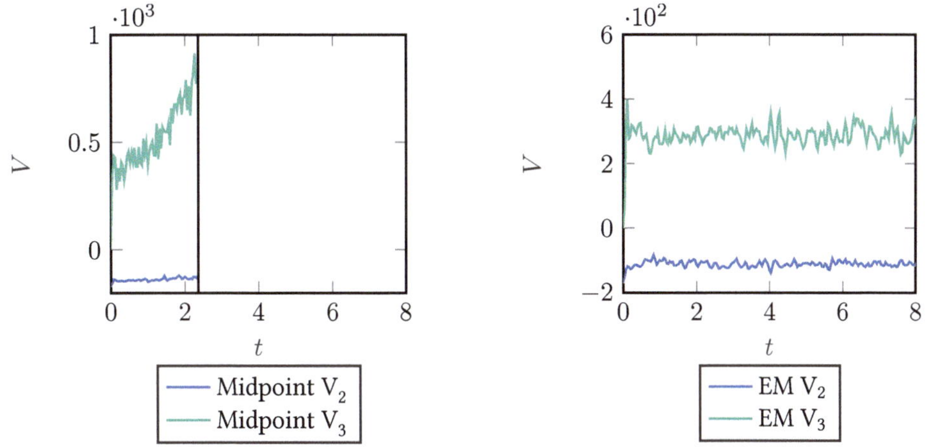

Figure 3.10.: Energy consistency study: Potential energies using the MP rule (left), Potential energies using the EM method (right). V_i refers to the $i-$body contribution to the potential

The largest contribution to the algorithmic energy error is due to the mid-point approximation of the forces generated from the three-body contribution of the Stillinger-Weber potential. It can be observed in Fig. 3.10 (left) that the potential energy of the two-body terms remains bounded, while the potential energy of the three-body terms increases unphysically causing the energy blow-up and, ultimately, the termination of the simulation.

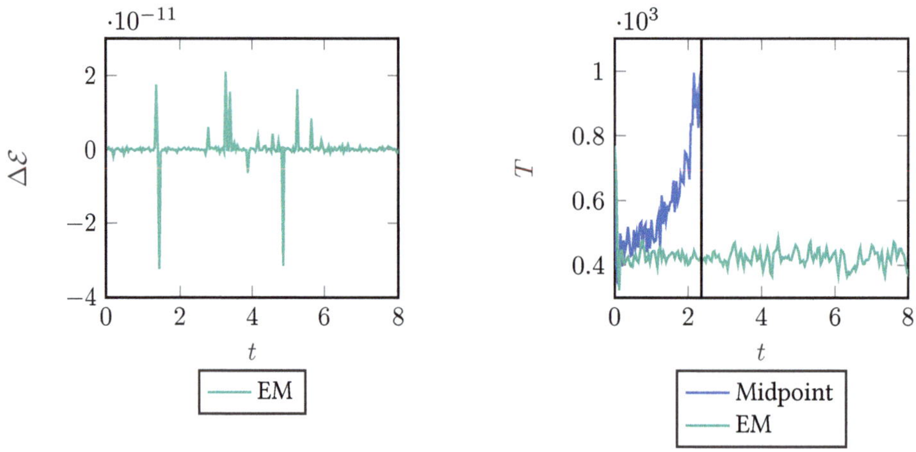

Figure 3.11.: Energy consistency study: Total energy difference (left), kinetic energy (right)

In contrast, using the EM method both contributions to the potential energy of the system remain bounded, see Fig. 3.10 (right). As expected, the EM method preserves the total energy up to round-off errors, see Fig. 3.11 (left). Eventually, as illustrated in Fig. 3.11 (right), the evolution of the kinetic energy reveals that the EM solution reaches thermodynamic equilibrium for the chosen time step size, in contrast with the mid-point rule.

3.5. Energy-momentum methods for periodic systems described by the embedded-atom method

In addition to three-body potentials of the type described in Section 3.4, more elaborate and accurate potentials include multi-body effects through an environment-dependent variable, resulting in effective potential that are extremely common, for example, in the simulation of metals. One of the most popular interatomic potentials of this class is the one employed in the embedded-atom method (EAM) [116–118], whose use in the context of conserving schemes is analyzed in this section. The interatomic potential in this case is of the form

$$V = \frac{1}{2} \sum_{\substack{a,b=1 \\ b \neq a}}^{N} \tilde{V}\left(d_T(\mathbf{x}^a, \mathbf{x}^b)\right) + \sum_{a=1}^{N} \tilde{F}\left(\bar{\rho}^a\right) . \tag{3.78}$$

The EAM potential consists of a pair potential contribution and electronic energies $\tilde{F}(\bar{\rho}^a)$. The latter is due to the embedding of a-th atom in a homogeneous electron gas of density

$\bar{\rho}^a$ [120]. The background electron density function $\bar{\rho}^a$ is a linear superposition of contributions from each neighbor atom such that the electronic energy $\tilde{F}(\bar{\rho}^a)$ of the a-th atom depends on the interatomic distance $d_T(\mathbf{x}^a, \mathbf{x}^b)$ to each neighbor.

In the case of metals, the environment of each atom is a nearly uniform electron gas and therefore the embedded-atom approximation is reasonable. Since we already investigated pair potentials in Section 3.3, we now investigate the discretization of the forces due to the electronic energy, noting that the resultant forces in an EAM potential must include the forces due to the pair potential, as well.

3.5.1. Equation of motion

We consider a system of N particles in a periodic box $\mathcal{B}$ of side L with equations of motion (3.16) and an effective potential that depends only on the electronic energy of each particle, that is

$$V = \sum_{a=1}^{N} \tilde{F}\left(\sum_{\substack{b=1\\ b\neq a}}^{N} g_b\left(d_T(\mathbf{x}^a, \mathbf{x}^b)\right)\right) , \tag{3.79}$$

where $\tilde{V} : \mathbb{R}^+ \cup \{0\} \mapsto \mathbb{R}$. Here, $g_a : \mathbb{R}^+ \cup \{0\} \mapsto \mathbb{R}$ is a function of the relative interatomic distance which represents a spherical electron density field around the isolated a-th particle [120]. Using the definitions in eq. (3.26), the background energy density is then given by

$$\bar{\rho}^a = \sum_{\substack{b=1\\ b\neq a}}^{N} g_b\left(\bar{r}^{ab}\right) . \tag{3.80}$$

The forces acting on the particles are defined by eq. (3.17) and have the standard form

$$\mathbf{f}^a = \sum_{\substack{b=1\\ a\neq b}}^{N} \mathbf{f}^{ab} , \qquad \text{with} \qquad \mathbf{f}^{ab} := \varphi^{ab} \frac{\bar{\mathbf{r}}^{ab}}{\bar{r}^{ab}} , \tag{3.81}$$

where the strength of the force has now the structure

$$\varphi^{ab} = \tilde{F}'\left(\bar{\rho}^a\right) g_b'\left(\bar{r}^{ab}\right) + \tilde{F}'(\bar{\rho}^b) g_a'\left(\bar{r}^{ab}\right) . \tag{3.82}$$

We observe that, for this potential contribution, the direction of the interatomic force depends only on the distance between the a-th and the b-th particle, while its strength is further determined by the background electron density at the a-th and b-th particle.

3.5.2. Time discretization

We consider now the integration in time of the equations of motion (3.16) of a system in the periodic box $\mathcal{B}$ and effective potential (3.78) and employ the same time integration strategy as outlined in Section 3.3.2.

3.5.2.1. Mid-point scheme

The canonical mid-point rule approximates the equation of motion by the implicit formula (3.29), where the mid-point approximation of the force acting on the a-th particle in the direction of the b-th particle is given by

$$\begin{aligned} \mathbf{f}_{\text{MP}}^{ab} &= \varphi_{\text{MP}}^{ab} \frac{\bar{\mathbf{r}}_{n+1/2}^{ab}}{|\bar{\mathbf{r}}_{n+1/2}^{ab}|} , \\ \varphi_{\text{MP}}^{ab} &= \tilde{F}' \left(\bar{\rho}_{\text{MP}}^{a}\right) g_b' \left(\bar{r}_{n+1/2}^{ab}\right) + \tilde{F}'(\bar{\rho}_{\text{MP}}^{b}) g_a' \left(\bar{r}_{n+1/2}^{ab}\right) , \end{aligned} \tag{3.83}$$

and the mid-point approximation of the background energy density reads

$$\bar{\rho}_{\text{MP}}^{a} = \sum_{\substack{b=1 \\ b\neq a}}^{N} g_b \left(\bar{r}_{n+1/2}^{ab}\right) . \tag{3.84}$$

As in the continuous case, the weak law of action and reaction is satisfied and responsible for the conservation of total linear momentum of the system, see Section 3.3.2.1.

3.5.2.2. Energy and momentum conserving discretization

As illustrated in previous sections, it is possible to construct a perturbation of the mid-point rule which, in addition to preserving the total linear momentum, preserves the total energy. For that, we follow once more the steps presented in Section 3.3.2.2. The partitioned discrete gradient will be used again with a slightly different structure due to the nature of the embedded function. The partitioned discrete gradient assumes the form

$$\mathsf{D}_{\mathbf{x}^a} V = - \sum_{\substack{a,b=1 \\ a\neq b}}^{N} \mathbf{f}_{\text{algo}}^{ab} = - \sum_{\substack{a,b=1 \\ a\neq b}}^{N} \frac{1}{2} \left(\mathbf{f}_{n,n+1}^{ab} + \mathbf{f}_{n+1,n}^{ab}\right) , \tag{3.85}$$

with the contributions

$$\begin{aligned} \mathbf{f}_{n,n+1}^{ab} &= \mathbf{f}_{\text{MP}}^{ab} + \frac{\Delta \tilde{F}_{n,n+1}^{ab} - \mathbf{f}_{\text{MP}}^{ab} \cdot (\mathbf{r}_{n+1}^{ab} - \mathbf{r}_{n}^{ab})}{|\mathbf{r}_{n+1}^{ab} - \mathbf{r}_{n}^{ab}|} \mathbf{n} , \\ \mathbf{f}_{n+1,n}^{ab} &= \mathbf{f}_{\text{MP}}^{ab} + \frac{\Delta \tilde{F}_{n+1,n}^{ab} - \mathbf{f}_{\text{MP}}^{ab} \cdot (\mathbf{r}_{n+1}^{ab} - \mathbf{r}_{n}^{ab})}{|\mathbf{r}_{n+1}^{ab} - \mathbf{r}_{n}^{ab}|} \mathbf{n} , \end{aligned} \tag{3.86}$$

for which we introduced the compact notation

$$\begin{aligned}\Delta\tilde{F}^{ab}_{n,n+1} &= \tilde{F}(\bar{\rho}^{a}_{n,n+1}(\bar{r}^{ab}_{n+1})) - \tilde{F}(\bar{\rho}^{a}_{n,n+1}(\bar{r}^{ab}_{n})) \\ &\quad + \tilde{F}(\bar{\rho}^{b}_{n,n+1}(\bar{r}^{ab}_{n+1})) - \tilde{F}(\bar{\rho}^{b}_{n,n+1}(\bar{r}^{ab}_{n}))\,, \\ \Delta\tilde{F}^{ab}_{n+1,n} &= \tilde{F}(\bar{\rho}^{a}_{n+1,n}(\bar{r}^{ab}_{n+1})) - \tilde{F}(\bar{\rho}^{a}_{n+1,n}(\bar{r}^{ab}_{n})) \\ &\quad + \tilde{F}(\bar{\rho}^{b}_{n+1,n}(\bar{r}^{ab}_{n+1})) - \tilde{F}(\bar{\rho}^{b}_{n+1,n}(\bar{r}^{ab}_{n}))\,,\end{aligned} \tag{3.87}$$

and

$$\begin{aligned}\bar{\rho}^{a}_{n,n+1}(\bar{r}^{ab}) &= \sum_{d=1}^{b-1} g_a(\bar{r}^{ad}_{n}) + g_a(\bar{r}^{ab}) + \sum_{e=b+1}^{N} g_a(\bar{r}^{ae}_{n+1})\,, \\ \bar{\rho}^{a}_{n+1,n}(\bar{r}^{ab}) &= \sum_{d=1}^{b-1} g_a(\bar{r}^{ad}_{n+1}) + g_a(\bar{r}^{ab}) + \sum_{e=b+1}^{N} g_a(\bar{r}^{ae}_{n})\,.\end{aligned} \tag{3.88}$$

The densities $\bar{\rho}^{b}_{n,n+1}(\bar{r}^{ab})$ and $\bar{\rho}^{b}_{n+1,n}(\bar{r}^{ab})$ are defined similarly.

To show that the integrator exactly preserves the total momentum, it suffices to follow the proof in Section 3.3.2.2. The critical condition that a conserving scheme must satisfy reads now

$$\sum_{\substack{b=1\\a<b}}^{N} \mathbf{f}^{ab}_{\text{algo}} \cdot (\mathbf{r}^{ab}_{n+1} - \mathbf{r}^{ab}_{n}) = \sum_{a=1}^{N} \left(\tilde{F}(\bar{\rho}^{a}_{n+1}) - \tilde{F}(\bar{\rho}^{a}_{n})\right)\,, \tag{3.89}$$

which is indeed satisfied by the proposed method.

3.5.3. Interatomic potential

We consider for our subsequent analysis the Lennard-Jones-Baskes (LJB) EAM model [118, 133], which is the extension of the Lennard-Jones material model into the many-body regime of the EAM formalism [134]. The total potential energy of the LJB model is given by

$$\tilde{V} = \frac{1}{2}\sum_{\substack{a,b=1\\a\neq b}}^{N} V(\bar{r}^{ab}) + \sum_{a=1}^{N} \tilde{F}(\bar{\rho}^{a})\,. \tag{3.90}$$

The two body part has been introduced in Section 3.3.3. For the EAM contribution we further define

$$\begin{aligned}
\tilde{F}(\bar{\rho}^a) &= \frac{1}{2}\varepsilon A Z_1 \bar{\rho}^a \left(\ln\left(\bar{\rho}^a\right) - 1\right), \\
\bar{\rho}^a &= \frac{1}{Z_1}\sum_{\substack{b=1\\ b\neq a}}^{N} g_b\left(|\bar{\mathbf{r}}^{ab}|\right), \\
g_b\left(|\bar{\mathbf{r}}^{ab}|\right) &= \begin{cases} \exp\left(-\beta\left(\sigma^{-1}|\bar{\mathbf{r}}^{ab}| - 1\right)\right) & \text{if } |\bar{\mathbf{r}}^{ab}| < r_c\,, \\ 0 & \text{otherwise}\,, \end{cases}
\end{aligned} \tag{3.91}$$

where ε, σ, β, A, r_c and Z_1 are material parameters.

3.5.4. Numerical evaluation

Since we already investigated pair potentials in Section 3.3, we focus on the EAM part of the LJB material model for the following numerical evaluations.

3.5.4.1. Accuracy study

A three-dimensional box $[-L/2, L/2]^3$ is now considered with 5 particles inside it. The particles form a unit body-centered cubic (BCC) cell of side 2, centered within the box. Starting from rest, the system builds up kinetic energy due to the fact that the particles in the exterior of the BCC crystal are not in equilibrium. Further parameters of the simulation can be found in Table 3.5.

Material parameters	ε	2	Initial configuration
	A	1	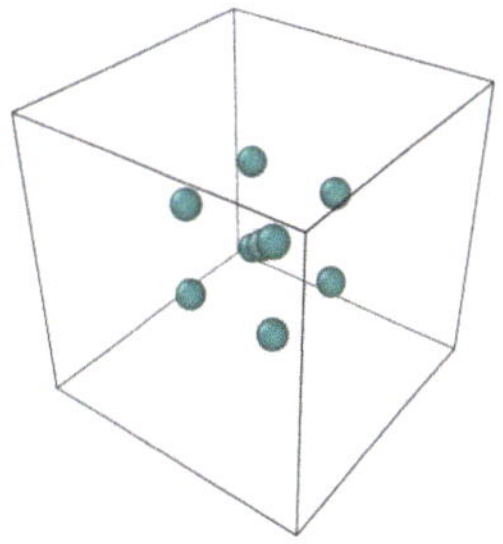
	σ	1	
	β	4	
	Z_1	12	
	r_c	3	
Mass	m^a	1	
Side length	L	6	
Newton tolerance	-	10^{-6}	
Simulation duration	T	0.5	
Reference			
time step size	Δt_{ref}	0.0001	
Time step size	Δt	0.0005,0.001,0.022	
		0.0025,0.004,0.005	
		0.01,0.02,0.025	

Table 3.5.: Accuracy study: Data used in the simulation

Proceeding as in Section 3.3.4.1, we perform an accuracy analysis of the EM integrator using the mid-point rule as the reference solution. To study the convergence of the numerical solutions, we employ ten time steps of decreasing size and the error measures defined in (3.56). Fig. 3.12 confirms again that both the mid-point rule and the EM method are second order accurate approximations.

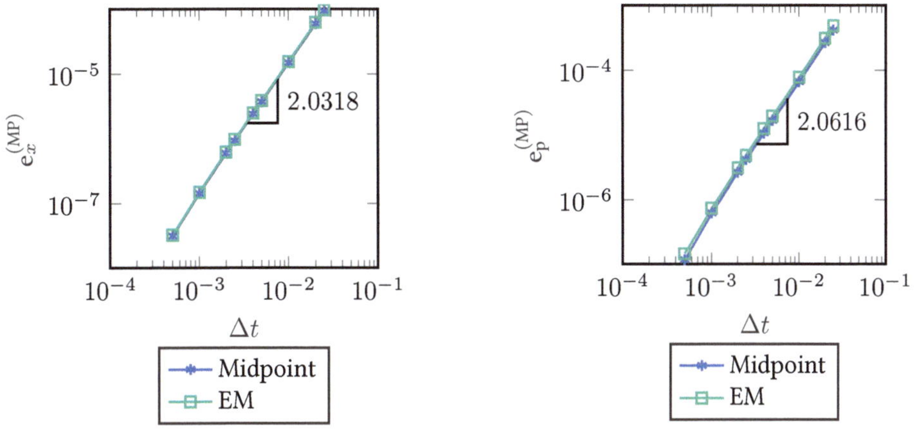

Figure 3.12.: Accuracy study: Relative error in the position w.r.t. mid-point rule (left), Relative error in the momentum w.r.t. mid-point rule (right)

3.5.5. Energy consistency study

In this example we consider 108 atoms inside a three-dimensional box $[-L/2, L/2]^3$, where the atoms are arranged in a perfect face-centered cubic (FCC) lattice structure. This FCC lattice is perturbed by the initial velocities of the atoms which are imparted in such a way that the total initial kinetic energy is approximately 3.251.

Material parameters	ε	2	Initial configuration
	A	1	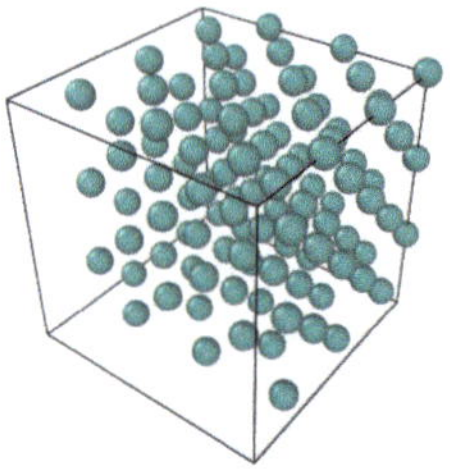
	σ	1	
	β	4	
	Z_1	12	
	r_c	3	
Mass	m^a	1	
Side length	L	6	
Newton tolerance	-	10^{-8}	
Simulation duration	T	25	
Time step size	Δt	0.1	

Table 3.6.: Energy consistency study: Data used in the simulation

Further parameters of the simulation can be found in Table 3.6.

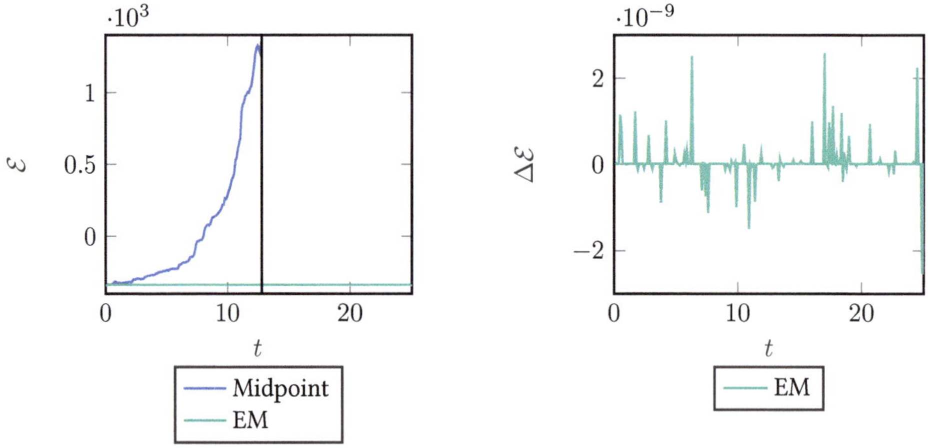

Figure 3.13.: Energy consistency study: Total energy (left), Total energy difference (right)

For the time step selected, the mid-point rule clearly violates the conservation of total energy, see Fig. 3.13 (left). As in previous examples, this leads to an energy blow-up and finally to a termination of the simulation, indicated by the black line on the same figure. In contrast, the EM method preserves the total energy up to round off errors, see Fig. 3.13 (right). Finally, and as in the previous examples, Fig. 3.14 shows that the EM, but not the mid-point rule, is able to evolve the system of particles up to thermodynamic equilibrium.

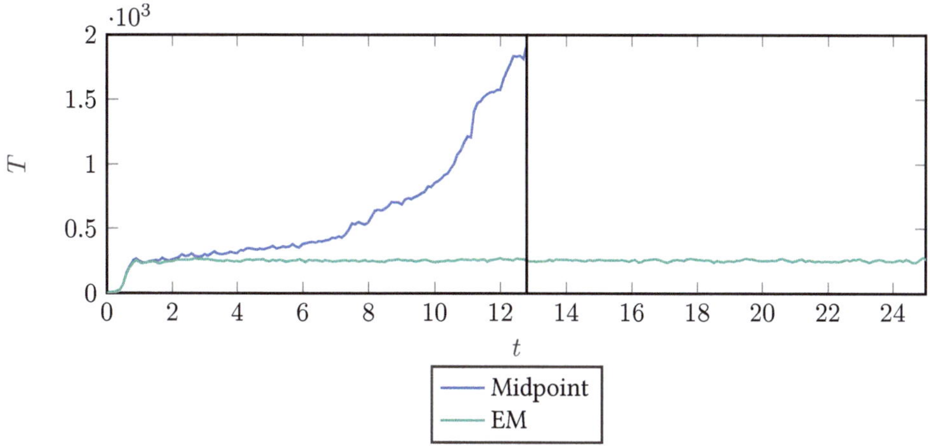

Figure 3.14.: Energy consistency study: Kinetic energy

4. The GENERIC framework for large-strain thermo-elasticity[1]

In this chapter we study the motion of large-strain thermo-elastic solids. Due the presence of heat conduction irreversible effects have to be considered. The continuum based description will resort to the GENERIC framework for infinitesmial dimensional systems, where the time-evolution of an arbitrary functional $\mathcal{A}$ can be expressed as

$$\frac{\mathrm{d}\mathcal{A}}{\mathrm{d}t} = \{\mathcal{A}, \mathcal{E}\} + [\mathcal{A}, \mathcal{S}] \,. \tag{4.1}$$

This equation represents a 2-generator formalism in which the reversible part is generated by the total energy $\mathcal{E}$ of the system via the Poisson bracket $\{\cdot, \cdot\}$, while the irreversible part is generated by the total entropy $\mathcal{S}$ of the system via the dissipative bracket $[\cdot, \cdot]$. In contrast to Chapters 2 and 3 the brackets now imply an integration over continues labels rather then summation over discrete indices, and functional derivatives are considered rather then partial derivatives. A concise definition of these quantities will be presented in the sequel within the context of a Lagrangian description of large-strain thermo-elasticity.

We propose a new variational formulation for finite-strain thermo-elastodynamics, which emanantes from the GENERIC formalism and makes the free choice of the thermodynamic state variable possible. In particular, one may choose the absolute temperature, the internal energy density or the entropy density as thermodynamic state variable. To solve initial boundary value problems, the GENERIC formalism is extended to open systems. The discretization in time makes use of the standard mid-point rule. Depending on the choice of the thermodynamic state variable, partially structure-preserving schemes are obtained. For example, choosing the internal energy as state variable yields a new energy-momentum consistent scheme. Thus the newly developed GENERIC-based weak form is particularly well suited for the design of (fully) structure-preserving methods as will be shown in Chapter 5. Furthermore, numerical investigations are presented which confirm the theoretical findings and shed light on the numerical stability of the newly developed schemes.

The first part of the present chapter deals with the description of the GENERIC framework which was originally developed for isolated systems. In the case of isolated (or closed) systems boundary effects are considered to be irrelevant.

---

[1] This Chapter is based on [1]

4.1. The reversible part: Isothermal elastodynamics

We start with the formulation of large-strain elastodynamics. Since elastic deformations are reversible the GENERIC evolution equation (4.1) reduces to the Hamiltonian form which is solely based on the Poisson bracket.

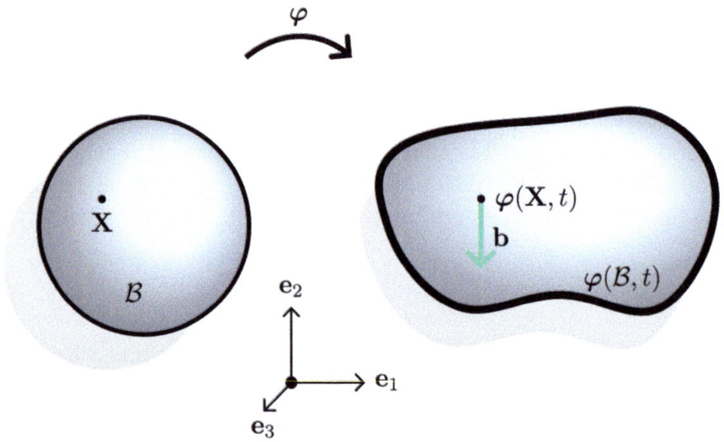

Figure 4.1.: Reference configuration $\mathcal{B}$ and deformed configuration $\varphi(\mathcal{B}, t)$ at time t. In Section 4.1 the focus is on the motion of isolated elastic solids. That is, external tractions acting on the boundary are disregarded.

Consider a continuum body with material points $\mathbf{X} = X^A \mathbf{e}_A$ in the reference configuration $\mathcal{B} \subset \mathbb{R}^3$ (Fig. 4.1). Here and in the sequel the summation convention applies to repeated indices. Moreover, $\mathbf{e}_A$ denote the canonical base vectors in $\mathbb{R}^3$. Within the Lagrangian description of continuum mechanics the deformed configuration of the body at time t is characterized by the deformation map $\varphi : \mathcal{B} \times \mathcal{I} \mapsto \mathbb{R}^3$, where $\mathcal{I} = [0, T]$ is the time interval of interest. Accordingly, the placement at time t of the material point $\mathbf{X} \in \mathcal{B}$ is given by $\mathbf{x} = \varphi(\mathbf{X}, t)$. Its velocity is defined by $\mathbf{v} = \dot{\varphi}$, where a superposed dot denotes the material time derivative.

The deformation gradient corresponds to the Jacobian of the deformation map

$$\mathbf{F} = \partial_{\mathbf{X}} \varphi \,. \tag{4.2}$$

The Hamiltonian formulation of nonlinear elastodynamics is based on the introduction of the momentum conjugate to φ. To this end we consider the Lagrangian functional

$$\mathcal{L}(\varphi, \dot{\varphi}) = \int_{\mathcal{B}} l(\varphi, \mathbf{F}, \dot{\varphi}) \, \mathrm{d}V \,, \tag{4.3}$$

which corresponds to the difference between the kinetic energy and the potential energy. The associated Lagrangian density function is given by

$$l(\boldsymbol{\varphi}, \mathbf{F}, \dot{\boldsymbol{\varphi}}) = \frac{1}{2}\rho\dot{\boldsymbol{\varphi}} \cdot \dot{\boldsymbol{\varphi}} - (W(\mathbf{F}) - \mathbf{b} \cdot \boldsymbol{\varphi}) \ , \tag{4.4}$$

where $\rho : \mathcal{B} \mapsto \mathbb{R}_+$ is the mass density in the reference configuration, $W : \mathbb{R}^{3\times 3} \mapsto \mathbb{R}$ is a stored energy function which characterizes hyperelastic material behavior, and $\mathbf{b} : \mathcal{B} \mapsto \mathbb{R}^3$ represents prescribed body forces, which are assumed to be dead loads. The conjugate momentum density is now defined by

$$\mathbf{p} = \partial_{\dot{\boldsymbol{\varphi}}} l(\boldsymbol{\varphi}, \mathbf{F}, \dot{\boldsymbol{\varphi}}) = \rho\dot{\boldsymbol{\varphi}} \ . \tag{4.5}$$

Moreover, the Hamiltonian density function results from a Legendre tranformation of the Lagrangian density function (4.4) with respect to the velocity leading to

$$h(\boldsymbol{\varphi}, \mathbf{F}, \mathbf{p}) = \mathbf{p} \cdot \dot{\boldsymbol{\varphi}} - l(\boldsymbol{\varphi}, \mathbf{F}, \dot{\boldsymbol{\varphi}}) \ , \tag{4.6}$$

subject to relation (4.5). With regard to (4.4), the Hamiltonian density function reads

$$h(\boldsymbol{\varphi}, \mathbf{F}, \mathbf{p}) = \frac{1}{2}\rho^{-1}\mathbf{p} \cdot \mathbf{p} + W(\mathbf{F}) - \mathbf{b} \cdot \boldsymbol{\varphi} \ . \tag{4.7}$$

The equations of motion can now be obtained by applying Hamilton's phase space principle. Accordingly, we consider the stationarity of the functional

$$A(\boldsymbol{\varphi}, \mathbf{p}) = \int_0^T \int_{\mathcal{B}} (\mathbf{p} \cdot \dot{\boldsymbol{\varphi}} - h(\boldsymbol{\varphi}, \mathbf{F}, \mathbf{p})) \ \mathrm{d}V \mathrm{d}t \ , \tag{4.8}$$

with fixed endpoints in time ($t = 0$ and $t = T > 0$). The first stationary condition is given by

$$\begin{aligned} DA(\boldsymbol{\varphi}, \mathbf{p}) \cdot \delta\boldsymbol{\varphi} &= \frac{d}{d\varepsilon} \int_0^T \int_{\mathcal{B}} (\mathbf{p} \cdot (\dot{\boldsymbol{\varphi}} + \varepsilon\delta\dot{\boldsymbol{\varphi}}) - h(\boldsymbol{\varphi} + \varepsilon\delta\boldsymbol{\varphi}, \mathbf{F} + \varepsilon\nabla\delta\boldsymbol{\varphi}, \mathbf{p})) \ \mathrm{d}V \mathrm{d}t \bigg|_{\varepsilon=0} \\ &= \int_0^T \left(\int_{\mathcal{B}} \mathbf{p} \cdot \delta\dot{\boldsymbol{\varphi}} \, \mathrm{d}V - \frac{d}{d\varepsilon} \int_{\mathcal{B}} h(\boldsymbol{\varphi} + \varepsilon\delta\boldsymbol{\varphi}, \mathbf{F} + \varepsilon\nabla\delta\boldsymbol{\varphi}, \mathbf{p}) \, \mathrm{d}V \bigg|_{\varepsilon=0} \right) \mathrm{d}t \\ &= 0 \ , \end{aligned} \tag{4.9}$$

where the variations $\delta\boldsymbol{\varphi} : \mathcal{B} \mapsto \mathbb{R}^3$ are subject to the endpoint conditions in time, $\delta\boldsymbol{\varphi}_0 = \mathbf{0}$ and $\delta\boldsymbol{\varphi}_T = \mathbf{0}$. The second integral on the right-hand side of (4.9) can be recast in the form

$$\begin{aligned}\frac{d}{d\varepsilon} \int_{\mathcal{B}} h(\boldsymbol{\varphi} + \varepsilon\delta\boldsymbol{\varphi}, \mathbf{F} + \varepsilon\nabla\delta\boldsymbol{\varphi}, \mathbf{p})\,\mathrm{d}V\Big|_{\varepsilon=0} &= \int_{\mathcal{B}} (\partial_{\boldsymbol{\varphi}} h \cdot \delta\boldsymbol{\varphi} + \partial_{\mathbf{F}} h : \nabla\delta\boldsymbol{\varphi})\;\mathrm{d}V \\ &= \int_{\mathcal{B}} \delta\boldsymbol{\varphi} \cdot (\partial_{\boldsymbol{\varphi}} h - \mathrm{Div}(\partial_{\mathbf{F}} h))\;\mathrm{d}V \\ &= \int_{\mathcal{B}} \delta\boldsymbol{\varphi} \cdot \delta_{\boldsymbol{\varphi}} h\,\mathrm{d}V\,,\end{aligned}$$

where $\delta_{\boldsymbol{\varphi}} h$ denotes the functional derivative (sometimes also called the Volterra derivative [135]) of the Hamiltonian density with respect to $\boldsymbol{\varphi}$. Note that in the above equation integration by parts has been applied along with the neglect of the boundary terms. As has been mentioned above, the neglect of boundary integrals is related to the fact that the focus in this section is on closed systems in which boundary effects are irrelevant. Performing integration by parts with respect to time on the first integral on the right-hand side of (4.9), the first stationary condition eventually yields

$$\int_0^T \int_{\mathcal{B}} \delta\boldsymbol{\varphi} \cdot (\dot{\mathbf{p}} + \delta_{\boldsymbol{\varphi}} h)\;\mathrm{d}V\mathrm{d}t = 0\,. \tag{4.10}$$

The second stationary condition is given by

$$\begin{aligned}DA(\boldsymbol{\varphi}, \mathbf{p}) \cdot \delta\mathbf{p} &= \frac{d}{d\varepsilon} \int_0^T \int_{\mathcal{B}} ((\mathbf{p} + \varepsilon\delta\mathbf{p}) \cdot \dot{\boldsymbol{\varphi}} - h(\boldsymbol{\varphi}, \mathbf{F}, \mathbf{p} + \varepsilon\delta\mathbf{p}))\;\mathrm{d}V\mathrm{d}t\Big|_{\varepsilon=0} \\ &= \int_0^T \int_{\mathcal{B}} \delta\mathbf{p} \cdot (\dot{\boldsymbol{\varphi}} - \delta_{\mathbf{p}} h)\;\mathrm{d}V\mathrm{d}t \\ &= 0\,.\end{aligned} \tag{4.11}$$

Since the variations $\delta\mathbf{p}$ and $\delta\boldsymbol{\varphi}$ are arbitrary (apart from the endpoint conditions on $\delta\boldsymbol{\varphi}$), the stationary conditions (4.10) and (4.11) give rise to the local form of the equations of motion

$$\begin{aligned}\dot{\boldsymbol{\varphi}} &= \delta_{\mathbf{p}} h\,, \\ \dot{\mathbf{p}} &= -\delta_{\boldsymbol{\varphi}} h\,,\end{aligned} \tag{4.12}$$

for all $\mathbf{X} \in \mathcal{B}$ and $t \in [0, T]$. Taking into account the specific form of the Hamiltonian density function (4.7), the functional derivatives on the right-hand side of (4.12) are given by

$$\delta_{\boldsymbol{\varphi}} h = \partial_{\boldsymbol{\varphi}} h - \mathrm{Div}(\partial_{\mathbf{F}} h) = -\mathbf{b} - \mathrm{Div}\left(\partial_{\mathbf{F}} W\right)\,,$$

and

$$\delta_{\mathbf{p}} h = \partial_{\mathbf{p}} h = \rho^{-1}\mathbf{p}\,.$$

Upon introduction of the first Piola-Kirchhoff stress tensor $\mathbf{P} = \partial_{\mathbf{F}} W$, the local form of the equations of motion (4.12) can be recast as

$$\rho \ddot{\varphi} = \operatorname{Div}(\mathbf{P}) + \mathbf{b} \, .$$

This is the familiar Lagrangian form of the balance law for linear momentum (see, e.g., [136]). To translate the equations of motion (4.12) into the GENERIC bracket form (4.1), we now consider functionals of the form

$$\mathcal{A}(\varphi, \mathbf{p}) = \int_{\mathcal{B}} a(\varphi, \mathbf{F}, \mathbf{p}) \, \mathrm{d}V \, . \tag{4.13}$$

Differentiation with respect to time yields

$$\begin{aligned}
\frac{\mathrm{d}\mathcal{A}}{\mathrm{d}t} &= \int_{\mathcal{B}} \frac{\mathrm{d}}{\mathrm{d}t} a(\varphi, \mathbf{F}, \mathbf{p}) \, \mathrm{d}V \\
&= \int_{\mathcal{B}} \left(\partial_{\varphi} a \cdot \dot{\varphi} + \partial_{\mathbf{F}} a : \dot{\mathbf{F}} + \partial_{\mathbf{p}} a \cdot \dot{\mathbf{p}} \right) \mathrm{d}V \\
&= \int_{\mathcal{B}} \left(\delta_{\varphi} a \cdot \dot{\varphi} + \delta_{\mathbf{p}} a \cdot \dot{\mathbf{p}} \right) \mathrm{d}V \\
&= \int_{\mathcal{B}} \left(\delta_{\varphi} a \cdot \delta_{\mathbf{p}} h - \delta_{\mathbf{p}} a \cdot \delta_{\varphi} h \right) \mathrm{d}V =: \{\mathcal{A}, \mathcal{H}\} \, .
\end{aligned}$$

Note that integration by parts has been applied. In this connection the boundary terms have been neglected again. The resulting equation

$$\frac{\mathrm{d}\mathcal{A}}{\mathrm{d}t} = \{\mathcal{A}, \mathcal{H}\} \, , \tag{4.14}$$

holds for arbitrary functionals of the form (4.13) and is sometimes referred to as the equation of motion in Poisson bracket form (see, e.g., [24]). Note that this equation represents the reversible part of the GENERIC evolution equation (4.1). The Poisson bracket introduced above assumes the canonical form

$$\{\mathcal{A}, \mathcal{B}\} = \int_{\mathcal{B}} \left(\delta_{\varphi} a \cdot \delta_{\mathbf{p}} b - \delta_{\mathbf{p}} a \cdot \delta_{\varphi} b \right) \mathrm{d}V \, , \tag{4.15}$$

being the infinitesimal version of the classical discrete Poisson bracket eq. (3.3). Hereby are $\mathcal{A}$ and $\mathcal{B}$ two arbitrary functionals of the form (4.13). It can be easily observed that the canonical Poisson bracket (4.15) is skew-symmetric, that is

$$\{\mathcal{A}, \mathcal{B}\} = -\{\mathcal{B}, \mathcal{A}\} \, . \tag{4.16}$$

Note that conservation of the Hamiltonian is an immediate consequence of property (4.16) implying $\{\mathcal{H}, \mathcal{H}\} = 0$, so that (4.14) yields $\mathrm{d}\mathcal{H}/\mathrm{d}t = 0$. Moreover, it is well-known that the Poisson bracket satisfies the Jacobi identity which is of paramount importance for the time-structure invariance of reversible dynamics (see, e.g., [94]).

4.2. Extension to dissipative systems: Thermo-elastodynamics

Departing from the isothermal problem outlined above, we now aim at the inclusion of thermal effects. In particular, we focus on the dynamics of thermoelastic solids with heat conduction (see [136, Sec. 9.1] for a concise definition of a thermoelastic solid). We stress that we still focus on closed (or isolated) systems.

In thermal solid mechanics the selection of the independent thermodynamic variable is a crucial point. From the viewpoint of constitutive modeling the temperature is a natural choice. The temperature field $\theta : \mathcal{B} \times \mathcal{I} \mapsto \mathbb{R}_+$ typically enters the definition of the Helmholtz free energy density

$$\psi = \overline{\psi}(\mathbf{F}, \theta) \, . \tag{4.17}$$

Alternatively, the entropy density field $\eta : \mathcal{B} \times \mathcal{I} \mapsto \mathbb{R}$ defined by

$$\eta = \overline{\eta}(\mathbf{F}, \theta) = -\partial_\theta \overline{\psi}(\mathbf{F}, \theta) \, , \tag{4.18}$$

could be used as thermodynamic variable. Then, for example, the temperature can be expressed in terms of the entropy density by inverting the last equation leading to

$$\theta = \widetilde{\Theta}(\mathbf{F}, \eta) \, . \tag{4.19}$$

Yet another viable alternative is to use the internal energy density field $u : \mathcal{B} \times \mathcal{I} \mapsto \mathbb{R}$, which follows from a Legendre transformation of the free energy density with respect to the temperature. Accordingly,

$$u = \overline{u}(\mathbf{F}, \theta) = \theta \, \overline{\eta}(\mathbf{F}, \theta) + \overline{\psi}(\mathbf{F}, \theta) \, . \tag{4.20}$$

As usual, we assume that $\partial_\theta \overline{u} > 0$ and $\partial_\theta \overline{\eta} > 0$ for all $(\mathbf{F}, \theta)$ with $\det(\mathbf{F}) > 0$ and $\theta > 0$. Using the internal energy density as independent thermodynamic variable, the temperature can be obtained by inverting the last equation leading to $\theta = \widehat{\Theta}(\mathbf{F}, u)$. Making use of this relation, the entropy density (4.18) can be expressed in terms of the internal energy density via

$$\eta = \overline{\eta}(\mathbf{F}, \widehat{\Theta}(\mathbf{F}, u)) = \widehat{\eta}(\mathbf{F}, u) \, . \tag{4.21}$$

Furthermore, inserting (4.19) into (4.20), the internal energy density can be expressed in terms of the entropy density, i.e.

$$u = \widetilde{u}(\mathbf{F}, \eta) \, . \tag{4.22}$$

As a further consequence of the Legendre transform (4.20) we get the following formula for the temperature in terms of the entropy density:

$$\theta = \widetilde{\Theta}(\mathbf{F}, \eta) = \partial_\eta \widetilde{u}(\mathbf{F}, \eta) \, . \tag{4.23}$$

A clear exposition of further interrelationships between the alternative thermodynamic variables can be found in [100, Sec. 2.3].

4.2.1. Free choice of the thermodynamic variable

Guided by [100], in what follows, we allow for the free choice of the independent thermodynamic state variable by introducing the variable $\tau : \mathcal{B} \times \mathcal{I} \mapsto \mathbb{R}$, which can be selected among the three aforementioned alternatives, i.e. $\tau \in \{\theta, \eta, u\}$. In particular, in view of (4.19) and (4.22), the thermodynamic state variable τ can be expressed in terms of the entropy density via

$$\tau = \widetilde{\tau}(\mathbf{F}, \eta) \,. \tag{4.24}$$

On the other hand, the entropy density can also be expressed in tems of τ, as can be observed from formulas (4.18) and (4.21). Accordingly, we introduce the map η' such that

$$\eta = \eta'(\mathbf{F}, \tau) \,, \tag{4.25}$$

for $\tau \in \{\theta, \eta, u\}$. The following relations will be needed in the sequel:

$$\begin{aligned} \partial_\eta \widetilde{\tau} &= \left(\partial_\tau \eta'\right)^{-1} \,, \\ \partial_{\mathbf{F}} \widetilde{\tau} &= -\left(\partial_\tau \eta'\right)^{-1} \partial_{\mathbf{F}} \eta' \,. \end{aligned} \tag{4.26}$$

To proof these relations we consider the identity $\eta = \eta'(\mathbf{F}, \widetilde{\tau}(\mathbf{F}, \eta))$ which follows from (4.24) and (4.25). Since $\partial_\eta \eta = 1$, we get $1 = \partial_\tau \eta' \partial_\eta \widetilde{\tau}$, from which follows $(4.26)_1$. Similarly, since $\partial_{\mathbf{F}} \eta = \mathbf{0}$, we have $\mathbf{0} = \partial_{\mathbf{F}} \eta' + \partial_\tau \eta' \partial_{\mathbf{F}} \widetilde{\tau}$, from which follows $(4.26)_2$. Eventually, we recall a very important formula for the temperature that is given by (see also [94] or [100])

$$\Theta'(\mathbf{F}, \tau) = \frac{\partial_\tau u'(\mathbf{F}, \tau)}{\partial_\tau \eta'(\mathbf{F}, \tau)} \,. \tag{4.27}$$

Note that if the entropy density is chosen as thermodynamic variable (i.e. $\tau = \eta$), formula (4.23) is recovered. In the case $\tau = \theta$, formula (4.27) yields the relationship $\theta = \partial_\theta \overline{u} / \partial_\theta \overline{\eta}$ which is an immediate consequence of the Legendre transform (4.20).

4.2.2. Entropy-based brackets

It is well-known that both the Poisson and the dissipative brackets are invariant under changes of the independent variables (see, e.g., [94] or Section 2.2, where this property has been verified for discrete systems). The Poisson bracket assumes a particularly simple form when the entropy density is chosen as the thermodynamic state variable. This is quite obvious since the reversible dynamics does not affect the entropy. Similarly, the dissipative bracket assumes its simplest form when the internal energy density is chosen as the thermodynamic state variable. This has been observed in [137] in the context of a compressible, non-isothermal fluid. Moreover, this observation lies at the heart of the *special form of GENERIC* proposed in [100].

In the present work we start with the entropy-based formulation of thermo-elastodynamics and subsequently apply a transformation of variables to obtain the expressions for the brackets depending on the choice of the thermodynamic state variable $\tau \in \{\theta, \eta, u\}$. In the entropy-based formulation we consider functionals of the form

$$\mathcal{A} = \widetilde{\mathcal{A}}(\boldsymbol{\varphi}, \mathbf{p}, \eta) = \int_{\mathcal{B}} \widetilde{a}(\boldsymbol{\varphi}, \mathbf{F}, \mathbf{p}, \eta) \, \mathrm{d}V \,. \tag{4.28}$$

Remarkably, in the entropy-based formulation the Poisson bracket retains the canonical form (4.15), i.e.

$$\{\widetilde{\mathcal{A}}, \widetilde{\mathcal{B}}\} = \int_{\mathcal{B}} \left(\delta_{\boldsymbol{\varphi}} \widetilde{a} \cdot \delta_{\mathbf{p}} \widetilde{b} - \delta_{\mathbf{p}} \widetilde{a} \cdot \delta_{\boldsymbol{\varphi}} \widetilde{b} \right) \, \mathrm{d}V \,, \tag{4.29}$$

while the dissipative bracket featuring in the GENERIC evolution equation (4.1) is given by (see, e.g., [105])

$$[\widetilde{\mathcal{A}}, \widetilde{\mathcal{B}}] = \int_{\mathcal{B}} \nabla \left(\widetilde{\Theta}^{-1} \delta_{\eta} \widetilde{a} \right) \cdot \widetilde{\Theta}^{2} \mathbf{K} \nabla \left(\widetilde{\Theta}^{-1} \delta_{\eta} \widetilde{b} \right) \, \mathrm{d}V \,. \tag{4.30}$$

Here $\mathbf{K} = \widetilde{\mathbf{K}}(\mathbf{F}, \eta)$ is a positive semi-definite and symmetric second-order material conductivity tensor. Note that in the entropy-based formulation the temperature is given by (4.23). That is, in (4.30), $\widetilde{\Theta}(\mathbf{F}, \eta) = \partial_{\eta} \widetilde{u}(\mathbf{F}, \eta)$. It can be easily observed that the dissipative bracket (4.30) is symmetric, i.e.

$$[\widetilde{\mathcal{A}}, \widetilde{\mathcal{B}}] = [\widetilde{\mathcal{B}}, \widetilde{\mathcal{A}}] \,, \tag{4.31}$$

and positive semi-definite, i.e. $[\widetilde{\mathcal{A}}, \widetilde{\mathcal{A}}] \geq 0$.

4.2.3. Change of the thermodynamic variable

We next aim at a uniform description of the Poisson bracket and the dissipative bracket in terms of the thermodynamic state variable $\tau \in \{\theta, \eta, u\}$. To this end we consider functionals of the form

$$\mathcal{A} = \mathcal{A}'(\boldsymbol{\varphi}, \mathbf{p}, \tau) = \int_{\mathcal{B}} a'(\boldsymbol{\varphi}, \mathbf{F}, \mathbf{p}, \tau) \, \mathrm{d}V \,, \tag{4.32}$$

and perform a change of the thermodynamic state variable. As has been outlined above the thermodynamic state variable τ can be expressed in terms of the entropy density through relation (4.24), i.e. $\tau = \widetilde{\tau}(\mathbf{F}, \eta)$. Consequently, the two alternative expressions for the functional $\mathcal{A}$ given by (4.28) and (4.32) can be linked to obtain

$$\widetilde{\mathcal{A}}(\boldsymbol{\varphi}, \mathbf{p}, \eta) = \int_{\mathcal{B}} \widetilde{a}(\boldsymbol{\varphi}, \mathbf{F}, \mathbf{p}, \eta) \, \mathrm{d}V = \int_{\mathcal{B}} a'(\boldsymbol{\varphi}, \mathbf{F}, \mathbf{p}, \widetilde{\tau}(\mathbf{F}, \eta)) \, \mathrm{d}V \,.$$

To get the functional derivative of $\widetilde{\mathcal{A}}$ with respect to φ we consider

$$\begin{aligned}
&\frac{d}{d\varepsilon}\int_{\mathcal{B}} a'(\varphi+\varepsilon\delta\varphi,\mathbf{F}+\varepsilon\nabla\delta\varphi,\mathbf{p},\widetilde{\tau}(\mathbf{F}+\varepsilon\nabla\delta\varphi,\eta))\,\mathrm{d}V\bigg|_{\varepsilon=0}\\
&=\int_{\mathcal{B}}(\partial_{\varphi}a'\cdot\delta\varphi+(\partial_{\mathbf{F}}a'+\partial_{\tau}a'\partial_{\mathbf{F}}\widetilde{\tau}):\nabla\delta\varphi)\,\mathrm{d}V\\
&=\int_{\mathcal{B}}\delta\varphi\cdot(\partial_{\varphi}a'-\mathrm{Div}(\partial_{\mathbf{F}}a'+\partial_{\tau}a'\partial_{\mathbf{F}}\widetilde{\tau}))\,\mathrm{d}V\\
&=\int_{\mathcal{B}}\delta\varphi\cdot\delta_{\varphi}\widetilde{a}\,\mathrm{d}V\,.
\end{aligned}$$

Here, again integration by parts and neglect of the boundary terms has been applied. Similarly, we obtain

$$\begin{aligned}
&\frac{d}{d\varepsilon}\int_{\mathcal{B}} a'(\varphi,\mathbf{F},\mathbf{p},\widetilde{\tau}(\mathbf{F},\eta+\varepsilon\delta\eta))\,\mathrm{d}V\bigg|_{\varepsilon=0}\\
&=\int_{\mathcal{B}}\delta\eta\partial_{\tau}a'\partial_{\eta}\widetilde{\tau}\,\mathrm{d}V\\
&=\int_{\mathcal{B}}\delta\eta\delta_{\eta}\widetilde{a}\,\mathrm{d}V\,.
\end{aligned}$$

To summarize, we have the following relationships

$$\begin{aligned}
\delta_{\varphi}\widetilde{a}&=\delta_{\varphi}a'-\mathrm{Div}(\delta_{\tau}a'\partial_{\mathbf{F}}\widetilde{\tau})\,,\\
\delta_{\mathbf{p}}\widetilde{a}&=\delta_{\mathbf{p}}a'\,,\\
\delta_{\eta}\widetilde{a}&=\delta_{\tau}a'\partial_{\eta}\widetilde{\tau}\,,
\end{aligned}$$

where, in view of (4.32), $\delta_{\tau}a'=\partial_{\tau}a'$. The derivatives $\partial_{\eta}\widetilde{\tau}$ and $\partial_{\mathbf{F}}\widetilde{\tau}$ in the above equation can be substituted from (4.26) to obtain

$$\begin{aligned}
\delta_{\varphi}\widetilde{a}&=\delta_{\varphi}a'+\mathrm{Div}\left(\frac{\delta_{\tau}a'}{\partial_{\tau}\eta'}\partial_{\mathbf{F}}\eta'\right)\,,\\
\delta_{\mathbf{p}}\widetilde{a}&=\delta_{\mathbf{p}}a'\,,\\
\delta_{\eta}\widetilde{a}&=\frac{\delta_{\tau}a'}{\partial_{\tau}\eta'}\,.
\end{aligned}\tag{4.33}$$

These relationships can now be applied in the transformation of the brackets. Inserting $(4.33)_{1,2}$ into the entropy-based expression for the Poisson bracket (4.29), we arrive at the Poisson bracket formulated in terms of $\tau\in\{\theta,\eta,u\}$:

$$\begin{aligned}
\{\mathcal{A}',\mathcal{B}'\}=&\int_{\mathcal{B}}(\delta_{\varphi}a'\cdot\delta_{\mathbf{p}}b'-\delta_{\mathbf{p}}a'\cdot\delta_{\varphi}b')\,\mathrm{d}V\\
&+\int_{\mathcal{B}}\left(\mathrm{Div}\left(\frac{\delta_{\tau}a'}{\partial_{\tau}\eta'}\partial_{\mathbf{F}}\eta'\right)\cdot\delta_{\mathbf{p}}b'-\delta_{\mathbf{p}}a'\cdot\mathrm{Div}\left(\frac{\delta_{\tau}b'}{\partial_{\tau}\eta'}\partial_{\mathbf{F}}\eta'\right)\right)\mathrm{d}V\,.
\end{aligned}\tag{4.34}$$

τ	η'	Θ'	u'	$\partial_\tau \eta'$	$\partial_\tau u'$	$\partial_{\mathbf{F}} \eta'$	$\partial_{\mathbf{F}} u'$
η	η	$\partial_\eta \widetilde{u}(\mathbf{F},\eta)$	$\widetilde{u}(\mathbf{F},\eta)$	1	$\partial_\eta \widetilde{u}(\mathbf{F},\eta)$	$\mathbf{0}$	$\partial_{\mathbf{F}} \widetilde{u}(\mathbf{F},\eta)$
θ	$-\partial_\theta \overline{\psi}(\mathbf{F},\theta)$	θ	$\overline{u}(\mathbf{F},\theta)$	$\partial_\theta \overline{\eta}(\mathbf{F},\theta)$	$\partial_\theta \overline{u}(\mathbf{F},\theta)$	$\partial_{\mathbf{F}} \overline{\eta}(\mathbf{F},\theta)$	$\partial_{\mathbf{F}} \overline{u}(\mathbf{F},\theta)$
u	$\widehat{\eta}(\mathbf{F},u)$	$[\partial_u \widehat{\eta}(\mathbf{F},u)]^{-1}$	u	$\partial_u \widehat{\eta}(\mathbf{F},u)$	1	$\partial_{\mathbf{F}} \widehat{\eta}(\mathbf{F},u)$	$\mathbf{0}$

Table 4.1.: Specific relationships needed for the evaluation of the uniform expressions for the Poisson bracket (4.34) and the dissipative bracket (4.35), depending on the choice for the thermodynamic state variable $\tau \in \{\theta, \eta, u\}$.

Obviously, symmetry property (4.16) still holds. That is, $\{\mathcal{A}', \mathcal{B}'\} = -\{\mathcal{B}', \mathcal{A}'\}$. Substituting (4.33)$_3$ into expression (4.30) for the entropy-based dissipative bracket we get the dissipative bracket in terms of $\tau \in \{\theta, \eta, u\}$:

$$[\mathcal{A}', \mathcal{B}'] = \int_{\mathcal{B}} \nabla \left(\frac{\delta_\tau a'}{\partial_\tau u'} \right) \cdot (\Theta')^2 \mathbf{K} \nabla \left(\frac{\delta_\tau b'}{\partial_\tau u'} \right) \mathrm{d}V \, . \tag{4.35}$$

Note that in the last equation Θ' is given by formula (4.27) for the temperature and the material conductivity tensor is given by $\mathbf{K} = \overline{\mathbf{K}}(\mathbf{F}, \Theta') = \mathbf{K}'(\mathbf{F}, \tau)$. Moreover, symmetry property (4.31) is still satisfied being $[\mathcal{A}', \mathcal{B}'] = [\mathcal{B}', \mathcal{A}']$.

Depending on the choice for $\tau \in \{\theta, \eta, u\}$, specific expressions for the respective brackets can be obtained from (4.34) and (4.35). In this connection Table 4.1 provides a summary of specific formulas. With regard to Table 4.1 it can be easily concluded that the Poisson bracket (4.34) assumes its simplest form in the entropy-based formulation, while the dissipative bracket (4.35) assumes its simplest form in the formulation based on the internal energy density.

Example: The Helmholtz free energy density (4.17) that will be used in the numerical investigations (see Section 4.5 and, for the following Chapters, Sections 5.4 and 6.5) is given by

$$\overline{\psi}(\mathbf{F}, \theta) = \psi_1(\mathbf{F}) + \psi_2(\theta) - (\theta - \theta_0)\psi_3(J) \, , \tag{4.36}$$

where

$$\begin{aligned} \psi_1(\mathbf{F}) &= \frac{\mu}{2} \left(\mathbf{F} : \mathbf{F} - 3 - 2 \log J - \frac{2}{3} (J-1)^2 \right) + W_{\text{vol}}(J) \, , \\ \psi_2(\theta) &= c \left(\theta - \theta_0 - \theta \log \left(\frac{\theta}{\theta_0} \right) \right) \, , \\ \psi_3(J) &= 3 \beta W'_{\text{vol}}(J) \, , \\ W_{\text{vol}}(J) &= \frac{\lambda + \frac{2}{3}\mu}{4} \left((\log J)^2 + (J-1)^2 \right) \, . \end{aligned} \tag{4.37}$$

Here, $J = \det(\mathbf{F})$ is the determinant of the deformation gradient and μ, λ are prescribed parameters, $c > 0$ is the specific heat capacity at constant deformation, β is the coefficient of thermal expansion, and θ_0 is the reference temperature. We refer to [74, App. C] for a detailed investigation of the specific Helmholtz free energy density (4.36). It is worth noting that the Helmholtz free energy density (4.36) is consistent with the special case of linear thermo-elasticity (see Appendix A.1). It is now a straightforward exercise to calculate the quantities needed in Table 4.1. In particular, the *temperature-based formulation* (cf. θ-row in Table 4.1) yields

$$
\begin{aligned}
\overline{\eta}(\mathbf{F},\theta) &= c\log\left(\frac{\theta}{\theta_0}\right) + \psi_3(J)\,,\\
\overline{u}(\mathbf{F},\theta) &= \psi_1(\mathbf{F}) + c(\theta-\theta_0) + \theta_0\psi_3(J)\,,\\
\partial_\theta\overline{\eta}(\mathbf{F},\theta) &= \frac{c}{\theta}\,,\\
\partial_\theta\overline{u}(\mathbf{F},\theta) &= c\,,\\
\partial_{\mathbf{F}}\overline{\eta}(\mathbf{F},\theta) &= \psi_3'(J)\mathrm{cof}(\mathbf{F})\,,\\
\partial_{\mathbf{F}}\overline{u}(\mathbf{F},\theta) &= \partial_{\mathbf{F}}\psi_1(\mathbf{F}) + \theta_0\psi_3'(J)\mathrm{cof}(\mathbf{F})\,.
\end{aligned}
$$

Here, $\mathrm{cof}(\mathbf{F}) = J(\mathbf{F})^{-\mathrm{T}}$ denotes the cofactor of the deformation gradient. The formulation based on the *entropy density* (cf. η-row in Table 4.1) gives

$$
\begin{aligned}
\widetilde{u}(\mathbf{F},\eta) &= \psi_1(\mathbf{F}) + c\left(\theta_0 e^{\frac{\eta-\psi_3(J)}{c}} - \theta_0\right) + \theta_0\psi_3(J)\,,\\
\partial_\eta\widetilde{u}(\mathbf{F},\eta) &= \theta_0 e^{\frac{\eta-\psi_3(J)}{c}} = \widetilde{\Theta}(\mathbf{F},\eta)\\
\partial_{\mathbf{F}}\widetilde{u}(\mathbf{F},\eta) &= \partial_{\mathbf{F}}\psi_1(\mathbf{F}) - \left(\widetilde{\Theta}(\mathbf{F},\eta) - \theta_0\right)\psi_3'(J)\mathrm{cof}(\mathbf{F})\,.
\end{aligned}
$$

Moreover, the formulation based on the *internal energy density* (cf. u-row in Table 4.1) leads to

$$
\begin{aligned}
\widehat{\eta}(\mathbf{F},u) &= c\log\left(\frac{u + c\theta_0 - \psi_1(\mathbf{F}) - \theta_0\psi_3(J)}{c}\right) + \psi_3(J)\,,\\
\partial_u\widehat{\eta}(\mathbf{F},u) &= \left(\frac{u + c\theta_0 - \psi_1(\mathbf{F}) - \theta_0\psi_3(J)}{c}\right)^{-1} = \left[\widehat{\Theta}(\mathbf{F},u)\right]^{-1}\,,\\
\partial_{\mathbf{F}}\widehat{\eta}(\mathbf{F},u) &= -\left[\widehat{\Theta}(\mathbf{F},u)\right]^{-1}\left(\partial_{\mathbf{F}}\psi_1(\mathbf{F}) + \theta_0\psi_3'(J)\mathrm{cof}(\mathbf{F})\right) + \psi_3'(J)\mathrm{cof}(\mathbf{F})\,.
\end{aligned}
$$

4.2.4. GENERIC evolution equation

The GENERIC evolution equation (4.1) can now be considered in more detail by using the previously derived Poisson bracket (4.34) along with the dissipative bracket (4.35). Accordingly, we obtain

$$\frac{\mathrm{d}\mathcal{A}'}{\mathrm{d}t} = \{\mathcal{A}', \mathcal{E}'\} + [\mathcal{A}', \mathcal{S}'] \, , \tag{4.38}$$

for functionals $\mathcal{A}'$ of the form (4.32). Note that the evolution equation (4.38) represents a uniform description of thermo-elastodynamics in terms of the thermodynamic state variable $\tau \in \{\theta, \eta, u\}$. In (4.38), the total energy of the system is given by the functional

$$\mathcal{E}'(\boldsymbol{\varphi}, \mathbf{p}, \tau) = \int_{\mathcal{B}} e'(\boldsymbol{\varphi}, \mathbf{F}, \mathbf{p}, \tau) \, \mathrm{d}V \, , \tag{4.39}$$

with associated density function

$$e'(\boldsymbol{\varphi}, \mathbf{F}, \mathbf{p}, \tau) = \frac{1}{2} \rho^{-1} \mathbf{p} \cdot \mathbf{p} + u'(\mathbf{F}, \tau) - \mathbf{b} \cdot \boldsymbol{\varphi} \, . \tag{4.40}$$

Note that in the isothermal case the energy density (4.40) reduces to the Hamiltonian density (4.7). In addition to the total energy, the total entropy of the system acts as second generator in the GENERIC evolution equation (4.38) and is given by the functional

$$\mathcal{S}'(\boldsymbol{\varphi}, \tau) = \int_{\mathcal{B}} \eta'(\mathbf{F}, \tau) \, \mathrm{d}V \, . \tag{4.41}$$

In the last equation, $\eta : \mathcal{B} \times \mathcal{I} \mapsto \mathbb{R}$ is the entropy density which has been introduced in (4.25). For later use we mention that the following functional derivatives follow from the definition of the total energy (4.39) and the total entropy (4.41), respectively:

$$\begin{aligned} \delta_{\boldsymbol{\varphi}} e' &= \partial_{\boldsymbol{\varphi}} e' - \mathrm{Div} \partial_{\mathbf{F}} e' = -\mathbf{b} - \mathrm{Div} \partial_{\mathbf{F}} u' \, , \\ \delta_{\mathbf{p}} e' &= \partial_{\mathbf{p}} e' = \rho^{-1} \mathbf{p} \, , \\ \delta_{\tau} e' &= \partial_{\tau} e' = \partial_{\tau} u' \, , \end{aligned} \tag{4.42}$$

and

$$\begin{aligned} \delta_{\boldsymbol{\varphi}} \eta' &= -\mathrm{Div} \partial_{\mathbf{F}} \eta' \, , \\ \delta_{\mathbf{p}} \eta' &= \mathbf{0} \, , \\ \delta_{\tau} \eta' &= \partial_{\tau} \eta' \, . \end{aligned} \tag{4.43}$$

Now the left-hand side of (4.38) can be written as

$$\begin{aligned} \frac{\mathrm{d}}{\mathrm{d}t} \mathcal{A}' &= \int_{\mathcal{B}} \frac{\mathrm{d}}{\mathrm{d}t} a'(\boldsymbol{\varphi}, \mathbf{F}, \mathbf{p}, \tau) \, \mathrm{d}V \\ &= \int_{\mathcal{B}} \left(\partial_{\boldsymbol{\varphi}} a' \cdot \dot{\boldsymbol{\varphi}} + \partial_{\mathbf{F}} a' : \dot{\mathbf{F}} + \partial_{\mathbf{p}} a' \cdot \dot{\mathbf{p}} + \partial_{\tau} a' \dot{\tau} \right) \mathrm{d}V \\ &= \int_{\mathcal{B}} \left(\delta_{\boldsymbol{\varphi}} a' \cdot \dot{\boldsymbol{\varphi}} + \delta_{\mathbf{p}} a' \cdot \dot{\mathbf{p}} + \delta_{\tau} a' \dot{\tau} \right) \mathrm{d}V \, . \end{aligned} \tag{4.44}$$

Note that again integration by parts along with the neglect of boundary terms has been applied. The Poisson bracket on the right-hand side of (4.38) follows from (4.34) together with (4.40). Thus we obtain

$$\begin{aligned}\{\mathcal{A}',\mathcal{E}'\} &= \int_{\mathcal{B}} (\delta_{\boldsymbol{\varphi}} a' \cdot \partial_{\mathbf{p}} e' - \delta_{\mathbf{p}} a' \cdot \delta_{\boldsymbol{\varphi}} e') \, \mathrm{d}V \\ &+ \int_{\mathcal{B}} \left(\mathrm{Div}\left(\frac{\delta_\tau a'}{\partial_\tau \eta'} \partial_{\mathbf{F}} \eta' \right) \cdot \partial_{\mathbf{p}} e' - \delta_{\mathbf{p}} a' \cdot \mathrm{Div}\left(\frac{\partial_\tau u'}{\partial_\tau \eta'} \partial_{\mathbf{F}} \eta' \right) \right) \mathrm{d}V \, .\end{aligned} \tag{4.45}$$

Similarly, the dissipative bracket on the right-hand side of (4.38) follows from inserting (4.41) into (4.35). Accordingly,

$$\left[\mathcal{A}',\mathcal{S}'\right] = \int_{\mathcal{B}} \nabla\left(\frac{\delta_\tau a'}{\partial_\tau u'}\right) \cdot (\Theta')^2 \mathbf{K} \nabla \left(\frac{\partial_\tau \eta'}{\partial_\tau u'}\right) \mathrm{d}V \, . \tag{4.46}$$

Next, we introduce the material heat flux vector $\mathbf{Q} : \mathcal{B} \times \mathcal{I} \mapsto \mathbb{R}^3$ through $\mathbf{Q} = \mathbf{Q}'(\mathbf{F},\tau)$, where

$$\begin{aligned}\mathbf{Q}' &= (\Theta')^2 \mathbf{K} \nabla \left(\frac{\partial_\tau \eta'}{\partial_\tau u'}\right) \\ &= (\Theta')^2 \mathbf{K} \nabla \left(\frac{1}{\Theta'}\right) \\ &= -\mathbf{K} \nabla \Theta' \, .\end{aligned} \tag{4.47}$$

Note that in the above equation use has been made of relationship (4.27) for the temperature field $\Theta'(\mathbf{F},\tau)$. Now, the bracket (4.46) can be recast in the form

$$\left[\mathcal{A}',\mathcal{S}'\right] = \int_{\mathcal{B}} \nabla\left(\frac{\delta_\tau a'}{\partial_\tau u'}\right) \cdot \mathbf{Q}' \, \mathrm{d}V \, . \tag{4.48}$$

It is important to emphasize that the present brackets need to satisfy the fundamental degeneracy requirements of the GENERIC framework

$$\begin{aligned}\{\mathcal{A}',\mathcal{S}'\} &= 0 \, , \\ \left[\mathcal{A}',\mathcal{E}'\right] &= 0 \, .\end{aligned} \tag{4.49}$$

The validity of these equations for arbitrary functionals $\mathcal{A}'$ can be easily verified. The degeneracy requirements (4.49) ensure the thermodynamical consistency of the GENERIC evolution equation (4.38) in the sense that $\mathrm{d}\mathcal{E}'/\mathrm{d}t = 0$ and $\mathrm{d}\mathcal{S}'/\mathrm{d}t = \left[\mathcal{S}',\mathcal{S}'\right] \geq 0$. Note that the last inequality follows from the above expression for the dissipative bracket together with the positive semi-definiteness of the material conductivity tensor $\mathbf{K}$. Accordingly, the total entropy is non-decreasing in the present case of isolated systems in which the net heating vanishes.

Taking into account relations (4.44), (4.45) and (4.48), the GENERIC evolution equation (4.38) can be recast in the form

$$\begin{aligned}0 =& \int_{\mathcal{B}} \delta_{\boldsymbol{\varphi}} a' \cdot (\dot{\boldsymbol{\varphi}} - \partial_{\mathbf{p}} e')\, \mathrm{d}V \\ &+ \int_{\mathcal{B}} \delta_{\mathbf{p}} a' \cdot \left(\dot{\mathbf{p}} + \delta_{\boldsymbol{\varphi}} e' + \mathrm{Div}\left(\frac{\partial_\tau u'}{\partial_\tau \eta'} \partial_{\mathbf{F}} \eta' \right) \right) \mathrm{d}V \\ &+ \int_{\mathcal{B}} \left(\delta_\tau a' \dot{\tau} - \mathrm{Div}\left(\frac{\delta_\tau a'}{\partial_\tau \eta'} \partial_{\mathbf{F}} \eta' \right) \cdot \partial_{\mathbf{p}} e' - \nabla \left(\frac{\delta_\tau a'}{\partial_\tau u'} \right) \cdot \mathbf{Q}' \right) \mathrm{d}V\,.\end{aligned} \tag{4.50}$$

This equation has to hold for arbitrary functionals $\mathcal{A}'$ of the form (4.32). Consequently, the first two integrals in the last equation yield the local equations

$$\begin{aligned}\dot{\boldsymbol{\varphi}} &= \rho^{-1} \mathbf{p}\,, \\ \dot{\mathbf{p}} &= \mathbf{b} + \mathrm{Div}\mathbf{P}\,,\end{aligned} \tag{4.51}$$

that must be satisfied for all $\mathbf{X} \in \mathcal{B}$ and $t > 0$. In $(4.51)_2$, the first Piola-Kirchhoff stress tensor $\mathbf{P} = \mathbf{P}'(\mathbf{F}, \tau)$ can be identified as

$$\mathbf{P}' = \partial_{\mathbf{F}} u' - \Theta' \partial_{\mathbf{F}} \eta'\,, \tag{4.52}$$

where Θ' is given by eq. (4.27). Concerning the third integral in (4.50), integration by parts and neglect of the boundary terms yields

$$\int_{\mathcal{B}} \delta_\tau a' \left(\dot{\tau} + \frac{1}{\partial_\tau \eta'} \partial_{\mathbf{F}} \eta' : \nabla(\rho^{-1} \mathbf{p}) + \frac{1}{\partial_\tau u'} \mathrm{Div}\mathbf{Q}' \right) \mathrm{d}V = 0\,.$$

Correspondingly, we arrive at the local form of the evolution of the thermodynamic variable $\tau \in \{\theta, \eta, u\}$, given by

$$\dot{\tau} = -\frac{1}{\partial_\tau \eta'} \partial_{\mathbf{F}} \eta' : \nabla(\rho^{-1} \mathbf{p}) - \frac{1}{\partial_\tau u'} \mathrm{Div}\mathbf{Q}'\,, \tag{4.53}$$

which has to hold for all $\mathbf{X} \in \mathcal{B}$ and $t > 0$.

Next, we summarize main ingredients of the present GENERIC-based formulation resulting from the specific choice of the thermodynamic state variable. Choosing $\tau = \theta$ leads to the *temperature-based formulation*. With regard to (4.52), the first Piola-Kirchhoff stress tensor assumes the form

$$\overline{\mathbf{P}} = \partial_{\mathbf{F}} \overline{u} - \theta \partial_{\mathbf{F}} \overline{\eta}\,.$$

Furthermore, (4.53) yields the temperature evolution equation given by

$$\dot{\theta} = -\frac{1}{\partial_\theta \overline{\eta}} \partial_{\mathbf{F}} \overline{\eta} : \nabla(\rho^{-1} \mathbf{p}) - \frac{1}{\partial_\theta \overline{u}} \mathrm{Div}\overline{\mathbf{Q}}\,.$$

The formulation in terms of the *entropy density* results from the choice $\tau = \eta$ leading to the first Piola-Kirchhoff stress tensor

$$\widetilde{\mathbf{P}} = \partial_{\mathbf{F}} \widetilde{u} \, .$$

The evolution of the specific entropy is governed by

$$\dot{\eta} = -\frac{1}{\widetilde{\Theta}} \mathrm{Div} \widetilde{\mathbf{Q}} \, .$$

Choosing $\tau = u$ yields the formulation based on the *internal energy density* with associated first Piola-Kirchhoff stress tensor

$$\widehat{\mathbf{P}} = -\widehat{\Theta} \partial_{\mathbf{F}} \widehat{\eta} \, .$$

Furthermore, the evolution of the internal energy density follows from

$$\dot{u} = -\widehat{\Theta} \partial_{\mathbf{F}} \widehat{\eta} : \nabla(\rho^{-1} \mathbf{p}) - \mathrm{Div} \widehat{\mathbf{Q}} \, .$$

Remark 3. Apart from the dead loads $\mathbf{b} = -\partial_{\varphi} e'$, non-potential material body forces per unit of reference volume, $\mathbf{f} : \mathcal{B} \times \mathcal{I} \mapsto \mathbb{R}^3$, and a heat supply field per unit of reference volume, $R : \mathcal{B} \times \mathcal{I} \mapsto \mathbb{R}$, can also be included into the present framework. To this end the supply term

$$(\mathcal{A}', \mathcal{F}') = \int_{\mathcal{B}} \left(\delta_{\mathbf{p}} a' \cdot \mathbf{f} + \frac{\delta_{\tau} a'}{\partial_{\tau} u'} R \right) \mathrm{d}V \, , \tag{4.54}$$

needs be appended to the right-hand side of the GENERIC evolution equation (4.38).

4.3. GENERIC for open systems

As pointed out in [94], there is no need to pay special attention to the boundary conditions if one is interested in the local field equations within an isolated system. Indeed, the majority of works dealing with the GENERIC formalism have been conducted in the context of closed systems. However, when it comes to numerical simulation, the boundary conditions are obviously of paramount importance. Consequently, we now aim at the inclusion of boundary effects (Fig. 4.2) into the GENERIC-based framework for finite-strain thermo-elastodynamics developed above. To this end, we build on ideas presented in [138] and apply them to the present framework for thermoelastic solids. Accordingly, the Poisson bracket (4.34) and the dissipative bracket (4.35) are now viewed as full brackets that can be decomposed according to

$$\begin{aligned} \{\mathcal{A}', \mathcal{B}'\} &= \{\mathcal{A}', \mathcal{B}'\}_{\text{bulk}} + \{\mathcal{A}', \mathcal{B}'\}_{\text{boun}} \, , \\ [\mathcal{A}', \mathcal{B}'] &= [\mathcal{A}', \mathcal{B}']_{\text{bulk}} + [\mathcal{A}', \mathcal{B}']_{\text{boun}} \, . \end{aligned} \tag{4.55}$$

That is, the full brackets are split additively into bulk and boundary parts, respectively.

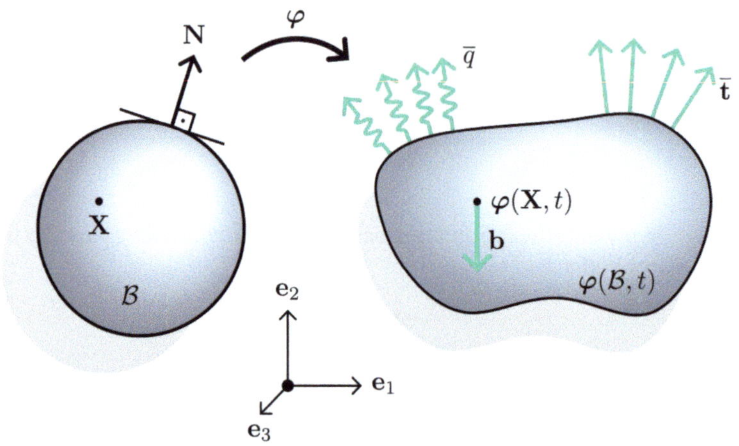

Figure 4.2.: Reference configuration $\mathcal{B}$ with boundary $\partial\mathcal{B}$ and current configuration $\varphi(\mathcal{B}, t)$ at time t. External tractions $\overline{\mathbf{t}} = \mathbf{PN}$ act on the boundary of the current configuration. In addition to that, the heat flux across the current boundary is denoted by $\overline{q} = \mathbf{Q} \cdot \mathbf{N}$.

This split can be accomplished by applying integration by parts to the full brackets. Following [138], the proper interpretation of the GENERIC evolution equation (4.38) is now given by

$$\frac{\mathrm{d}\mathcal{A}'}{\mathrm{d}t} = \{\mathcal{A}', \mathcal{E}'\}_{\text{bulk}} + [\mathcal{A}', \mathcal{S}']_{\text{bulk}} \,. \tag{4.56}$$

The degeneracy requirements (4.49) for the full brackets transfer to the corresponding degeneracy requirements for the bulk brackets

$$\begin{aligned} \{\mathcal{A}', \mathcal{S}'\}_{\text{bulk}} &= 0 \,, \\ [\mathcal{A}', \mathcal{E}']_{\text{bulk}} &= 0 \,, \end{aligned} \tag{4.57}$$

for arbitrary functionals $\mathcal{A}'$.

Invoking the symmetry properties of the full brackets, i.e. $\{\mathcal{A}', \mathcal{B}'\} = -\{\mathcal{B}', \mathcal{A}'\}$ and $[\mathcal{A}', \mathcal{B}'] = [\mathcal{B}', \mathcal{A}']$, (4.55) leads to

$$\begin{aligned} \{\mathcal{A}', \mathcal{B}'\}_{\text{bulk}} &= -\{\mathcal{B}', \mathcal{A}'\}_{\text{bulk}} - \left(\{\mathcal{A}', \mathcal{B}'\}_{\text{boun}} + \{\mathcal{B}', \mathcal{A}'\}_{\text{boun}}\right) \,, \\ [\mathcal{A}', \mathcal{B}']_{\text{bulk}} &= \;\; [\mathcal{B}', \mathcal{A}']_{\text{bulk}} - \left([\mathcal{A}', \mathcal{B}']_{\text{boun}} - [\mathcal{B}', \mathcal{A}']_{\text{boun}}\right) \,. \end{aligned} \tag{4.58}$$

Substitution from (4.58) into (4.56) yields the GENERIC evolution equation for open systems

$$\begin{aligned}\frac{\mathrm{d}\mathcal{A}'}{\mathrm{d}t} &= -\{\mathcal{E}',\mathcal{A}'\}_{\text{bulk}} + [\mathcal{S}',\mathcal{A}']_{\text{bulk}} \\ &\quad - \left(\{\mathcal{A}',\mathcal{E}'\}_{\text{boun}} + \{\mathcal{E}',\mathcal{A}'\}_{\text{boun}} + [\mathcal{A}',\mathcal{S}']_{\text{boun}} - [\mathcal{S}',\mathcal{A}']_{\text{boun}}\right) .\end{aligned} \tag{4.59}$$

In the sequel, the above equation will be used to formulate important balance laws and to derive the variational formulation of large-strain thermo-elastodynamics. It is worth mentioning that Remark 3 also applies to evolution equation (4.59).

4.3.1. Specific brackets

Next, we provide the specific brackets featuring in the GENERIC evolution equation (4.59) pertaining to thermoelastic solids. Proceeding along the lines of [138], we start from the full brackets derived above and apply integration by parts. In this connection, the goal is to get the derivatives $(\delta_{\varphi}a', \delta_{\mathbf{p}}a', \delta_{\tau}a')$ free of any spatial derivatives. To this end, we rewrite the full Poisson bracket (4.34) as

$$\begin{aligned}\{\mathcal{A}',\mathcal{B}'\} &= \int_{\mathcal{B}} \left(\partial_{\varphi}a' \cdot \partial_{\mathbf{p}}b' - \partial_{\mathbf{p}}a' \cdot \left(\partial_{\varphi}b' - \operatorname{Div}\left(\partial_{\mathbf{F}}b' - \frac{\partial_{\tau}b'}{\partial_{\tau}\eta'}\partial_{\mathbf{F}}\eta'\right)\right)\right) \mathrm{d}V \\ &\quad - \int_{\mathcal{B}} \operatorname{Div}\left(\partial_{\mathbf{F}}a' - \frac{\partial_{\tau}a'}{\partial_{\tau}\eta'}\partial_{\mathbf{F}}\eta'\right) \cdot \partial_{\mathbf{p}}b' \, \mathrm{d}V .\end{aligned} \tag{4.60}$$

Applying integration by parts to the second integral on the right-hand side of (4.60), and taking into account the additive decomposition $(4.55)_1$, we obtain

$$\begin{aligned}\{\mathcal{A}',\mathcal{B}'\}_{\text{bulk}} &= \int_{\mathcal{B}} \left(\partial_{\varphi}a' \cdot \partial_{\mathbf{p}}b' - \partial_{\mathbf{p}}a' \cdot \left(\partial_{\varphi}b' - \operatorname{Div}\left(\partial_{\mathbf{F}}b' - \frac{\partial_{\tau}b'}{\partial_{\tau}\eta'}\partial_{\mathbf{F}}\eta'\right)\right)\right) \mathrm{d}V \\ &\quad + \int_{\mathcal{B}} \left(\partial_{\mathbf{F}}a' - \frac{\partial_{\tau}a'}{\partial_{\tau}\eta'}\partial_{\mathbf{F}}\eta'\right) : \nabla\partial_{\mathbf{p}}b' \, \mathrm{d}V ,\end{aligned} \tag{4.61}$$

and

$$\{\mathcal{A}',\mathcal{B}'\}_{\text{boun}} = -\int_{\partial\mathcal{B}} \partial_{\mathbf{p}}b' \cdot \left(\partial_{\mathbf{F}}a' - \frac{\partial_{\tau}a'}{\partial_{\tau}\eta'}\partial_{\mathbf{F}}\eta'\right) \mathbf{N} \, \mathrm{d}A . \tag{4.62}$$

Here, the vector $\mathbf{N}$ denotes the unit outward normal field on the boundary $\partial\mathcal{B}$ of the reference configuration (Fig. 4.2). Similarly, applying integration by parts to the full dissipative bracket (4.35), yields the corresponding bulk bracket

$$[\mathcal{A}',\mathcal{B}']_{\text{bulk}} = -\int_{\mathcal{B}} \frac{\partial_{\tau}a'}{\partial_{\tau}u'} \operatorname{Div}\left((\Theta')^2 \mathbf{K} \nabla\left(\frac{\partial_{\tau}b'}{\partial_{\tau}u'}\right)\right) \mathrm{d}V , \tag{4.63}$$

along with the boundary bracket

$$[\mathcal{A}', \mathcal{B}']_{\text{boun}} = \int_{\partial\mathcal{B}} \frac{\partial_\tau a'}{\partial_\tau u'} \mathbf{N} \cdot (\Theta')^2 \mathbf{K} \nabla \left(\frac{\partial_\tau b'}{\partial_\tau u'} \right) \mathrm{d}A \,. \tag{4.64}$$

It can be easily verified that the degeneracy conditions (4.57) are satisfied by the bulk brackets (4.61) and (4.63), respectively. Table 4.2 contains a summary of the specific brackets appearing in the GENERIC evolution equations (4.59) for open systems. The brackets in Table 4.2 can be easily obtained from expressions (4.61) through (4.64) by taking into account the generators in (4.39) and (4.41).

Table 4.2.: Summary of the brackets featuring in the GENERIC evolution equations (4.59) for open systems.

$$\begin{aligned}
\{\mathcal{E}', \mathcal{A}'\}_{\text{bulk}} &= -\int_{\mathcal{B}} \left(\mathbf{b} \cdot \partial_{\mathbf{p}} a' + \rho^{-1} \mathbf{p} \cdot \left(\partial_{\varphi} a' - \operatorname{Div} \left(\partial_{\mathbf{F}} a' - \frac{\partial_\tau a'}{\partial_\tau \eta'} \partial_{\mathbf{F}} \eta' \right) \right) \right) \mathrm{d}V \\
&\quad + \int_{\mathcal{B}} \mathbf{P}' : \nabla \partial_{\mathbf{p}} a' \, \mathrm{d}V \,, \\
[\mathcal{S}', \mathcal{A}']_{\text{bulk}} &= -\int_{\mathcal{B}} \frac{1}{\Theta'} \operatorname{Div} \left((\Theta')^2 \mathbf{K} \nabla \left(\frac{\partial_\tau a'}{\partial_\tau u'} \right) \right) \mathrm{d}V \,, \\
\{\mathcal{A}', \mathcal{E}'\}_{\text{boun}} &= -\int_{\partial\mathcal{B}} \rho^{-1} \mathbf{p} \cdot \left(\partial_{\mathbf{F}} a' - \frac{\partial_\tau a'}{\partial_\tau \eta'} \partial_{\mathbf{F}} \eta' \right) \mathbf{N} \, \mathrm{d}A \,, \\
\{\mathcal{E}', \mathcal{A}'\}_{\text{boun}} &= -\int_{\partial\mathcal{B}} \partial_{\mathbf{p}} a' \cdot \mathbf{P}' \mathbf{N} \, \mathrm{d}A \,, \\
[\mathcal{A}', \mathcal{S}']_{\text{boun}} &= \int_{\partial\mathcal{B}} \frac{\partial_\tau a'}{\partial_\tau u'} \mathbf{N} \cdot \mathbf{Q}' \, \mathrm{d}A \,, \\
[\mathcal{S}', \mathcal{A}']_{\text{boun}} &= \int_{\partial\mathcal{B}} \frac{1}{\Theta'} \mathbf{N} \cdot (\Theta')^2 \mathbf{K} \nabla \left(\frac{\partial_\tau a'}{\partial_\tau u'} \right) \mathrm{d}A \,.
\end{aligned}$$

4.3.2. Balance laws

Next, we deduce the balance laws in material (or Lagrangian) form from the GENERIC evolution equation (4.59). We start with the total linear momentum of the continuum body defined by $\mathbf{L} = \int_{\mathcal{B}} \mathbf{p} \, \mathrm{d}V$. In particular, we choose the density $a' = l'_{\xi}$, with $l'_{\xi} = \boldsymbol{\xi} \cdot \mathbf{p}$, where $\boldsymbol{\xi} \in \mathbb{R}^3$ is arbitrary and constant. This corresponds to the functional $\mathcal{L}'_{\xi} = \boldsymbol{\xi} \cdot \mathbf{L}$. Substituting $\mathcal{L}'_{\xi}$ for $\mathcal{A}'$ in (4.59), the non-zero contributions on the right-hand side of (4.59) are given by $\{\mathcal{E}', \mathcal{L}'_{\xi}\}_{\text{bulk}} = -\int_{\mathcal{B}} \mathbf{b} \cdot \boldsymbol{\xi} \, \mathrm{d}V$ and $\{\mathcal{E}', \mathcal{L}'_{\xi}\}_{\text{boun}} = -\int_{\partial\mathcal{B}} \boldsymbol{\xi} \cdot \mathbf{P}\mathbf{N} \, \mathrm{d}A$. Consequently, we obtain

$$\boldsymbol{\xi} \cdot \frac{\mathrm{d}\mathbf{L}}{\mathrm{d}t} = \boldsymbol{\xi} \cdot \left(\int_{\mathcal{B}} \mathbf{b} \, \mathrm{d}V + \int_{\partial\mathcal{B}} \mathbf{P}\mathbf{N} \, \mathrm{d}A \right) . \tag{4.65}$$

Due to the arbitrariness of $\boldsymbol{\xi} \in \mathbb{R}^3$, (4.65) coincides with the balance law for linear momentum. Note that the parentheses on the right-hand side of (4.65) contain the resultant external loads applied to the continuum body (see also Fig. 4.2).

The total angular momentum relative to the origin of the inertial frame is defined by $\mathbf{J} = \int_{\mathcal{B}} \boldsymbol{\varphi} \times \mathbf{p} \, \mathrm{d}V$. We choose $a' = j'_{\xi}$, where $j'_{\xi} = \boldsymbol{\xi} \cdot (\boldsymbol{\varphi} \times \mathbf{p})$. Correspondingly, we substitute the functional $\mathcal{J}'_{\xi} = \boldsymbol{\xi} \cdot \mathbf{J}$ for $\mathcal{A}'$ in (4.59). The non-zero terms emanating from the right-hand side of (4.59) are given by $\{\mathcal{E}', \mathcal{J}'_{\xi}\}_{\text{bulk}} = -\int_{\mathcal{B}} \mathbf{b} \cdot (\boldsymbol{\xi} \times \boldsymbol{\varphi}) \, \mathrm{d}V$ and $\{\mathcal{E}', \mathcal{J}'_{\xi}\}_{\text{boun}} = -\int_{\partial\mathcal{B}} (\boldsymbol{\xi} \times \boldsymbol{\varphi}) \cdot \mathbf{PN} \, \mathrm{d}A$. Note that in the bulk bracket use has been made of the symmetry condition $\mathbf{FP}^{\mathrm{T}} = \mathbf{PF}^{\mathrm{T}}$. Consequently, the GENERIC evolution equation (4.59) yields

$$\boldsymbol{\xi} \cdot \frac{\mathrm{d}\mathbf{J}}{\mathrm{d}t} = \boldsymbol{\xi} \cdot \left(\int_{\mathcal{B}} \boldsymbol{\varphi} \times \mathbf{b} \, \mathrm{d}V + \int_{\partial\mathcal{B}} \boldsymbol{\varphi} \times \mathbf{PN} \, \mathrm{d}A \right) . \tag{4.66}$$

Since $\boldsymbol{\xi} \in \mathbb{R}^3$ is arbitrary, the last equation corresponds to the balance of angular momentum. Note that the parentheses on the right-hand side of (4.66) contain the resultant external torque about the origin (see also Fig. 4.2).

Employing expression (4.39) for the total energy in the GENERIC evolution equation (4.59), we obtain

$$\begin{aligned}
\frac{\mathrm{d}\mathcal{E}'}{\mathrm{d}t} &= -\{\mathcal{E}', \mathcal{E}'\}_{\text{bulk}} + [\mathcal{S}', \mathcal{E}']_{\text{bulk}} \\
&\quad - \left(\{\mathcal{E}', \mathcal{E}'\}_{\text{boun}} + \{\mathcal{E}', \mathcal{E}'\}_{\text{boun}} + [\mathcal{E}', \mathcal{S}']_{\text{boun}} - [\mathcal{S}', \mathcal{E}']_{\text{boun}}\right) \\
&= -\{\mathcal{E}', \mathcal{E}'\} + [\mathcal{S}', \mathcal{E}'] - \{\mathcal{E}', \mathcal{E}'\}_{\text{boun}} - [\mathcal{E}', \mathcal{S}']_{\text{boun}} \\
&= -\left(\{\mathcal{E}', \mathcal{E}'\}_{\text{boun}} + [\mathcal{E}', \mathcal{S}']_{\text{boun}}\right) .
\end{aligned}$$

In the above equation use has been made of the decomposition of the full brackets in (4.55) along with the skew-symmetry of the full Poisson bracket and the degeneracy condition $(4.49)_2$. Taking into account the specific brackets $\{\mathcal{E}', \mathcal{E}'\}_{\text{boun}} = -\int_{\partial\mathcal{B}} \rho^{-1}\mathbf{p} \cdot \mathbf{PN} \, \mathrm{d}A$ and $[\mathcal{E}', \mathcal{S}']_{\text{boun}} = \int_{\partial\mathcal{B}} \mathbf{N} \cdot \mathbf{Q} \, \mathrm{d}A$, the above rate of change of the total energy can be recast in the form

$$\frac{\mathrm{d}}{\mathrm{d}t} \int_{\mathcal{B}} \left(\frac{1}{2} \rho^{-1} \mathbf{p} \cdot \mathbf{p} + u' \right) \mathrm{d}V = \int_{\mathcal{B}} \mathbf{b} \cdot \dot{\boldsymbol{\varphi}} \, \mathrm{d}V + \int_{\partial\mathcal{B}} \left(\rho^{-1}\mathbf{p} \cdot \mathbf{PN} - \mathbf{N} \cdot \mathbf{Q} \right) \mathrm{d}A \, . \tag{4.67}$$

This is the balance of energy in material form which is in line with the first law of thermodynamics (see, for example, [139, Sec. 2.3]).

Inserting expression (4.41) for the total entropy into the GENERIC evolution equation (4.59) yields

$$\begin{aligned}
\frac{\mathrm{d}\mathcal{S}'}{\mathrm{d}t} &= -\{\mathcal{E}', \mathcal{S}'\} + [\mathcal{S}', \mathcal{S}'] - \{\mathcal{S}', \mathcal{E}'\}_{\text{boun}} - [\mathcal{S}', \mathcal{S}']_{\text{boun}} \\
&= [\mathcal{S}', \mathcal{S}'] - [\mathcal{S}', \mathcal{S}']_{\text{boun}} ,
\end{aligned}$$

where again use has been made of the decomposition of the full brackets in (4.55), degeneracy condition $(4.49)_1$, and $\{\mathcal{S}', \mathcal{E}'\}_{\text{boun}} = 0$. Since, as has been shown above, $[\mathcal{S}', \mathcal{S}'] \geq 0$, and $[\mathcal{S}', \mathcal{S}']_{\text{boun}} = \int_{\partial\mathcal{B}} (\Theta')^{-1} \mathbf{N} \cdot \mathbf{Q} \, \mathrm{d}A$ the above equation implies

$$\frac{\mathrm{d}\mathcal{S}'}{\mathrm{d}t} \geq - \int_{\partial\mathcal{B}} \frac{1}{\Theta'} \mathbf{N} \cdot \mathbf{Q} \, \mathrm{d}A \,. \tag{4.68}$$

This corresponds to the Clausius-Duhem form of the second law of thermodynamics (see, for example, [136, Sec. 5]).

4.3.3. Initial boundary value problem

We next deal with the initial boundary value problem (IBVP) pertaining to large-strain thermo-elastodynamics. To this end, we decompose the boundary $\partial\mathcal{B}$ of the continuum body into a displacement boundary $\partial_\varphi\mathcal{B}$, on which $\varphi = \overline{\varphi}$, and a traction boundary $\partial_\sigma\mathcal{B}$, on which $\mathbf{PN} = \overline{\mathbf{t}}$, where $\overline{\varphi}$ and $\overline{\mathbf{t}}$ are prescribed functions for $t \geq 0$ (Fig. 4.3). Moreover, $\partial_\varphi\mathcal{B} \cup \partial_\sigma\mathcal{B} = \partial\mathcal{B}$ and $\partial_\varphi\mathcal{B} \cap \partial_\sigma\mathcal{B} = \emptyset$. Similarly, for the thermal part we consider the subsets $\partial_\tau\mathcal{B}$ and $\partial_q\mathcal{B}$, with the properties $\partial_\tau\mathcal{B} \cup \partial_q\mathcal{B} = \partial\mathcal{B}$ and $\partial_\tau\mathcal{B} \cap \partial_q\mathcal{B} = \emptyset$ (Fig. 4.4). Here, the thermodynamic state variable is prescribed on $\partial_\tau\mathcal{B}$, i.e. $\tau = \overline{\tau}$, whereas the heat flux is prescribed on $\partial_q\mathcal{B}$, i.e. $\mathbf{Q} \cdot \mathbf{N} = \overline{q}$.

The goal is now to determine the motion $\varphi : \mathcal{B} \times \mathcal{I} \mapsto \mathbb{R}^3$, the linear momentum $\mathbf{p} : \mathcal{B} \times \mathcal{I} \mapsto \mathbb{R}^3$, and the thermodynamic state variable $\tau : \mathcal{B} \times \mathcal{I} \mapsto \mathbb{R}$. The unknown fields are subject to initial conditions of the form $\varphi(\cdot, 0) = \mathbf{X}$, $\mathbf{p}(\cdot, 0) = \rho\mathbf{V}_0$, and $\tau(\cdot, 0) = \tau^{\text{ini}}$ in $\mathcal{B}$. Here, $\mathbf{V}_0$ is a prescribed material velocity field and τ^{ini} is a prescribed field of the thermodynamic state variable $\tau \in \{\theta, \eta, u\}$. The unknown fields are determined by the variational problem to be dealt with in the next section.

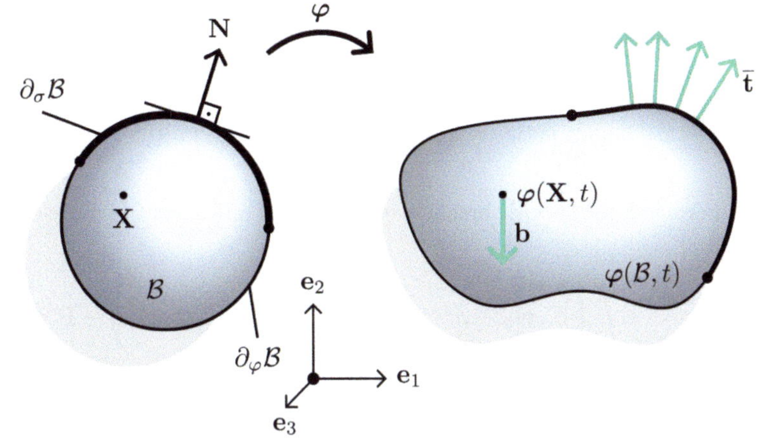

Figure 4.3.: Mechanical part of the IBVP. Note that $\bar{\mathbf{t}} = \mathbf{PN}$ denotes prescribed external Piola tractions acting on the current boundary expressed per unit area of the reference boundary $\partial_\sigma\mathcal{B}$.

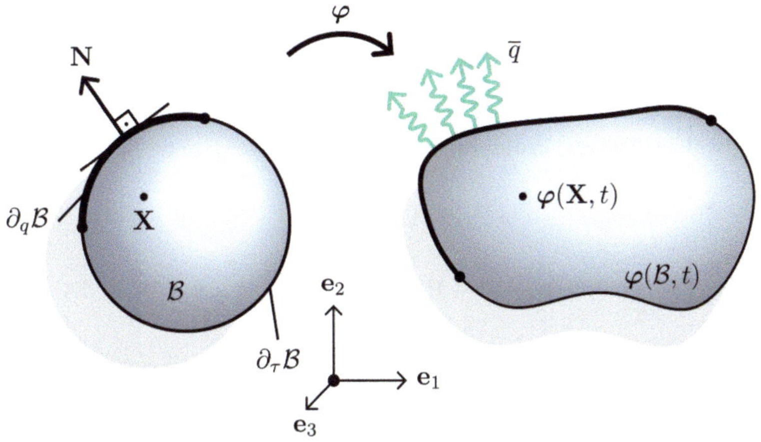

Figure 4.4.: Thermal part of the IBVP. Note that $\bar{q} = \mathbf{Q} \cdot \mathbf{N}$ is the prescribed rate of heat transfer accross the current boundary expressed per unit area of the reference boundary $\partial_q\mathcal{B}$.

4.3.3.1. Variational formulation

To deduce the variational formulation of the present IBVP from the GENERIC evolution equation (4.59), we choose the specific density function $a' = w'$, where

$$w' = \mathbf{w}_\varphi \cdot \boldsymbol{\varphi} + \mathbf{w}_\mathbf{p} \cdot \mathbf{p} + \mathrm{w}_\tau \tau \; . \tag{4.69}$$

Here, $\mathbf{w}_\varphi, \mathbf{w}_\mathbf{p} : \mathcal{B} \mapsto \mathbb{R}^3$ and $\mathrm{w}_\tau : \mathcal{B} \mapsto \mathbb{R}$ are test functions that have to satisfy the boundary conditions $\mathbf{w}_\varphi = \mathbf{0}$ and $\mathbf{w}_\mathbf{p} = \mathbf{0}$ on $\partial_\varphi \mathcal{B}$, and $\mathrm{w}_\tau = 0$ on $\partial_\tau \mathcal{B}$. With the choice (4.69), the left-hand side of the GENERIC evolution equation (4.59) yields

$$\frac{\mathrm{d}}{\mathrm{d}t}\mathcal{W}' = \int_\mathcal{B} \left(\mathbf{w}_\varphi \cdot \dot{\varphi} + \mathbf{w}_\mathbf{p} \cdot \dot{\mathbf{p}} + \mathrm{w}_\tau \dot{\tau}\right) \mathrm{d}V \,. \tag{4.70}$$

On the other hand, the specific brackets on the right-hand side of (4.59) assume the form (cf. Table 4.2)

$$\begin{aligned}
\{\mathcal{E}', \mathcal{W}'\}_{\text{bulk}} &= -\int_\mathcal{B} \left(\mathbf{b} \cdot \mathbf{w}_\mathbf{p} + \rho^{-1}\mathbf{p} \cdot \left(\mathbf{w}_\varphi + \mathrm{Div}\left(\frac{\mathrm{w}_\tau}{\partial_\tau \eta'}\partial_\mathbf{F}\eta'\right)\right) - \mathbf{P}' : \nabla \mathbf{w}_\mathbf{p}\right) \mathrm{d}V \,, \\
[\mathcal{S}', \mathcal{W}']_{\text{bulk}} &= -\int_\mathcal{B} \frac{1}{\Theta'}\mathrm{Div}\left((\Theta')^2 \mathbf{K} \nabla\left(\frac{\mathrm{w}_\tau}{\partial_\tau u'}\right)\right) \mathrm{d}V \,, \\
\{\mathcal{W}', \mathcal{E}'\}_{\text{boun}} &= -\int_{\partial\mathcal{B}} \rho^{-1}\mathbf{p} \cdot \left(-\frac{\mathrm{w}_\tau}{\partial_\tau \eta'}\partial_\mathbf{F}\eta'\right) \mathbf{N} \,\mathrm{d}A \,, \\
\{\mathcal{E}', \mathcal{W}'\}_{\text{boun}} &= -\int_{\partial\mathcal{B}} \mathbf{w}_\mathbf{p} \cdot \mathbf{P}'\mathbf{N} \,\mathrm{d}A \,, \\
[\mathcal{W}', \mathcal{S}']_{\text{boun}} &= \int_{\partial\mathcal{B}} \frac{\mathrm{w}_\tau}{\partial_\tau u'} \mathbf{N} \cdot \mathbf{Q}' \,\mathrm{d}A \,, \\
[\mathcal{S}', \mathcal{W}']_{\text{boun}} &= \int_{\partial\mathcal{B}} \frac{1}{\Theta'} \mathbf{N} \cdot (\Theta')^2 \mathbf{K} \nabla\left(\frac{\mathrm{w}_\tau}{\partial_\tau u'}\right) \mathrm{d}A \,.
\end{aligned}$$

Accordingly, the GENERIC evolution equation (4.59) gives rise to the equations

$$\begin{aligned}
0 &= \int_\mathcal{B} \mathbf{w}_\varphi \cdot \left(\dot{\varphi} - \rho^{-1}\mathbf{p}\right) \mathrm{d}V \,, \\
0 &= \int_\mathcal{B} \left(\mathbf{w}_\mathbf{p} \cdot (\dot{\mathbf{p}} - \mathbf{b}) + \mathbf{P}' : \nabla \mathbf{w}_\mathbf{p}\right) \mathrm{d}V - \int_{\partial\mathcal{B}} \mathbf{w}_\mathbf{p} \cdot \mathbf{P}'\mathbf{N} \,\mathrm{d}A \,,
\end{aligned} \tag{4.71}$$

along with

$$\begin{aligned}
0 = &\int_\mathcal{B} \left(\mathrm{w}_\tau \dot{\tau} - \mathrm{Div}\left(\frac{\mathrm{w}_\tau}{\partial_\tau \eta'}\partial_\mathbf{F}\eta'\right) \cdot \rho^{-1}\mathbf{p} + \frac{1}{\Theta'}\mathrm{Div}\left((\Theta')^2 \mathbf{K} \nabla\left(\frac{\mathrm{w}_\tau}{\partial_\tau u'}\right)\right)\right) \mathrm{d}V \\
&+ \int_{\partial\mathcal{B}} \left(\rho^{-1}\mathbf{p} \cdot \left(\frac{\mathrm{w}_\tau}{\partial_\tau \eta'}\partial_\mathbf{F}\eta'\right) \mathbf{N} - \frac{1}{\Theta'} \mathbf{N} \cdot (\Theta')^2 \mathbf{K} \nabla\left(\frac{\mathrm{w}_\tau}{\partial_\tau u'}\right) + \frac{\mathrm{w}_\tau}{\partial_\tau u'} \mathbf{N} \cdot \mathbf{Q}'\right) \mathrm{d}A \,.
\end{aligned}$$

Integrating by parts twice, we arrive at an alternative representation of the last equation given by

$$\begin{aligned}
0 = &\int_\mathcal{B} \left(\mathrm{w}_\tau \dot{\tau} + \nabla(\rho^{-1}\mathbf{p}) : \left(\frac{\mathrm{w}_\tau}{\partial_\tau \eta'}\partial_\mathbf{F}\eta'\right) - \nabla\left(\frac{1}{\Theta'}\right) \cdot (\Theta')^2 \mathbf{K} \nabla\left(\frac{\mathrm{w}_\tau}{\partial_\tau u'}\right)\right) \mathrm{d}V \\
&+ \int_{\partial\mathcal{B}} \frac{\mathrm{w}_\tau}{\partial_\tau u'} \mathbf{N} \cdot \mathbf{Q}' \,\mathrm{d}A \,.
\end{aligned}$$

Taking into account expression (4.47) for the material heat flux, and the symmetry property $\mathbf{K} = \mathbf{K}^{\mathrm{T}}$, the above equation can be recast in the form

$$0 = \int_{\mathcal{B}} \left(\mathrm{w}_\tau \dot{\tau} + \nabla(\rho^{-1}\mathbf{p}) : \left(\frac{\mathrm{w}_\tau}{\partial_\tau \eta'} \partial_{\mathbf{F}} \eta' \right) - \nabla \left(\frac{\mathrm{w}_\tau}{\partial_\tau u'} \right) \cdot \mathbf{Q}' \right) \mathrm{d}V + \int_{\partial\mathcal{B}} \frac{\mathrm{w}_\tau}{\partial_\tau u'} \mathbf{N} \cdot \mathbf{Q}' \, \mathrm{d}A \,. \tag{4.72}$$

Taking into account the above stated boundary conditions, (4.71) and (4.72) give rise to the following variational formulation of the present IBVP:

$$\begin{aligned}
0 &= \int_{\mathcal{B}} \mathbf{w}_\varphi \cdot (\dot{\varphi} - \rho^{-1}\mathbf{p}) \, \mathrm{d}V \,, \\
0 &= \int_{\mathcal{B}} \left(\mathbf{w}_\mathbf{p} \cdot (\dot{\mathbf{p}} - \mathbf{b}) + \mathbf{P}' : \nabla \mathbf{w}_\mathbf{p} \right) \mathrm{d}V - \int_{\partial_\sigma \mathcal{B}} \mathbf{w}_\mathbf{p} \cdot \bar{\mathbf{t}} \, \mathrm{d}A \,, \\
0 &= \int_{\mathcal{B}} \left(\mathrm{w}_\tau \dot{\tau} + \nabla(\rho^{-1}\mathbf{p}) : \left(\frac{\mathrm{w}_\tau}{\partial_\tau \eta'} \partial_{\mathbf{F}} \eta' \right) - \nabla \left(\frac{\mathrm{w}_\tau}{\partial_\tau u'} \right) \cdot \mathbf{Q}' \right) \mathrm{d}V + \int_{\partial_q \mathcal{B}} \frac{\mathrm{w}_\tau}{\partial_\tau u'} \bar{q} \, \mathrm{d}A \,.
\end{aligned} \tag{4.73}$$

These equations have to hold for all times $t \geq 0$ and for arbitrary test functions subject to the above mentioned boundary conditions.

Remark 4. The variational formulation (4.73) can be easily extended to account for non-potential body forces and a heat supply field. To this end, the supply term introduced in Remark 3 just needs be included in the above derivation of the variational formulation.

Remark 5. The majority of previously developed numerical methods for finite-strain thermomechanics rely on the use of the temperature as the thermodynamic state variable. Choosing $\tau = \theta$ in the GENERIC-based weak form $(4.73)_3$, leads to the variational equation for the temperature evolution given by

$$\int_{\mathcal{B}} \left(\mathrm{w}_\theta \dot{\theta} + \nabla(\rho^{-1}\mathbf{p}) : \left(\frac{\mathrm{w}_\theta}{\partial_\theta \overline{\eta}} \partial_{\mathbf{F}} \overline{\eta} \right) - \nabla \left(\frac{\mathrm{w}_\theta}{\partial_\theta \overline{u}} \right) \cdot \overline{\mathbf{Q}} \right) \mathrm{d}V + \int_{\partial_q \mathcal{B}} \frac{\mathrm{w}_\theta}{\partial_\theta \overline{u}} \bar{q} \, \mathrm{d}A = 0 \,. \tag{4.74}$$

The last equation can be recast in the form

$$\int_{\mathcal{B}} \left\{ \frac{\mathrm{w}_\theta}{\partial_\theta \overline{u}} \left(\partial_\theta \overline{u} \dot{\theta} + \frac{\partial_\theta \overline{u}}{\partial_\theta \overline{\eta}} \nabla(\rho^{-1}\mathbf{p}) : \partial_{\mathbf{F}} \overline{\eta} \right) - \nabla \left(\frac{\mathrm{w}_\theta}{\partial_\theta \overline{u}} \right) \cdot \overline{\mathbf{Q}} \right\} \mathrm{d}V \,, + \int_{\partial_q \mathcal{B}} \frac{\mathrm{w}_\theta}{\partial_\theta \overline{u}} \bar{q} \, \mathrm{d}A = 0 \,,$$

or

$$\int_{\mathcal{B}} \left\{ v_\theta \left(\partial_\theta \overline{u} \dot{\theta} + \theta \nabla(\rho^{-1}\mathbf{p}) : \partial_{\mathbf{F}} \overline{\eta} \right) + \nabla v_\theta \cdot \mathbf{K} \nabla \theta \right\} \mathrm{d}V + \int_{\partial_q \mathcal{B}} v_\theta \bar{q} \, \mathrm{d}A = 0 \,. \tag{4.75}$$

In the last equation the modified test function $v_\theta = \frac{\mathrm{w}_\theta}{\partial_\theta \overline{u}}$ has been introduced. In addition to that, temperature formula (4.27) has been used along with constitutive equation (4.47) for the material heat flux vector. In essence, weak form (4.75) of the temperature evolution equation lies at the heart of numerous works on finite-strain thermomechanics such as, for example, [140–142]. Alternatively, differentiating (4.20) with respect to time, one arrives at the relation

$$\partial_\theta \overline{u} \dot{\theta} = \theta \dot{\overline{\eta}} - \theta \partial_{\mathbf{F}} \overline{\eta} : \dot{\mathbf{F}} \, .$$

Inserting the last equation into (4.75), and making use of the identity $\dot{\varphi} = \rho^{-1}\mathbf{p}$, we obtain the entropy form of (4.75) given by

$$\int_{\mathcal{B}} \left\{ v_\theta \theta \dot{\overline{\eta}} + \nabla v_\theta \cdot \mathbf{K} \nabla \theta \right\} \mathrm{d}V + \int_{\partial_q \mathcal{B}} v_\theta \overline{q} \, \mathrm{d}A = 0 \, . \tag{4.76}$$

In essence, variational formulation (4.76) lies at the origin of alternative temperature-based numerical methods such as, for example, [143, 144]. It is worth mentioning that in (4.76), $\dot{\overline{\eta}} = \frac{\mathrm{d}}{\mathrm{d}t} \overline{\eta}(\mathbf{F}, \theta)$ is the time derivative of the "constitutive entropy" $\overline{\eta}(\mathbf{F}, \theta)$. This is to be contrasted with the entropy-based formulation emanating from the GENERIC-based weak form $(4.73)_3$. Specifically, choosing $\tau = \eta$, $(4.73)_3$ yields

$$\int_{\mathcal{B}} \left\{ \mathrm{w}_\eta \dot{\eta} - \nabla \left(\mathrm{w}_\eta \widetilde{\Theta}^{-1} \right) \cdot \widetilde{\Theta}^2 \mathbf{K} \nabla \left(\widetilde{\Theta}^{-1} \right) \right\} \mathrm{d}V + \int_{\partial_q \mathcal{B}} \frac{\mathrm{w}_\eta}{\widetilde{\Theta}} \overline{q} \, \mathrm{d}A = 0 \, , \tag{4.77}$$

where $\widetilde{\Theta} = \partial_\eta \widetilde{u}(\mathbf{F}, \eta)$.

4.3.3.2. Balance laws revisited

The balance laws dealt with in Section 4.3.2 can also be recovered from the variational formulation (4.73), by choosing specific test functions. For that purpose we confine our attention in this section to the pure Neumann problem (i.e. $\partial_\sigma \mathcal{B} = \partial_q \mathcal{B} = \partial \mathcal{B}$).

Consider $\mathbf{w_p} = \partial_{\mathbf{p}} l'_{\xi}$, where $l'_{\xi} = \boldsymbol{\xi} \cdot \mathbf{p}$ has been introduced in Section 4.3.2. Correspondingly, inserting $\mathbf{w_p} = \boldsymbol{\xi}$ into $(4.73)_2$, we directly recover the balance law (4.65) for linear momentum.

Next, we choose $\mathbf{w}_\varphi = \partial_\varphi j'_\xi$ and $\mathbf{w_p} = \partial_{\mathbf{p}} j'_\xi$, where $j'_\xi = \boldsymbol{\xi} \cdot (\varphi \times \mathbf{p})$ has been introduced in Section 4.3.2. Accordingly, inserting $\mathbf{w}_\varphi = \mathbf{p} \times \boldsymbol{\xi}$ and $\mathbf{w_p} = \boldsymbol{\xi} \times \varphi$ into $(4.73)_{1,2}$, and subsequently adding both equations, we recover the balance law (4.66) for angular momentum. To verify the balance of energy we choose for the test functions in (4.73)

$$\begin{aligned} \mathbf{w}_\varphi &= \partial_\varphi e' \, , \\ \mathbf{w_p} &= \partial_{\mathbf{p}} e' \, , \\ \mathrm{w}_\tau &= \partial_\tau e' \, , \end{aligned} \qquad \text{or} \qquad \begin{aligned} \mathbf{w}_\varphi &= -\mathbf{b} \, , \\ \mathbf{w_p} &= \rho^{-1} \mathbf{p} \, , \\ \mathrm{w}_\tau &= \partial_\tau u' \, . \end{aligned}$$

Substituting these quantities into (4.73) and subsequently adding the three resulting equations, a straightforward calculation recovers the balance law (4.67) for energy. In this connection, the following relation is of importance:

$$\begin{aligned}
&\int_{\mathcal{B}} \left(\mathbf{P}' : \nabla(\rho^{-1}\mathbf{p}) + \partial_\tau u' \dot{\tau} + \nabla(\rho^{-1}\mathbf{p}) : \left(\frac{\partial_\tau u'}{\partial_\tau \eta'} \partial_{\mathbf{F}} \eta' \right) \right) \mathrm{d}V \\
&= \int_{\mathcal{B}} \Big(\partial_\tau u' \dot{\tau} + \underbrace{\Big(\partial_{\mathbf{F}} u' - \frac{\partial_\tau u'}{\partial_\tau \eta'} \partial_{\mathbf{F}} \eta'}_{=\mathbf{P}'} + \frac{\partial_\tau u'}{\partial_\tau \eta'} \partial_{\mathbf{F}} \eta' \Big) : \nabla(\rho^{-1}\mathbf{p}) \Big) \mathrm{d}V \\
&= \int_{\mathcal{B}} \left(\partial_\tau u' \dot{\tau} + \partial_{\mathbf{F}} u' : \dot{\mathbf{F}} \right) \mathrm{d}V \\
&= \frac{\mathrm{d}}{\mathrm{d}t} \int_{\mathcal{B}} u' \, \mathrm{d}V \,.
\end{aligned} \tag{4.78}$$

Here, formula (4.52) for the first Piola-Kirchhoff stress tensor has been used, along with the identity $\dot{\varphi} = \rho^{-1}\mathbf{p}$.

Concerning the balance of entropy, we insert $\mathrm{w}_\tau = \partial_\tau \eta'$ into $(4.73)_3$, to obtain

$$\begin{aligned}
0 &= \int_{\mathcal{B}} \left(\partial_\tau \eta' \dot{\tau} + \nabla(\rho^{-1}\mathbf{p}) : (\partial_{\mathbf{F}} \eta') - \nabla \left(\frac{\partial_\tau \eta'}{\partial_\tau u'} \right) \cdot \mathbf{Q}' \right) \mathrm{d}V + \int_{\partial\mathcal{B}} \frac{\partial_\tau \eta'}{\partial_\tau u'} \mathbf{Q}' \cdot \mathbf{N} \, \mathrm{d}A \\
&= \int_{\mathcal{B}} \left(\frac{\mathrm{d}\eta'}{\mathrm{d}t} - \nabla \left(\frac{1}{\Theta'} \right) \cdot (\Theta')^2 \mathbf{K} \nabla \left(\frac{1}{\Theta'} \right) \right) \mathrm{d}V + \int_{\partial\mathcal{B}} \frac{1}{\Theta'} \mathbf{Q}' \cdot \mathbf{N} \, \mathrm{d}A \,.
\end{aligned}$$

Here, use has been made of formula (4.27) for the temperature along with expression (4.47) for the material heat flux vector. Moreover, the identity $\dot{\varphi} = \rho^{-1}\mathbf{p}$ has again been taken into account. The above equation can be rewritten as

$$\frac{\mathrm{d}\mathcal{S}'}{\mathrm{d}t} = \underbrace{\int_{\mathcal{B}} \nabla \left(\frac{1}{\Theta'} \right) \cdot (\Theta')^2 \mathbf{K} \nabla \left(\frac{1}{\Theta'} \right) \mathrm{d}V}_{\geq 0} - \int_{\partial\mathcal{B}} \frac{1}{\Theta'} \mathbf{Q}' \cdot \mathbf{N} \, \mathrm{d}A \,, \tag{4.79}$$

which recovers the second law of thermodynamics in the form (4.68).

4.4. Discretization in time and space

4.4.1. Discretization in time

We first perform the discretization in time of the variational formulation (4.73). To this end we focus on a representative time interval $[t_\mathrm{n}, t_{\mathrm{n}+1}]$ with corresponding time-step size

$\Delta t = t_{n+1} - t_n$. The discrete approximations at times t_n and t_{n+1} of the continuous variable $(\bullet)_t$ will be denoted by $(\bullet)_n$ and $(\bullet)_{n+1}$, respectively. Moreover, the approximation of any state variable $(\bullet)_t$ at mid-point time $t_{n+\frac{1}{2}} = \frac{1}{2}(t_n + t_{n+1})$ is given by

$$(\bullet)_{n+\frac{1}{2}} = \frac{1}{2}\left((\bullet)_n + (\bullet)_{n+1}\right) .$$

Assume that the state variables $\varphi_n, \mathbf{p}_n : \mathcal{B} \mapsto \mathbb{R}^3$ and $\tau_n : \mathcal{B} \mapsto \mathbb{R}$, $\tau_n \in \{\theta_n, \eta_n, u_n\}$ are given. We aim at the determination of the corresponding state variables at time t_{n+1} $(\varphi_{n+1}, \mathbf{p}_{n+1}, \tau_{n+1})$ through the mid-point type discretization of the variational formulation (4.73) given by

$$\begin{aligned}
0 &= \int_{\mathcal{B}} \mathbf{w}_\varphi \cdot \left(\frac{\varphi_{n+1} - \varphi_n}{\Delta t} - \rho^{-1} \mathbf{p}_{n+\frac{1}{2}} \right) \mathrm{d}V , \\
0 &= \int_{\mathcal{B}} \left(\mathbf{w}_\mathbf{p} \cdot \frac{\mathbf{p}_{n+1} - \mathbf{p}_n}{\Delta t} - \mathbf{w}_\mathbf{p} \cdot \mathbf{b} + \nabla \mathbf{w}_\mathbf{p} : \mathbf{P}'\big|_{t_{n+\frac{1}{2}}} \right) \mathrm{d}V - \int_{\partial_\sigma \mathcal{B}} \mathbf{w}_\mathbf{p} \cdot \bar{\mathbf{t}}_{n+\frac{1}{2}} \, \mathrm{d}A , \\
0 &= \int_{\mathcal{B}} \mathrm{w}_\tau \left(\frac{\tau_{n+1} - \tau_n}{\Delta t} + \nabla(\rho^{-1} \mathbf{p}_{n+\frac{1}{2}}) : \left(\frac{\partial_\mathbf{F} \eta'}{\partial_\tau \eta'} \right)\bigg|_{t_{n+\frac{1}{2}}} \right) \mathrm{d}V \\
&\quad - \int_{\mathcal{B}} \nabla \left(\frac{\mathrm{w}_\tau}{\partial_\tau u'|_{t_{n+\frac{1}{2}}}} \right) \cdot \mathbf{Q}'\big|_{t_{n+\frac{1}{2}}} \mathrm{d}V + \int_{\partial_q \mathcal{B}} \frac{\mathrm{w}_\tau}{\partial_\tau u'|_{t_{n+\frac{1}{2}}}} \bar{q}_{n+\frac{1}{2}} \, \mathrm{d}A .
\end{aligned} \tag{4.80}$$

Note that depending on the choice for the thermodynamic state variable $\tau \in \{\theta, \eta, u\}$, three alternative semi-discrete formulations result from (4.80). With regard to (4.52), the time-discrete version of the first Piola-Kirchhoff stress tensor in $(4.80)_2$ assumes the form

$$\begin{aligned}
\mathbf{P}'\big|_{t_{n+\frac{1}{2}}} &= \mathbf{P}'(\mathbf{F}_{n+\frac{1}{2}}, \tau_{n+\frac{1}{2}}) \\
&= \partial_\mathbf{F} u'\big|_{t_{n+\frac{1}{2}}} - \left(\frac{\partial_\tau u'}{\partial_\tau \eta'} \partial_\mathbf{F} \eta' \right)\bigg|_{t_{n+\frac{1}{2}}} .
\end{aligned} \tag{4.81}$$

A frame-indifferent thermoelastic formulation requires that the functions u' and η' can be expressed as

$$\begin{aligned}
u'(\mathbf{F}, \tau) &= u''(\mathbf{C}, \tau) , \\
\eta'(\mathbf{F}, \tau) &= \eta''(\mathbf{C}, \tau) ,
\end{aligned} \tag{4.82}$$

where $\mathbf{C} = \mathbf{F}^{\mathrm{T}}\mathbf{F}$ is the right Cauchy-Green tensor written in terms of the deformation gradient $\mathbf{F}$. Taking into account (4.82), we obtain

$$\begin{aligned} \partial_{\mathbf{F}} u'|_{t_{\mathrm{n}+\frac{1}{2}}} &= 2\mathbf{F}_{\mathrm{n}+\frac{1}{2}} \partial_{\mathbf{C}} u''(\mathbf{C}|_{t_{\mathrm{n}+\frac{1}{2}}}, \tau_{\mathrm{n}+\frac{1}{2}}) , \\ \partial_{\mathbf{F}} \eta'|_{t_{\mathrm{n}+\frac{1}{2}}} &= 2\mathbf{F}_{\mathrm{n}+\frac{1}{2}} \partial_{\mathbf{C}} \eta''(\mathbf{C}|_{t_{\mathrm{n}+\frac{1}{2}}}, \tau_{\mathrm{n}+\frac{1}{2}}) , \end{aligned} \tag{4.83}$$

together with

$$\begin{aligned} \partial_{\tau} u'|_{t_{\mathrm{n}+\frac{1}{2}}} &= \partial_{\tau} u''(\mathbf{C}|_{t_{\mathrm{n}+\frac{1}{2}}}, \tau_{\mathrm{n}+\frac{1}{2}}) , \\ \partial_{\tau} \eta'|_{t_{\mathrm{n}+\frac{1}{2}}} &= \partial_{\tau} \eta''(\mathbf{C}|_{t_{\mathrm{n}+\frac{1}{2}}}, \tau_{\mathrm{n}+\frac{1}{2}}) , \end{aligned} \tag{4.84}$$

where $\mathbf{C}|_{t_{\mathrm{n}+\frac{1}{2}}}$ denotes the right Cauchy-Green tensor evaluated in the mid-point configuration. That is,

$$\mathbf{C}|_{t_{\mathrm{n}+\frac{1}{2}}} = \mathbf{F}^{\mathrm{T}}_{\mathrm{n}+\frac{1}{2}} \mathbf{F}_{\mathrm{n}+\frac{1}{2}} . \tag{4.85}$$

Moreover, in view of (4.47), the time-discrete version of the material heat flux vector $\mathbf{Q}'|_{t_{\mathrm{n}+\frac{1}{2}}}$ featuring in $(4.80)_3$ is given by

$$\begin{aligned} \mathbf{Q}'|_{t_{\mathrm{n}+\frac{1}{2}}} &= \mathbf{Q}'(\mathbf{F}_{\mathrm{n}+\frac{1}{2}}, \tau_{\mathrm{n}+\frac{1}{2}}) \\ &= -\,\mathbf{K}'|_{t_{\mathrm{n}+\frac{1}{2}}} \nabla \left(\frac{\partial_{\tau} u'}{\partial_{\tau} \eta'} \right) \bigg|_{t_{\mathrm{n}+\frac{1}{2}}} , \end{aligned} \tag{4.86}$$

where the time-discrete version of the material conductivity tensor $\mathbf{K}'|_{t_{\mathrm{n}+\frac{1}{2}}}$ follows from the relation

$$\mathbf{K}'|_{t_{\mathrm{n}+\frac{1}{2}}} = \mathbf{K}'(\mathbf{F}_{\mathrm{n}+\frac{1}{2}}, \tau_{\mathrm{n}+\frac{1}{2}}) = \mathbf{K}''(\mathbf{C}|_{t_{\mathrm{n}+\frac{1}{2}}}, \tau_{\mathrm{n}+\frac{1}{2}}) . \tag{4.87}$$

4.4.1.1. Semi-discrete balance laws

We next investigate to what extent the balance laws outlined in Sections 4.3.2 and 4.3.3.2 are inherited by the three alternative semi-discrete formulations developed in the last section. To this end we proceed along the lines of Section 4.3.3.2. We again confine our attention to the pure Neumann problem (i.e. $\partial_{\sigma}\mathcal{B} = \partial_{q}\mathcal{B} = \partial\mathcal{B}$). Inserting $\mathbf{w}_{\mathbf{p}} = \boldsymbol{\xi}$ into $(4.80)_2$, we directly obtain

$$\boldsymbol{\xi} \cdot \frac{\mathbf{L}_{\mathrm{n}+1} - \mathbf{L}_{\mathrm{n}}}{\Delta t} = \boldsymbol{\xi} \cdot \left(\int_{\mathcal{B}} \mathbf{b} \,\mathrm{d}V + \int_{\partial\mathcal{B}} \bar{\mathbf{t}}_{\mathrm{n}+\frac{1}{2}} \,\mathrm{d}A \right) ,$$

where $\bar{\mathbf{t}}_{\mathrm{n}+\frac{1}{2}} = \mathbf{P}'\big|_{t_{\mathrm{n}+\frac{1}{2}}}\mathbf{N}$. The last equation corresponds to the semi-discrete version of the balance of linear momentum (4.65). Inserting $\mathbf{w}_{\varphi} = \mathbf{p}_{\mathrm{n}+\frac{1}{2}} \times \boldsymbol{\xi}$ and $\mathbf{w}_{\mathbf{p}} = \boldsymbol{\xi} \times \boldsymbol{\varphi}_{\mathrm{n}+\frac{1}{2}}$ into $(4.80)_{1,2}$, and subsequently adding both equations, a straightforward calculation yields

$$\boldsymbol{\xi} \cdot \frac{\mathbf{J}_{\mathrm{n}+1} - \mathbf{J}_{\mathrm{n}}}{\Delta t} = \boldsymbol{\xi} \cdot \left(\int_{\mathcal{B}} \boldsymbol{\varphi}_{\mathrm{n}+\frac{1}{2}} \times \mathbf{b}\, \mathrm{d}V + \int_{\partial\mathcal{B}} \boldsymbol{\varphi}_{\mathrm{n}+\frac{1}{2}} \times \bar{\mathbf{t}}_{\mathrm{n}+\frac{1}{2}}\, \mathrm{d}A \right) ,$$

which corresponds to the semi-discrete version of the balance of angular momentum (4.66). Accordingly, independent of the choice for $\tau \in \{\theta, \eta, u\}$, the three alternative formulations at hand consistently reproduce the balance laws for linear and angular momentum for arbitrary time-steps Δt.

Inserting $\mathbf{w}_{\varphi} = -\mathbf{b}$, $\mathbf{w}_{\mathbf{p}} = \rho^{-1}\mathbf{p}_{\mathrm{n}+\frac{1}{2}}$, and $\mathrm{w}_{\tau} = \partial_{\tau}u'\big|_{t_{\mathrm{n}+\frac{1}{2}}}$ into the semi-discrete formulation (4.80), and subsequently adding the three equations, we arrive at

$$\begin{aligned} &\frac{1}{2}\int_{\mathcal{B}} \rho^{-1}\frac{\mathbf{p}_{\mathrm{n}+1}\cdot\mathbf{p}_{\mathrm{n}+1} - \mathbf{p}_{\mathrm{n}}\cdot\mathbf{p}_{\mathrm{n}}}{\Delta t} + \left(\partial_{\tau}u'\big|_{t_{\mathrm{n}+\frac{1}{2}}} \frac{\tau_{\mathrm{n}+1} - \tau_{\mathrm{n}}}{\Delta t} + \partial_{\mathbf{F}}u'\big|_{t_{\mathrm{n}+\frac{1}{2}}} : \nabla(\rho^{-1}\mathbf{p}_{\mathrm{n}+\frac{1}{2}}) \right) \mathrm{d}V \\ &= \int_{\mathcal{B}} \mathbf{b} \cdot \frac{\boldsymbol{\varphi}_{\mathrm{n}+1} - \boldsymbol{\varphi}_{\mathrm{n}}}{\Delta t}\, \mathrm{d}V + \int_{\partial\mathcal{B}} \left(\rho^{-1}\mathbf{p}_{\mathrm{n}+\frac{1}{2}} \cdot \bar{\mathbf{t}}_{\mathrm{n}+\frac{1}{2}} - \bar{q}_{\mathrm{n}+\frac{1}{2}} \right) \mathrm{d}A , \end{aligned}$$

where $\bar{q}_{\mathrm{n}+\frac{1}{2}} = \mathbf{Q}'\big|_{t_{\mathrm{n}+\frac{1}{2}}} \cdot \mathbf{N}$. The last equation can be viewed as semi-discrete version of the balance of energy (4.67) provided that the second integral on the left-hand side of the last equation is equal to $\frac{1}{\Delta t}\int_{\mathcal{B}} u'(\mathbf{F}_{\mathrm{n}+1}, \tau_{\mathrm{n}+1}) - u'(\mathbf{F}_{\mathrm{n}}, \tau_{\mathrm{n}})\, \mathrm{d}V$. However, for $\tau \in \{\theta, \eta\}$ this is generally not the case. In contrast to that, for $\tau = u$, the above equation reads

$$\begin{aligned} &\frac{1}{2}\int_{\mathcal{B}} \rho^{-1}\frac{\mathbf{p}_{\mathrm{n}+1}\cdot\mathbf{p}_{\mathrm{n}+1} - \mathbf{p}_{\mathrm{n}}\cdot\mathbf{p}_{\mathrm{n}}}{\Delta t}\, \mathrm{d}V + \int_{\mathcal{B}} \frac{u_{\mathrm{n}+1} - u_{\mathrm{n}}}{\Delta t}\, \mathrm{d}V \\ &\quad = \int_{\mathcal{B}} \mathbf{b} \cdot \frac{\boldsymbol{\varphi}_{\mathrm{n}+1} - \boldsymbol{\varphi}_{\mathrm{n}}}{\Delta t}\, \mathrm{d}V + \int_{\partial\mathcal{B}} \left(\rho^{-1}\mathbf{p}_{\mathrm{n}+\frac{1}{2}} \cdot \bar{\mathbf{t}}\big|_{t_{\mathrm{n}+\frac{1}{2}}} - \bar{q}_{\mathrm{n}+\frac{1}{2}} \right) \mathrm{d}A , \end{aligned} \tag{4.88}$$

which indeed corresponds to the semi-discrete counterpart of the balance of energy (4.67). Accordingly, the formulation in terms of the internal energy density leads to an energy-momentum scheme.

Eventually, inserting $\mathrm{w}_{\tau} = \partial_{\tau}\eta'\big|_{t_{\mathrm{n}+\frac{1}{2}}}$ into $(4.80)_3$, we obtain

$$\begin{aligned} &\int_{\mathcal{B}} \left(\partial_{\tau}\eta'\big|_{t_{\mathrm{n}+\frac{1}{2}}} \frac{\tau_{\mathrm{n}+1} - \tau_{\mathrm{n}}}{\Delta t} + \partial_{\mathbf{F}}\eta'\big|_{t_{\mathrm{n}+\frac{1}{2}}} : \nabla(\rho^{-1}\mathbf{p}_{\mathrm{n}+\frac{1}{2}}) \right) \mathrm{d}V \\ &= \underbrace{\int_{\mathcal{B}} \nabla\left(\Theta'\big|^{-1}_{t_{\mathrm{n}+\frac{1}{2}}} \right) \cdot \Theta'\big|^{2}_{t_{\mathrm{n}+\frac{1}{2}}} \mathbf{K}'\big|_{t_{\mathrm{n}+\frac{1}{2}}} \nabla\left(\Theta'\big|^{-1}_{t_{\mathrm{n}+\frac{1}{2}}} \right) \mathrm{d}V}_{\geq 0} - \int_{\partial\mathcal{B}} \Theta'\big|^{-1}_{t_{\mathrm{n}+\frac{1}{2}}} \bar{q}_{\mathrm{n}+\frac{1}{2}}\, \mathrm{d}A . \end{aligned}$$

The last equation does conform with the balance of entropy provided that the integral on the left-hand side of the last equation equals

$$\frac{1}{\Delta t}\int_{\mathcal{B}} \eta'(\mathbf{F}_{\mathrm{n+1}}, \tau_{\mathrm{n+1}}) - \eta'(\mathbf{F}_{\mathrm{n}}, \tau_{\mathrm{n}})\, \mathrm{d}V \,.$$

However, in general this is not true for $\tau \in \{\theta, u\}$. In contrast to that, for $\tau = \eta$, the last equation yields

$$\int_{\mathcal{B}} \frac{\eta_{\mathrm{n+1}} - \eta_{\mathrm{n}}}{\Delta t}\, \mathrm{d}V =$$
$$\underbrace{\int_{\mathcal{B}} \nabla\left(\widetilde{\Theta}\Big|_{t_{\mathrm{n}+\frac{1}{2}}}^{-1}\right) \cdot \widetilde{\Theta}\Big|_{t_{\mathrm{n}+\frac{1}{2}}}^{2} \widetilde{\mathbf{K}}\Big|_{t_{\mathrm{n}+\frac{1}{2}}} \nabla\left(\widetilde{\Theta}\Big|_{t_{\mathrm{n}+\frac{1}{2}}}^{-1}\right) \mathrm{d}V}_{\geq 0} - \int_{\partial\mathcal{B}} \widetilde{\Theta}\Big|_{t_{\mathrm{n}+\frac{1}{2}}}^{-1} \bar{q}_{\mathrm{n}+\frac{1}{2}}\, \mathrm{d}A,$$

which corresponds to the semi-discrete counterpart of the second law of thermodynamics in the form (4.68). Accordingly, the formulation in terms of the entropy density leads to an momentum-entropy scheme.

4.4.2. Discretization in space

We apply standard isoparametric finite elements (see, for example, [145]) based on finite-dimensional approximations of the state variables at time t, given by

$$\varphi_{\mathrm{t}}^{\mathrm{h}}(\mathbf{X}) = \sum_{a=1}^{n_{node}} N^a(\mathbf{X})\, \varphi_a(t)\,, \qquad \mathbf{p}_{\mathrm{t}}^{\mathrm{h}}(\mathbf{X}) = \sum_{a=1}^{n_{node}} N^a(\mathbf{X})\, \mathbf{p}_a(t)\,, \tag{4.89}$$

and

$$\tau_{\mathrm{t}}^{\mathrm{h}}(\mathbf{X}) = \sum_{a=1}^{n_{node}} N^a(\mathbf{X})\, \tau_a(t)\,. \tag{4.90}$$

Here, $N^a : \mathcal{B} \mapsto \mathbb{R}$ denote the nodal shape functions and $\varphi_a(t), \mathbf{p}_a(t) \in \mathbb{R}^3$, $\tau_a(t) \in \mathbb{R}$ are the respective nodal values at time t. Moreover, n_{node} denotes the total number of nodes in the finite element mesh. The standard (Bubnov) Galerkin approach relies on analogous approximations for the test functions in the variational equations (4.80), $\mathbf{w}_{\varphi}, \mathbf{w}_{\mathbf{p}}$ and w_{τ}, denoted by ${\mathbf{w}_{\varphi}}^{\mathrm{h}}, {\mathbf{w}_{\mathbf{p}}}^{\mathrm{h}}$ and w_{τ}^{h}.

It is worth noting that the semi-discrete balance laws investigated in the last section still hold for the fully discrete finite element formulation as can be easily verified by proceeding along the lines of the last section.

4.5. Numerical investigations

In this section, the three alternative mid-point type schemes newly developed in the present work are applied to representative numerical examples dealing with finite-strain thermo-elastodynamics. Depending on the specific choice of the thermodynamic state variable $\tau \in \{\theta, \eta, u\}$, the following methods are applied:

$\tau = u$	$(\mathrm{EM})_u$	Energy-Momentum consistent scheme
$\tau = \eta$	$(\mathrm{ME})_\eta$	Momentum-Entropy consistent scheme
$\tau = \theta$	$(\mathrm{M})_\theta$	Momentum consistent scheme

Concerning the constitutive equation (4.47) for the material heat flux vector, we assume thermally isotropic material, with material conductivity tensor given by

$$\mathbf{K} = kJ\mathbf{C}^{-1} . \tag{4.91}$$

Here, k is a prescribed coefficient of thermal conductivity and, as before, $J = \sqrt{\det(\mathbf{C})}$. Moreover, the Helmholtz free energy assumes the form (4.36).

In the numerical investigations we shall focus on momentum maps associated with symmetries of the mechanical system at hand, and the balance laws associated with the two fundamental laws of thermodynamics. In this connection we also consider the functional

$$\mathcal{L} = \mathcal{E} - \theta_0 \mathcal{S} . \tag{4.92}$$

According to [146] (cf. Appendix A.2), for certain types of environments, $\mathcal{L}$ is a natural Lyapunov function and thus qualifies as estimate for the numerical stability of the schemes under consideration. Concerning the numerical solution of the resulting algebraic system of nonlinear equations we apply Newton's method together with the convergence criteria summarized in Appendix A.3.

4.5.1. Flying L-shaped block

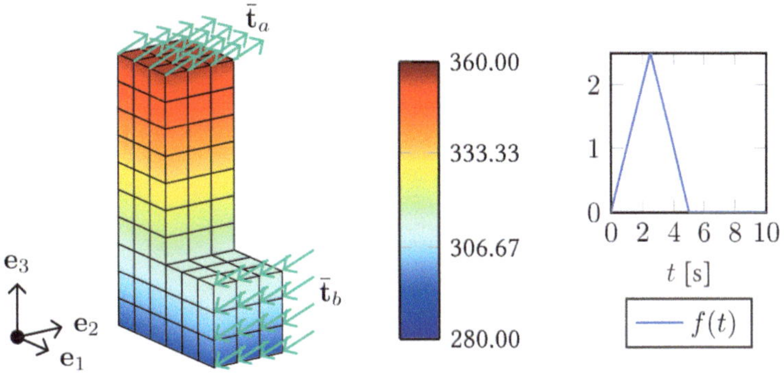

Figure 4.5.: L-shaped block: Discretised block with initial temperature distribution and mechanical boundary conditions (left), load function over time (right)

The first numerical example deals with the L-shaped block depicted in Fig. 4.5. The spatial discretization of the block relies on 117 tri-linear finite elements leading to 224 nodes. The initial temperature field is varying linearly over the height (x_3 direction) of the block. In particular, at $x_3 = 0$, the initial temperature is prescribed as θ_a, while at $x_3 = h$, the temperature is prescribed as θ_b. The whole block is assumed to be thermally insulated ($\bar{q} = 0$ on $\partial_q \mathcal{B} = \partial \mathcal{B}$). Starting at rest, Piola traction vectors $\bar{\mathbf{t}}_a$ and $\bar{\mathbf{t}}_b$ are acting on two parts of the boundary surface of the block (Fig. 4.5). The external loads are applied in the form of a hat function over time. In particular, the traction vectors are given by

$$\bar{\mathbf{t}}_a = -\bar{\mathbf{t}}_b = f(t) \begin{pmatrix} 256/9 \\ 512/9 \\ 768/9 \end{pmatrix} \text{Pa}\,, \quad f(t) = \begin{cases} t & \text{for} \quad 0\text{s} \le t \le 2.5\text{s}\,, \\ 5-t & \text{for} \quad 2.5\text{s} \le t \le 5\text{s}\,, \\ 0 & \text{for} \quad t > 5\text{s}\,. \end{cases} \tag{4.93}$$

Table 4.3 provides a summary of the data used in the simulations. No Dirichlet boundary conditions are applied.

Material parameters	μ	997.5	Pa
	λ	5209	Pa
Specific heat capacity	c	100	$JK^{-1}m^{-3}$
Expansion coefficient	β	$2.233 \cdot 10^{-4}$	K^{-1}
Thermal conductivity	k	10	$WK^{-1}m^{-1}$
Ref. temperature	θ_0	293.15	K
Mass density	ρ	100	kgm^{-3}
Initial temperature	θ_a	290	K
	θ_b	350	K
Geometry	h	10	m
	b	3	m
Newton tolerance	ε	10^{-8}	-
Simulation duration	T	250	s
Time steps	Δt	0.08	s
		0.2	s
		0.4	s

Geometry

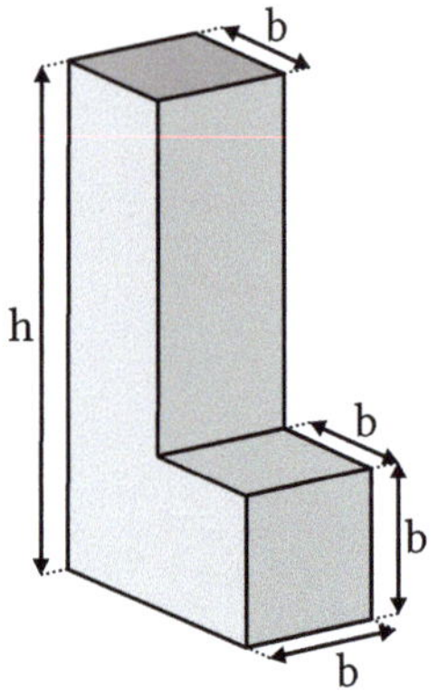

Table 4.3.: L-shaped block: Data used in the simulations

Since in the initial loading phase the external forces are equilibrated, the total linear momentum of the block is a conserved quantity. In addition to that, after the loading phase ($t > 5$s) no external torque is acting on the block.

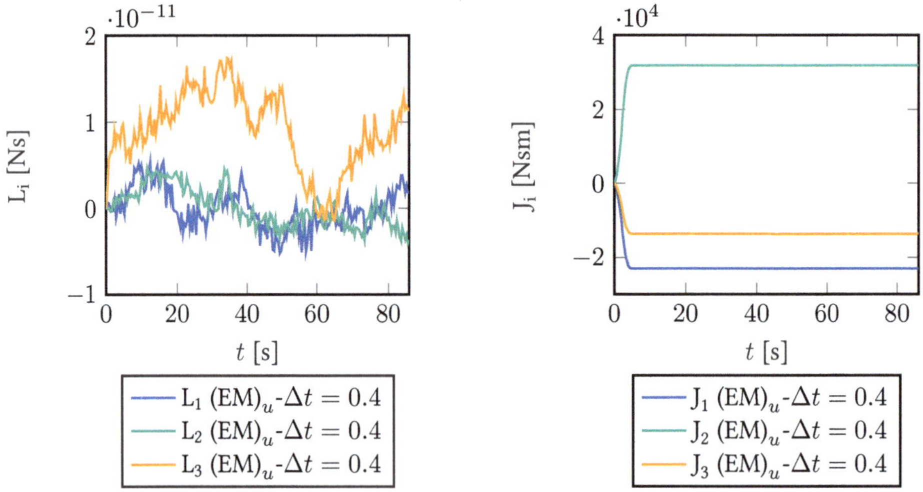

Figure 4.6.: L-shaped block: Total discrete linear momentum $(EM)_u$ scheme (left), Total discrete angular momentum $(EM)_u$ scheme (right)

Consequently, the total angular momentum is conserved as well. All of the integrators under consideration are capable to exactly conserve both momentum maps (up to numerical round-off), independent of the chosen time step. This can be observed from Fig. 4.6, where representative numerical results of the energy-momentum integrator $(\mathrm{EM})_u$ are shown. In the simulations three alternative constant time steps $\Delta t \in \{0.08, 0.2, 0.4\}$s have been used. After the loading phase the total energy must be a conserved quantity. As expected, the $(\mathrm{EM})_u$ scheme does exactly reproduce this conservation law (up to numerical round-off), see Fig. 4.7. In contrast to that, the schemes $(\mathrm{M})_\theta$ and $(\mathrm{ME})_\eta$ are not capable of conserving the total energy.

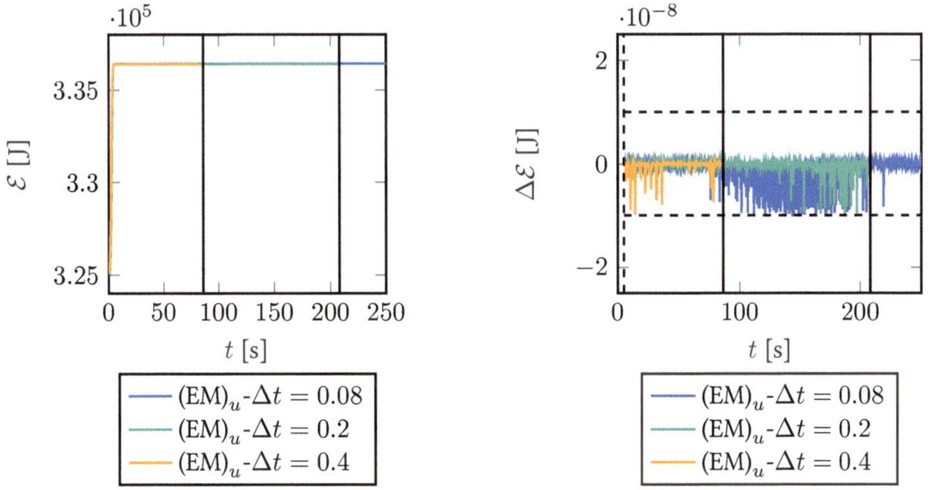

Figure 4.7.: L-shaped block: Total energy $(\mathrm{EM})_u$ scheme (left), Incremental change of total energy $(\mathrm{EM})_u$ scheme (right)

It can be observed from Fig. 4.8 that both schemes have a tendency to increase the energy, depending on the time step. Typically, such energy blow-ups eventually lead to a failure of the iterative (Newton-Raphson) solution procedure. In the diagrams the break down of the simulation is indicated by vertical lines. Despite the capability of the $(\mathrm{EM})_u$ scheme to conserve the total energy, the simulation typically breaks down at about the same point in time as the break down of the other two schemes occurs. The numerical instability of the $(\mathrm{EM})_u$ is accompanied by a nonphysical decrease of the total entropy as can be observed from Fig. 4.9. In fact, the total entropy ought to be a non-decreasing function of time. Only the $(\mathrm{ME})_\eta$ scheme is capable to correctly adhere to the second law of thermodynamics, independent of the time step (Fig. 4.10). Indeed, in each time step the total entropy does increase, as can be observed from Fig. 4.10 (right). The above investigations indicate that all of the three schemes under consideration are prone to numerical instability.

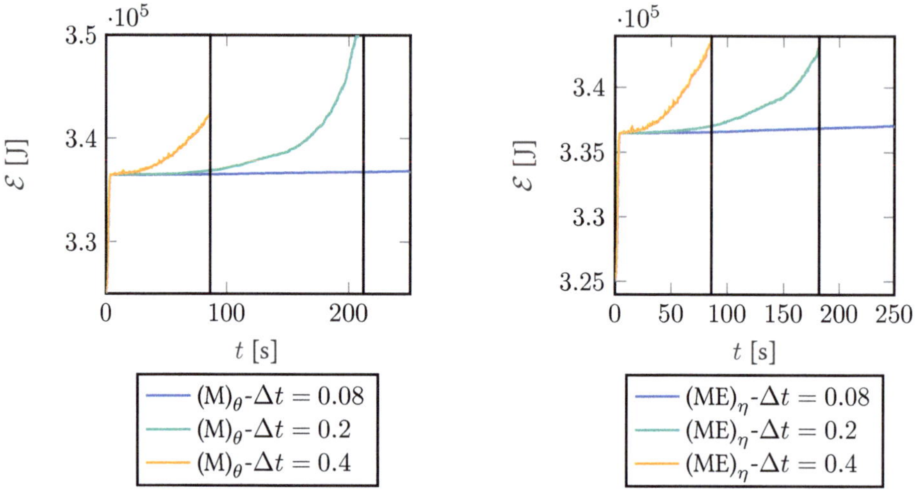

Figure 4.8.: L-shaped block: Total energy $(\mathrm{M})_\theta$ scheme (left), Total energy $(\mathrm{ME})_\eta$ scheme (right)

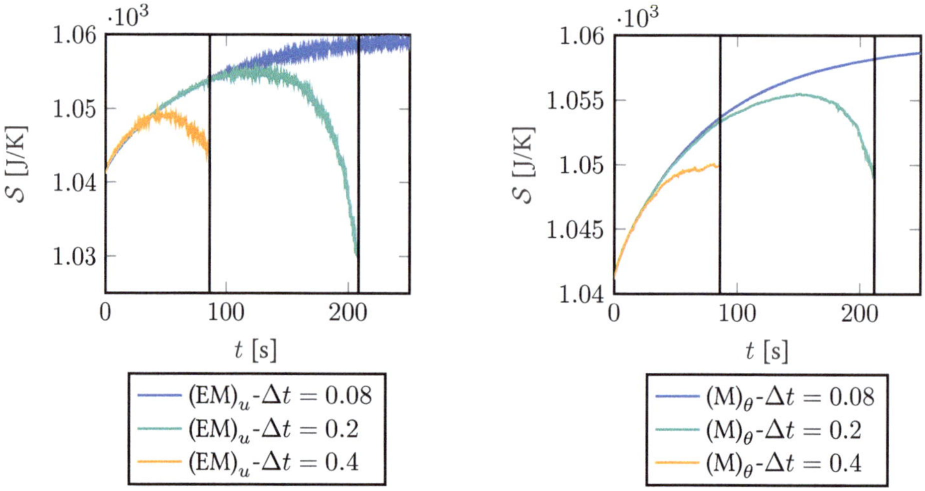

Figure 4.9.: L-shaped block: Total entropy $(\mathrm{EM})_u$ scheme (left), Total entropy $(\mathrm{M})_\theta$ scheme (right)

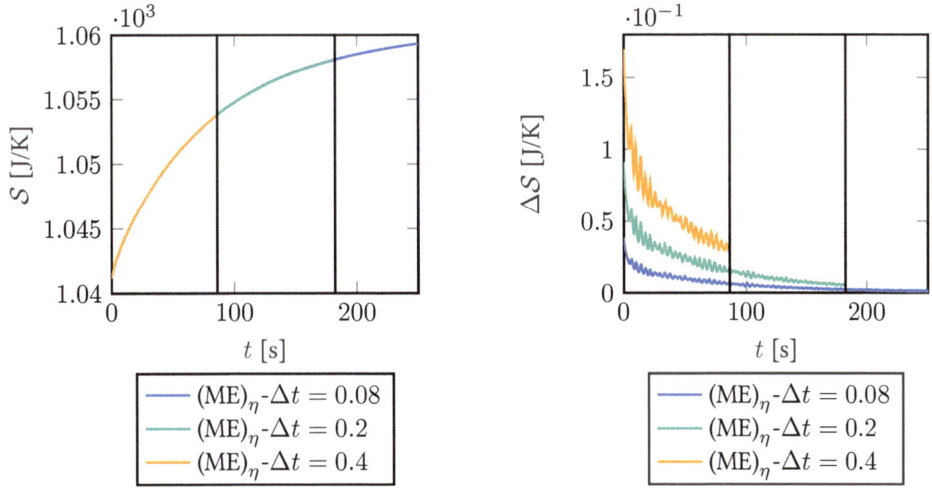

Figure 4.10.: L-shaped block: Total entropy (ME)$_\eta$ scheme (left), Incremental change of total entropy (ME)$_\eta$ scheme (right)

After the loading phase, the environment of the present example can be characterized as thermally perfect in the sense of [146] (cf. Appendix A.2). Thus $\mathcal{L}$ defined in (4.92) plays the role of a Lyapunov function that has to decrease with time. However, none of the schemes under investigation does correctly reproduce this behavior, as can be seen from Figs. 4.11 and 4.12. That is, depending on the time step and the duration of the simulation, all of the schemes inevitably exhibit numerical instabilities characterized by increasing values of $\mathcal{L}$.

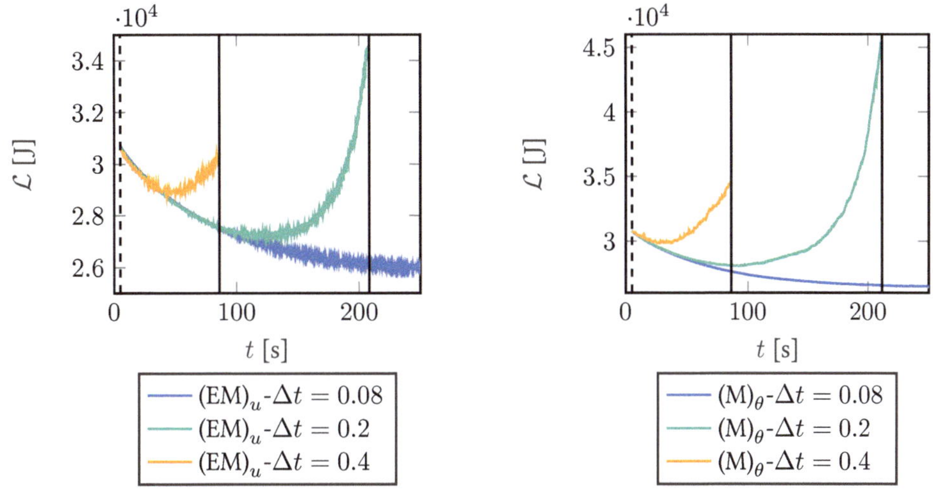

Figure 4.11.: L-shaped block: Lyapunov function (EM)$_u$ scheme (left), Lyapunov function (M)$_\theta$ scheme (right)

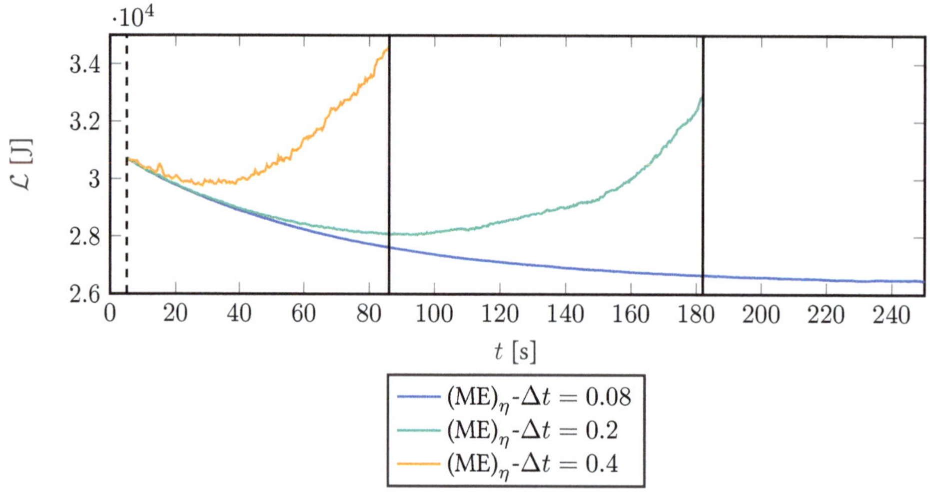

Figure 4.12.: L-shaped block: Lyapunov function $(\mathrm{ME})_\eta$ scheme

Eventually, the motion of the L-shaped block is illustrated in Figs. 4.13 and 4.14 with snapshots at successive points in time. In addition to that, the distribution of the temperature and $\det(\mathbf{F})$ over the block is shown.

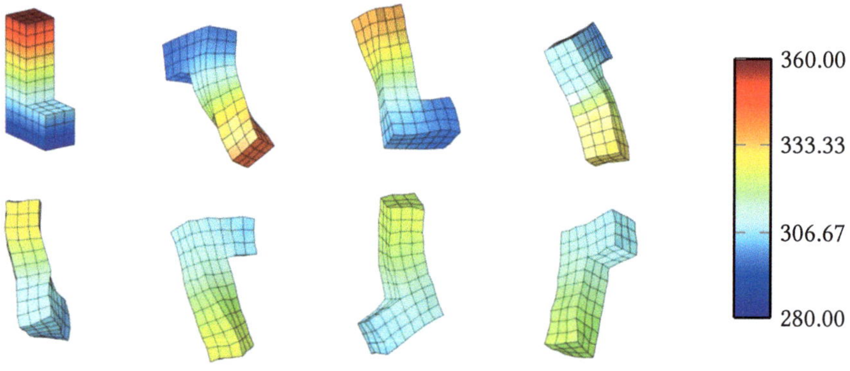

Figure 4.13.: L-shaped block: Snapshots of the motion along with the temperature distribution over the block at $t \in \{0, 32, 64, 96, 128, 160, 192, 224\}$s, obtained with the $(\mathrm{M})_\theta$ scheme and time step $\Delta t = 0.08$s

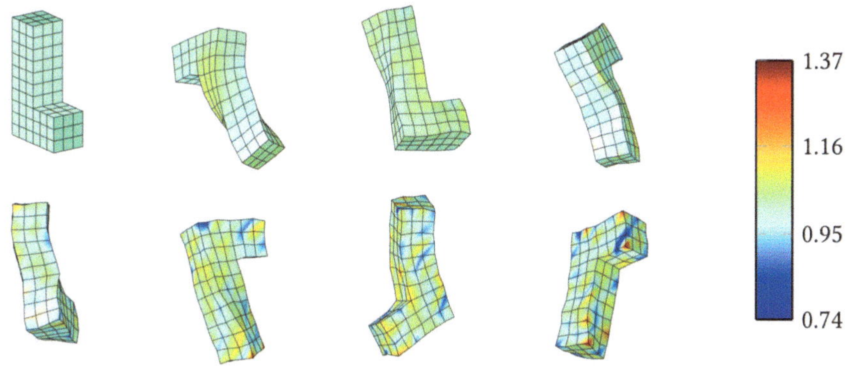

Figure 4.14.: L-shaped block: Snapshots of the motion along with the distribution of the determinant of the deformation gradient over the block at $t \in \{0, 32, 64, 96, 128, 160, 192, 224\}$s, obtained with the $(\mathrm{M})_\theta$ scheme and time step $\Delta t = 0.08$s

4.5.2. Rotating disc

The second example deals with a rotating disc subjected to prescribed heat flow over part of the boundary surface (Fig. 4.15). The spatial discretization of the disc is based on 200 tri-linear finite elements leading to a total of 360 nodes.

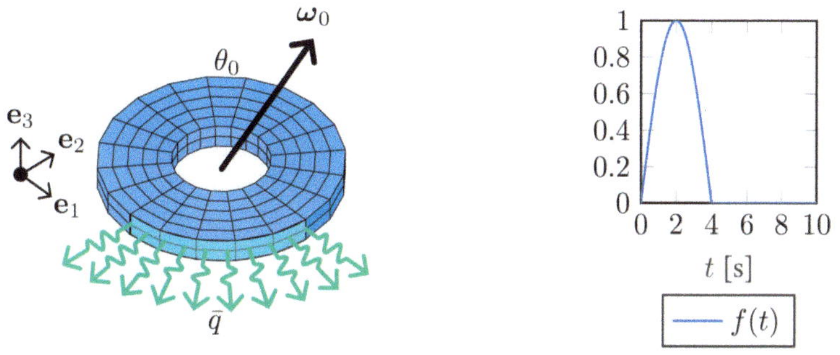

Figure 4.15.: Rotating disc: Initial configuration and thermal boundary conditions (left), function $f(t)$ for the prescribed heat flow over part of the boundary surface (right)

The initial velocity distribution over the disc results from a prescribed angular velocity $\boldsymbol{\omega}_0 \in \mathbb{R}^3$ and is given by

$$\mathbf{v}_0(\mathbf{X}) = \boldsymbol{\omega}_0 \times \mathbf{X}, \qquad \boldsymbol{\omega}_0 = \begin{pmatrix} 1 \\ 1 \\ 1 \end{pmatrix} \frac{1}{\mathrm{s}} \, .$$

The initial temperature of the disc is homogeneously distributed and equal to the reference temperature θ_0. In an initial period of time, $t \in [0, 4]$s, heat flow is prescribed over one quarter of the lateral boundary surface (Fig. 4.15). In particular, the heat flow into the disc is described by

$$\bar{q} = -\frac{2000W}{\pi \text{m}^2} f(t) \,, \qquad f(t) = \begin{cases} \sin(\frac{\pi}{4\text{s}} t) & \text{for} \quad 0 \leq t \leq 4\text{s} \,, \\ 0 & \text{for} \quad t > 4\text{s} \,. \end{cases}$$

A plot of function $f(t)$ can be found in Fig. 4.15. The rest of the boundary surface of the disc is assumed to be thermally insulated ($\bar{q} = 0$). Note that the prescribed heat flow vanishes after $t = 4$s. For $t > 4$s, the complete boundary surface of the disc is assumed to be thermally insulated ($\bar{q} = 0$ on $\partial_q \mathcal{B} = \partial \mathcal{B}$). Then, the environment of the disc is thermally perfect in the sense of [146] (cf. Appendix A.2). A summary of the data used in the simulation of the rotating disc can be found in Table 4.4.

Material parameters	λ	3000	Pa	Geometry
	μ	750	Pa	
Specific heat capacity	c	150	$\text{JK}^{-1}\text{m}^{-3}$	
Expansion coefficient	β	$1 \cdot 10^{-4}$	K^{-1}	
Thermal conductivity	k	20	$\text{WK}^{-1}\text{m}^{-1}$	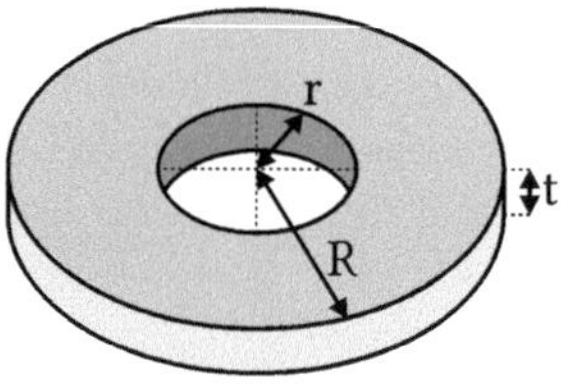
Ref. temperature	θ_0	300	K	
Mass density	ρ	8.93	kgm^{-3}	
Radius	r	0.8	m	
	R	2	m	
Thickness	t	0.4	m	
Newton tolerance	ε	10^{-8}	-	
Simulation time	T	30	s	
Time steps	Δt	0.04	s	
		0.08	s	
		0.1	s	

Table 4.4.: Rotating disc: Data used in the simulations

In the simulations constant time steps of size $\Delta t \in \{0.04, 0.08, 0.1\}$s are applied. Since neither external loads act on the disc, nor displacement boundary conditions are imposed, the mechanical system at hand has translational and rotational symmetry. Consequently, the corresponding momentum maps are first integrals of the motion.

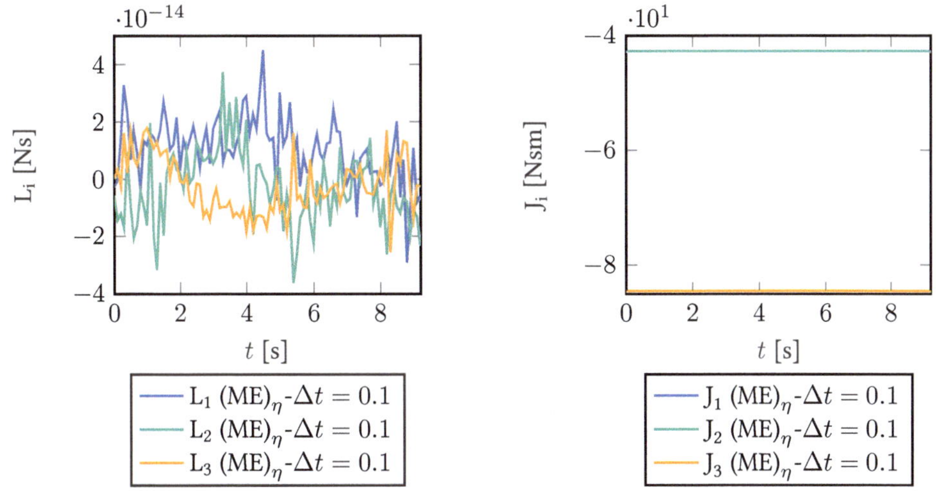

Figure 4.16.: Rotating disc: Total linear momentum $(\mathrm{ME})_\eta$ scheme (left), Total angular momentum $(\mathrm{ME})_\eta$ scheme (right)

All three integrators at hand are capable to conserve the respective momentum map. Representative numerical results are shown in Fig. 4.16.

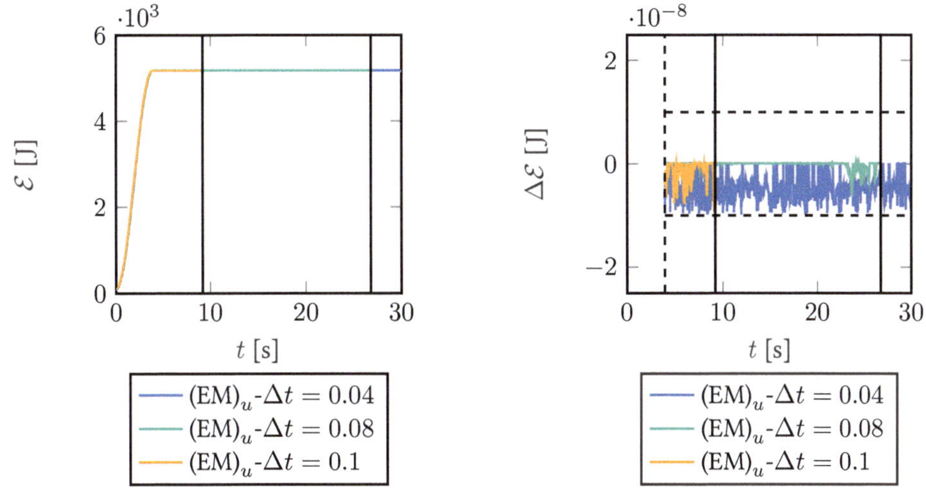

Figure 4.17.: Rotating disc: Total energy $(\mathrm{EM})_u$ scheme (left), Discrete change of total energy $(\mathrm{EM})_u$ scheme (right)

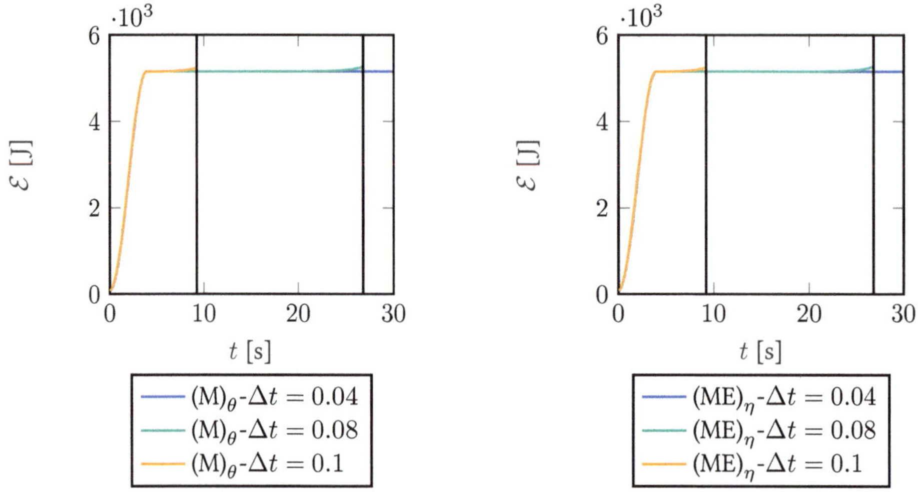

Figure 4.18.: Rotating disc: Total energy (M)$_\theta$ scheme (left), Total energy (ME)$_\eta$ scheme (right)

Since heat flow into the system is prescribed in the initial time period $[0, 4]$s, the total energy is expected to increase. For $t > 4$s, the system is closed and the total energy should stay constant. Again the (EM)$_u$ scheme is capable to correctly reproduce the first law (Fig. 4.17). However, despite the algorithmic energy conservation (for $t > 4$s), the (EM)$_u$ scheme gets unstable, depending on the time step. The corresponding point in time of the break down of the simulation is indicated with a vertical line in the diagrams. At about the same points in time the other two schemes break down as well (Fig. 4.18). For these schemes the break down is accompanied by a sudden increase of the total energy leading to the divergence of the Newton-Raphson iterations. For the considered duration of the simulation ($t \in [0, 30]$s), a time step of $\Delta t = 0.04$s is small enough to retain numerical stability of the three schemes at hand.

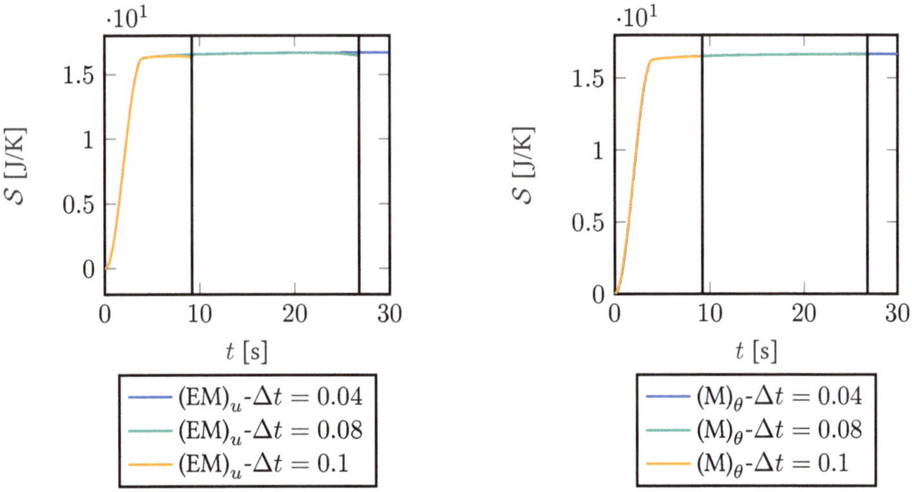

Figure 4.19.: Rotating disc: Total entropy $(\mathrm{EM})_u$ scheme (left), Total entropy $(\mathrm{M})_\theta$ scheme (right)

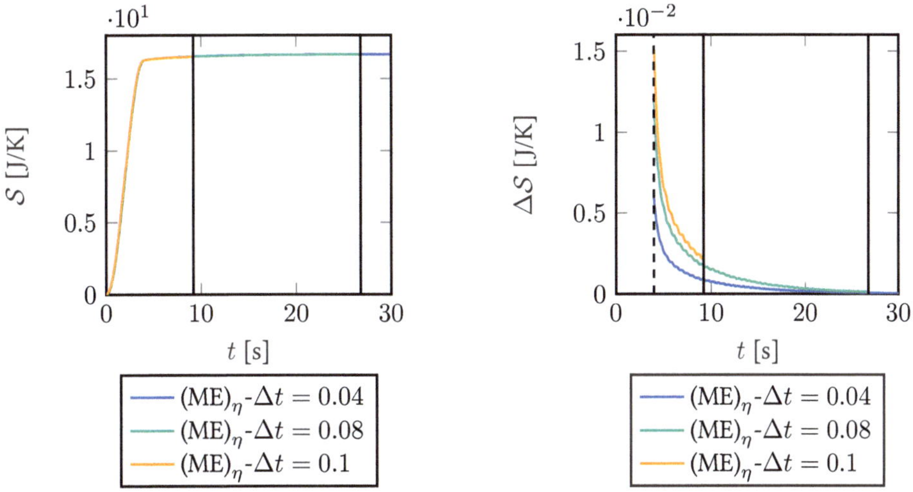

Figure 4.20.: Rotating disc: Total entropy $(\mathrm{ME})_\eta$ scheme (left), Discrete change of total entropy $(\mathrm{ME})_\eta$ scheme (right)

Due to the prescribed heat flow into the disc, the total entropy of the disc is expected to increase in the initial time period $[0, 4]$s. For $t > 4$s, the system is closed and the total entropy should be a non-decreasing function of time. As expected, the $(\mathrm{ME})_\eta$ scheme is capable to exactly satisfy the second law, independent of the time step. This can be observed from Fig. 4.20. In particular, Fig. 4.20 (right) confirms that the change per time step of the total entropy is always positive. The $(\mathrm{EM})_u$ scheme does not correctly reproduce the second law as can be seen from Fig. 4.19. Accordingly, the divergence of the iterative solution procedure is accompanied by an nonphysical decrease of the total entropy. The $(M)_\theta$ closely adheres to the second law as can be observed from Fig. 4.19.

To shed further light on the numerical stability of the present schemes, we consider the Lyapunov function $\mathcal{L}$ defined in (4.92). For $t > 4$s the system is closed and the function $\mathcal{L}$ should decrease with time (see also Appendix A.2). All three schemes fail to correctly reproduce this behavior, depending on the size of the time step and the duration of the simulation (Figs. 4.21 and 4.22). In particular, it can be seen that the smallest time step $\Delta t = 0.04$s yields a stable numerical simulation, at least in the considered time interval $[0, 30]$s. However, for larger time steps $\Delta t = 0.08$s and $\Delta t = 0.1$s, the numerical instability of each scheme becomes visible through the increase of $\mathcal{L}$.

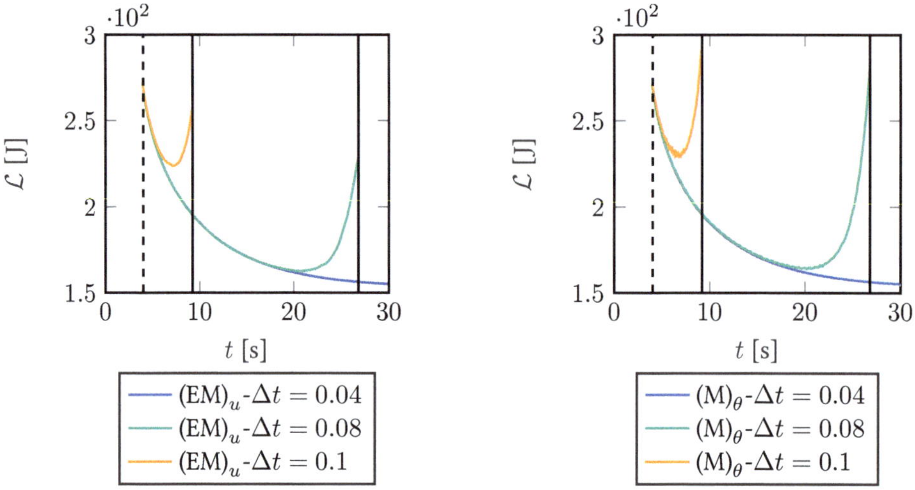

Figure 4.21.: Rotating disc: Lyapunov function $(EM)_u$ scheme (left), Lyapunov function $(M)_\theta$ scheme (right)

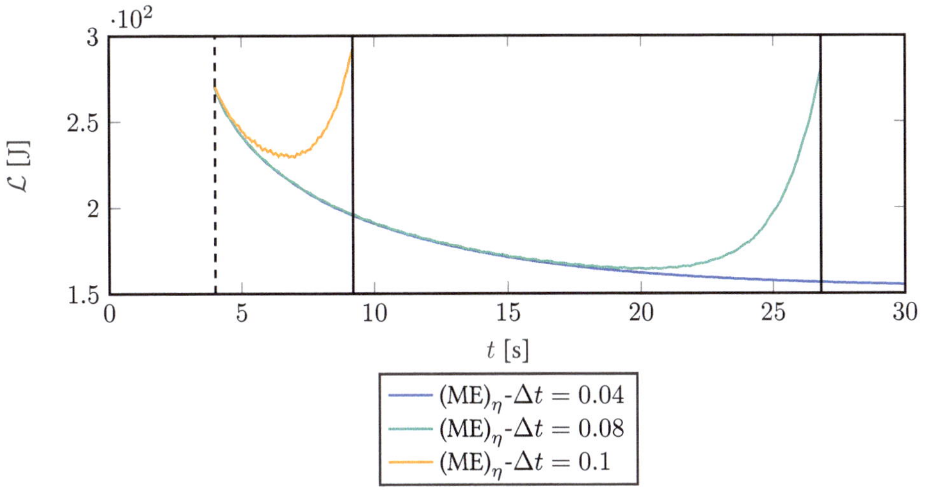

Figure 4.22.: Rotating disc: Lyapunov function (ME)$_\eta$ scheme

In addition we investigate the order of accuracy of the present schemes. We therefore introduce the relative error in the position, linear momentum density and absolute temperature defined by

$$\mathrm{e}_{\varphi} = \frac{||\boldsymbol{\varphi}-\boldsymbol{\varphi}_r||_{L_2}}{||\boldsymbol{\varphi}_r||_{L_2}}, \quad \mathrm{e}_{\mathrm{p}} = \frac{||\mathbf{p}-\mathbf{p}_r||_{L_2}}{||\mathbf{p}_r||_{L_2}}, \quad \mathrm{e}_{\theta} = \frac{||\theta-\theta_r||_{L_2}}{||\theta_r||_{L_2}}, \quad ||\bullet||_{L_2} = \left[\int_{\mathcal{B}} \langle \bullet, \bullet \rangle \,\mathrm{d}V\right]^{\frac{1}{2}},$$

where index r refers to the reference solution obtained with a very small time step. Moreover, $\langle \bullet, \bullet \rangle$ denotes the inner product. As can be observed from Figs. 4.23 and 4.24, all schemes are second order accurate in the position, linear momentum density and absolute temperature.

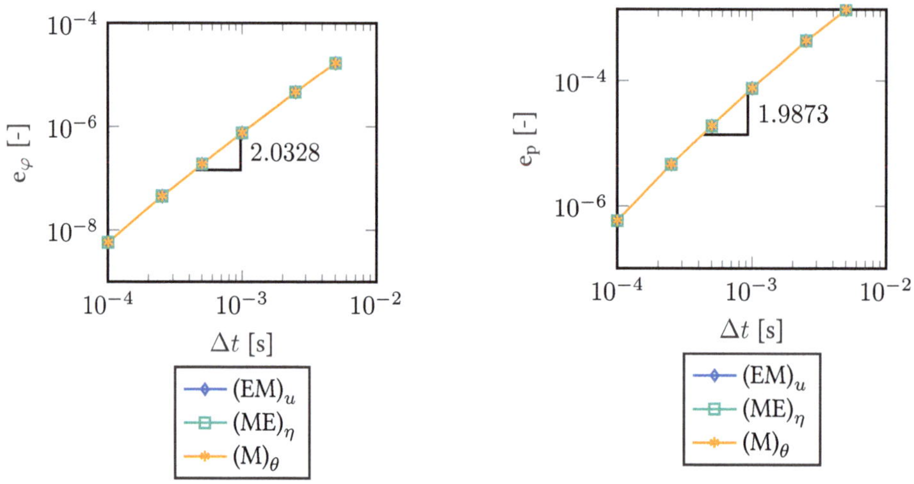

Figure 4.23.: Rotating disc: Error in the position (left), Error in the linear momentum density (right)

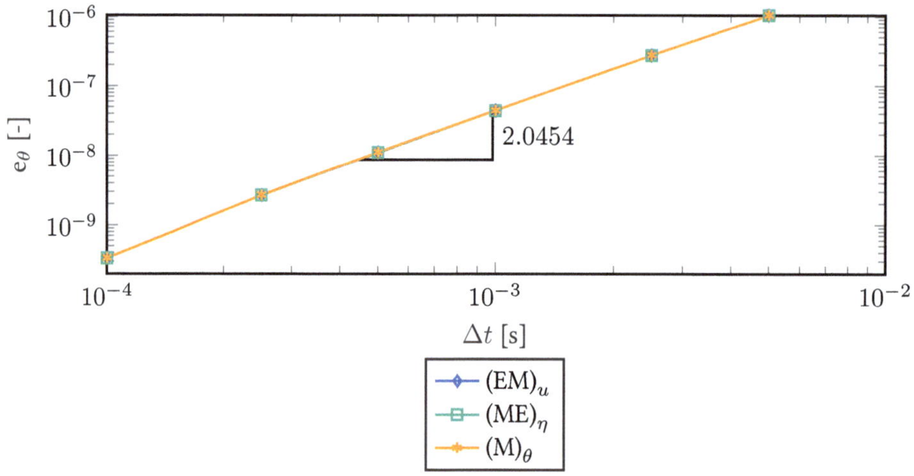

Figure 4.24.: Rotating disc: Error in the absolute temperature

Eventually, the motion of the disc is illustrated in Fig. 4.25 with snapshots at successive points in time. In addition to that, the distribution of the temperature over the disc is shown.

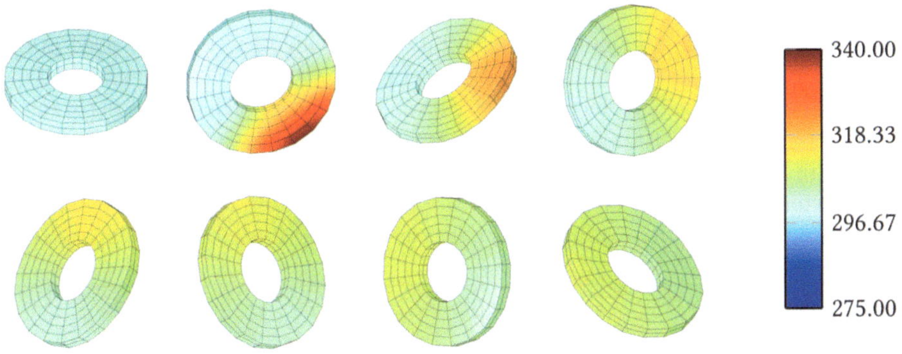

Figure 4.25.: Rotating disc: Snapshots of the motion at successive points in time $t \in \{0, 4, 8, 12, 16, 18, 24, 28\}$s, and corresponding temperature distribution, calculated with the $(\mathrm{M})_\theta$ scheme and $\Delta t = 0.04$s

4.5.3. Rotating disc in a thermally perfect environment

The last example of this Chapter deals with a slight modification of the previous example which renders a thermally perfect environment for the disc throughout the whole duration of the simulation (Fig. 4.26).

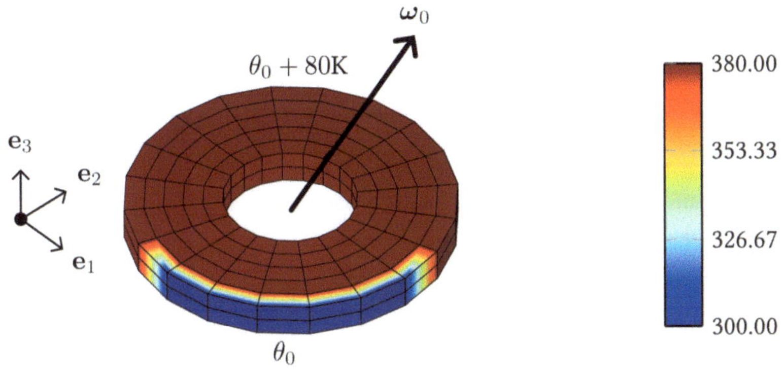

Figure 4.26.: Rotating disc in a thermally perfect environment: Initial configuration, temperature boundary conditions imposed on a quarter of the lateral surface of the disc and initial distribution of the temperature

Instead of prescribing the heat flow over a quarter of the lateral surface of the disc, we now impose temperature boundary conditions on the same portion of the lateral surface of the disc. In particular, the temperature at the finite element nodes located on the boundary in question is constrained to be equal to the reference temperature θ_0. The initial temperature at the remaining nodes of the finite element mesh is chosen to assume the value $\theta_0 + 80\text{K}$. The remaining simulation data are identical to those used in the previous example (see also Table 4.4). A summary of the data used in the simulation of the rotating disc in a thermally perfect environment can be found in Table 4.5. In this example we focus on the temperature-based $(\text{M})_\theta$ scheme, which makes possible to impose the temperature boundary conditions in a straightforward way through standard Dirichlet boundary conditions. Constant time steps of size $\Delta t \in \{0.04, 0.075, 0.0875\}$s are applied. Again the mechanical system at hand has both translational and rotational symmetry.

Material parameters	λ	3000	Pa	Geometry
	μ	750	Pa	
Specific heat capacity	c	150	$\text{JK}^{-1}\text{m}^{-3}$	
Expansion coefficient	β	$1 \cdot 10^{-4}$	K^{-1}	
Thermal conductivity	k	20	$\text{WK}^{-1}\text{m}^{-1}$	
Ref. temperature	θ_0	300	K	
Mass density	ρ	8.93	kgm^{-3}	
Radius	r	0.8	m	
	R	2	m	
Thickness	t	0.4	m	
Newton tolerance	ε	10^{-8}	-	
Simulation time	T	70	s	
Time steps	Δt	0.04	s	
		0.075	s	
		0.0875	s	

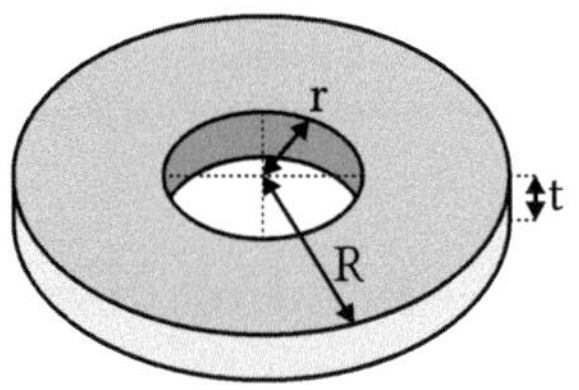

Table 4.5.: Rotating disc in a thermally perfect environment: Data used in the simulations

The corresponding momentum maps are exactly conserved by the $(\text{M})_\theta$ scheme as can be observed from Fig. 4.27.

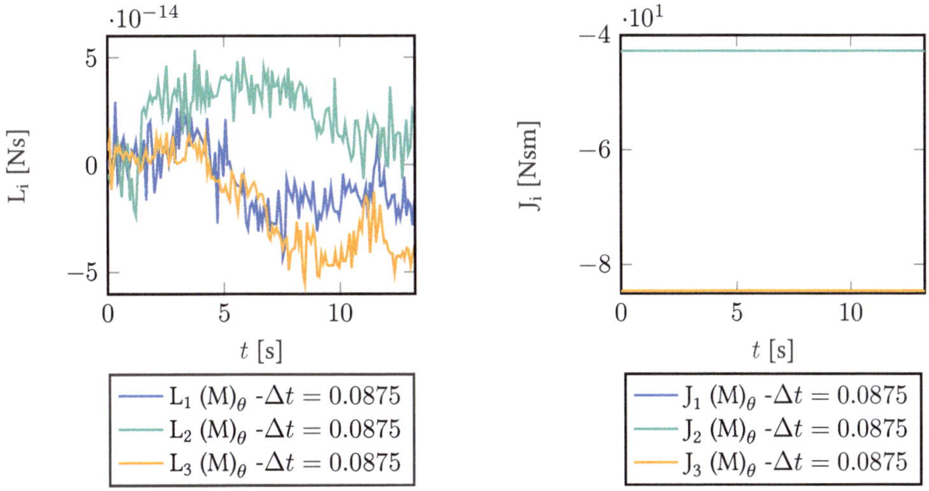

Figure 4.27.: Rotating disc in a thermally perfect environment: Total linear momentum $(M)_\theta$ scheme (left), Total angular momentum $(M)_\theta$ scheme (right)

Due to the temperature boundary condition heat is flowing out of the disc. The corresponding evolution of the total energy is depicted in Fig. 4.28 (left). In addition to that, the evolution of the total entropy is shown in Fig. 4.28 (right).

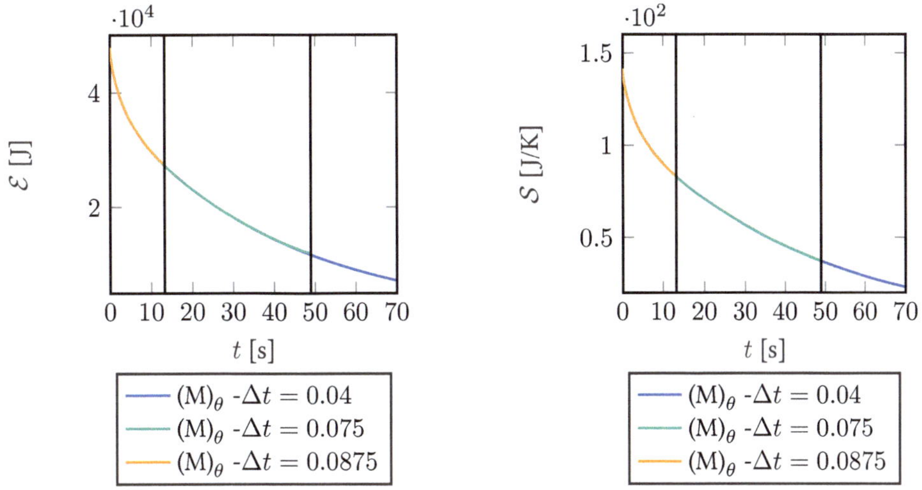

Figure 4.28.: Rotating disc in a thermally perfect environment: Total energy (left), Total entropy (right)

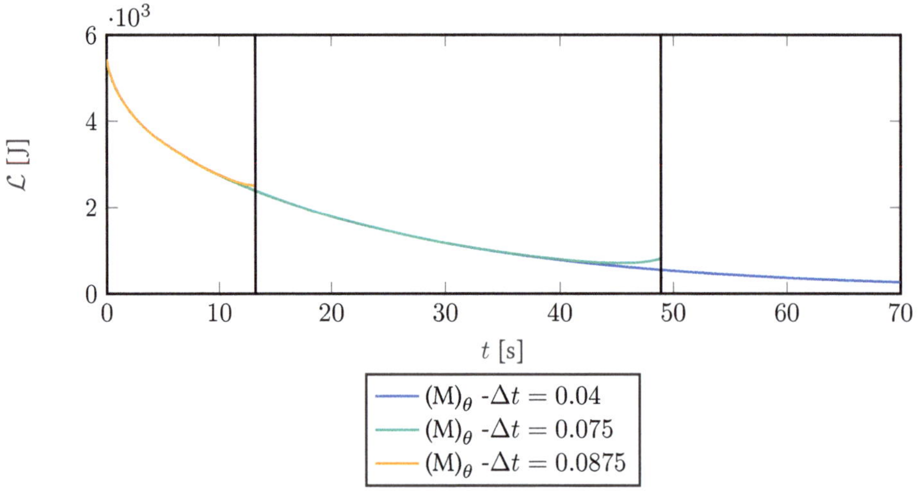

Figure 4.29.: Rotating disc in a thermally perfect environment: Lyapunov function

As before, the break down of a simulation due to numerical instability is indicated by vertical lines. Again the onset of numerical instability can be detected from the plot of the Lyapunov function in Fig. 4.29, although this time the nonphysical increase is not as pronounced as in the two previous examples. Eventually, both the motion and the evolution of the temperature distribution are shown in Fig. 4.30 with snapshots at successive points in time.

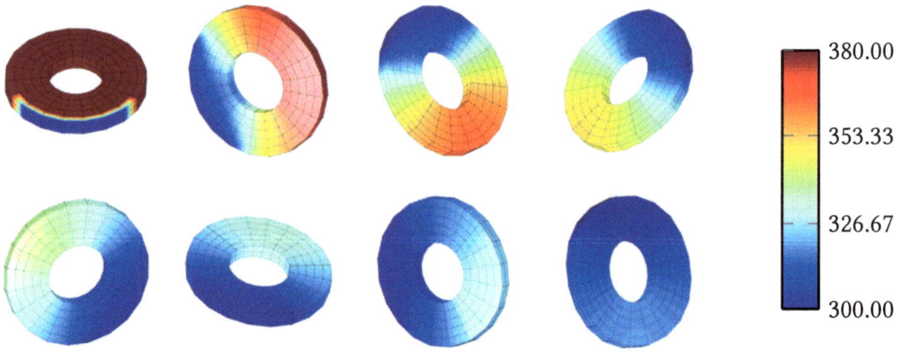

Figure 4.30.: Rotating disc in a thermally perfect environment: Snapshots of the motion at $t \in \{0, 10, 20, 30, 40, 50, 60, 70\}$s and corresponding temperature distribution, calculated with the $(M)_\theta$ scheme and $\Delta t = 0.04$s

5. Energy-momentum-entropy consistent numerical methods for large-strain thermo-elasticity based on the GENERIC formalism[1]

The results of Chapter 4 lead to the conclusion that all major balance laws should be maintained in the discrete setting in order to enhance the numerical stability of numerical methods for finite-strain thermo-elastodynamics. Therefore fully structure-preserving numerical methods are developed in the present chapter. We will resort to the GENERIC-based description of large-strain thermo-elasticity, which makes the free choice of the thermodynamic state variable possible. In particular, one may choose (i) the internal energy density, (ii) the entropy density, or (iii) the absolute temperature as the thermodynamic state variable. Three alternative energy-momentum-entropy (EME) schemes result from the present approach that make use of a *GENERIC-consistent space discretization.* These schemes are directly linked to the respective choice of the thermodynamic state variable. Numerical examples confirm the structure-preserving properties of the newly developed (EME) schemes, which exhibit superior numerical stability.

5.1. Large-strain thermoelasticty

In this section we briefly recapitulate the variational formulation of large-strain thermo-elasticity with heat conduction of Chapter 4 which lies at the heart of the proposed discretization in space and time. In contrast to Chapter 4 we introduce a frame-invariant formulation from the outset.

5.1.1. Underlying variational formulation

We consider a continuum body with material points $\mathbf{X} = X^A \mathbf{e}_A$ in the reference configuration $\mathcal{B} \subset \mathbb{R}^3$, see Fig. 5.1. Here and in the sequel the summation convention applies to repeated indices. Moreover, $\mathbf{e}_A$ denote the canonical base vectors in $\mathbb{R}^3$.

[1] This Chapter is based on [2]

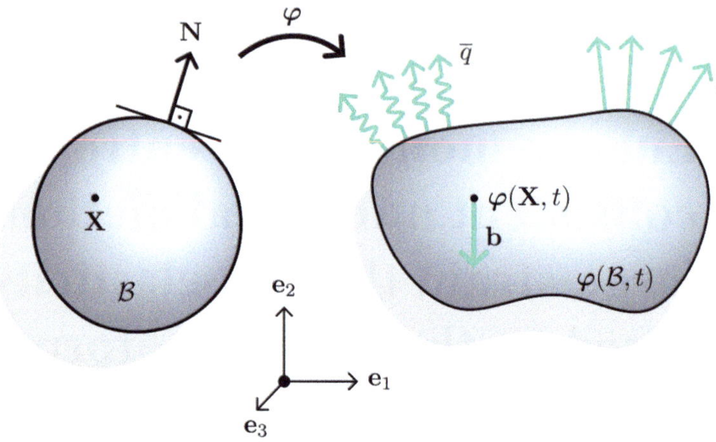

Figure 5.1.: Reference configuration $\mathcal{B}$ with boundary $\partial\mathcal{B}$ and current configuration $\varphi(\mathcal{B}, t)$ at time t. External tractions $\overline{\mathbf{t}} = \mathbf{FSN}$ act on the boundary of the current configuration. In addition to that, the heat flux across the current boundary is denoted by $\overline{q} = \mathbf{Q} \cdot \mathbf{N}$.

Within the Lagrangian description of continuum mechanics the deformed configuration of the body at time t is characterised by the deformation map $\varphi : \mathcal{B} \times \mathcal{I} \mapsto \mathbb{R}^3$, where $\mathcal{I} = [0, T]$ is the time interal of interest. The velocity of the material point $\mathbf{X} \in \mathcal{B}$ located at $\mathbf{x} = \varphi(\mathbf{X}, t)$ is given by $\mathbf{v} = \dot{\varphi}$, where a superposed dotes denotes the material time derivate. The deformation gradient is given by

$$\mathbf{F} = \partial_{\mathbf{X}}\varphi \; . \tag{5.1}$$

Main ingredients of the GENERIC framework are the internal energy and the entropy. In addition to that, the choice of the thermodynamic state variable plays an important role. Following Chapter 4 we allow for the free choice of the thermodynamic state variable $\tau : \mathcal{B} \times \mathcal{I} \mapsto \mathbb{R}$ from among three options $\tau \in \{\theta, \eta, u\}$. These options are: (i) the absolute temperature θ, (ii) the entropy density η, and (iii) the internal energy density u. Depending on the choice of the thermodynamic state variable, the absolute temperature can be written in the form (see also [94] and [100])

$$\Theta'(\mathbf{C}, \tau) = \frac{\partial_\tau u'(\mathbf{C}, \tau)}{\partial_\tau \eta'(\mathbf{C}, \tau)} \; . \tag{5.2}$$

Now the internal energy density and the entropy density, respectively, are given by

$$u = u'(\mathbf{C}, \tau) \quad \text{and} \quad \eta = \eta'(\mathbf{C}, \tau) \; . \tag{5.3}$$

In this connection, a frame-indifferent constitutive formulation for thermoelastic materials is based on the right Cauchy-Green tensor $\mathbf{C} = \mathbf{F}^{\mathrm{T}}\mathbf{F}$. The GENERIC-based weak form pertaining to large-strain thermo-elasticity can be written in the form (see Chapter 4)

$$
\begin{aligned}
0 &= \int_{\mathcal{B}} \mathbf{w}_{\varphi} \cdot (\dot{\boldsymbol{\varphi}} - \rho^{-1}\mathbf{p}) \, \mathrm{d}V \,, \\
0 &= \int_{\mathcal{B}} (\mathbf{w}_{\mathbf{p}} \cdot (\dot{\mathbf{p}} - \mathbf{b}) + \mathbf{FS} : \nabla \mathbf{w}_{\mathbf{p}}) \, \mathrm{d}V - \int_{\partial_\sigma \mathcal{B}} \mathbf{w}_{\mathbf{p}} \cdot \bar{\mathbf{t}} \, \mathrm{d}A \,, \\
0 &= \int_{\mathcal{B}} \left(\mathrm{w}_\tau \dot{\tau} + \nabla(\rho^{-1}\mathbf{p}) : \left(\frac{\mathrm{w}_\tau}{\partial_\tau \eta'} 2\mathbf{F} \partial_{\mathbf{C}} \eta' \right) - \nabla \left(\frac{\mathrm{w}_\tau}{\partial_\tau u'} \right) \cdot \mathbf{Q} \right) \mathrm{d}V + \int_{\partial_q \mathcal{B}} \frac{\mathrm{w}_\tau}{\partial_\tau u'} \bar{q} \, \mathrm{d}A \,,
\end{aligned}
\tag{5.4}
$$

where $\rho : \mathcal{B} \mapsto \mathbb{R}_+$ is the mass density in the reference configuration. Moreover, $\mathbf{p} : \mathcal{B} \times \mathcal{I} \mapsto \mathbb{R}^3$ is the linear momentum density and $\mathbf{b} : \mathcal{B} \mapsto \mathbb{R}^3$ represent prescribed body forces which, for simplicity, are assumed to be dead loads. The second Piola-Kirchhoff stress tensor is given by

$$
\mathbf{S} = \mathbf{S}'(\mathbf{C}, \tau) = 2\,(\partial_{\mathbf{C}} u' - \Theta' \partial_{\mathbf{C}} \eta') \,. \tag{5.5}
$$

Furthermore, the material heat flux vector assumes the form

$$
\mathbf{Q} = \mathbf{Q}'(\mathbf{C}, \tau) = (\Theta')^2 \mathbf{K}' \nabla \left(\frac{1}{\Theta'} \right) \,, \tag{5.6}
$$

where $\mathbf{K} = \mathbf{K}'(\mathbf{C}, \tau)$ is the positive semi-definite material conductivity tensor. The weak form needs be supplemented with initial and boundary conditions, respectively. For that purpose, the boundary $\partial\mathcal{B}$ of the continuum body is decomposed into a displacement boundary $\partial_\varphi\mathcal{B}$, on which $\varphi = \bar{\varphi}$ is prescribed, and a traction boundary $\partial_\sigma\mathcal{B}$, on which the external traction $\bar{\mathbf{t}}$ is prescribed such that $\mathbf{FSN} = \bar{\mathbf{t}}$ (Fig. 5.2). In this connection, the standard relations $\partial_\varphi\mathcal{B} \cup \partial_\sigma\mathcal{B} = \partial\mathcal{B}$ and $\partial_\varphi\mathcal{B} \cap \partial_\sigma\mathcal{B} = \emptyset$ hold. Similarly, for the thermal part we consider the subsets $\partial_\tau\mathcal{B}$ and $\partial_q\mathcal{B}$, with the properties $\partial_\tau\mathcal{B} \cup \partial_q\mathcal{B} = \partial\mathcal{B}$ and $\partial_\tau\mathcal{B} \cap \partial_q\mathcal{B} = \emptyset$ (Fig. 5.3). The thermodynamic state variable is prescribed on $\partial_\tau\mathcal{B}$, i.e. $\tau = \bar{\tau}$, whereas the heat flux is prescribed on $\partial_q\mathcal{B}$, i.e. $\mathbf{Q} \cdot \mathbf{N} = \bar{q}$.

The unknown fields are subject to initial conditions of the form $\varphi(\cdot, 0) = \mathbf{X}$, $\mathbf{p}(\cdot, 0) = \rho\mathbf{v}_0$, and $\tau(\cdot, 0) = \tau_0$ in $\mathcal{B}$. Here, $\mathbf{v}_0$ is a prescribed velocity field and τ_0 is a prescribed field of the thermodynamic state variable $\tau \in \{\theta, \eta, u\}$.

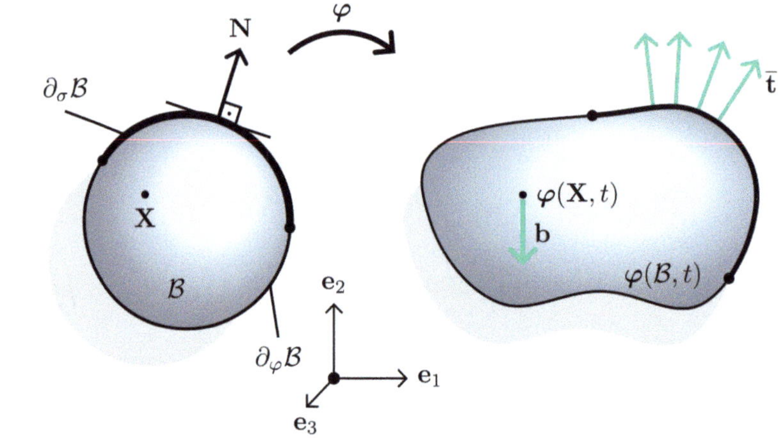

Figure 5.2.: Mechanical part of the IBVP. Note that $\bar{\mathbf{t}} = \mathbf{FSN}$ denotes prescribed external Piola tractions acting on the current boundary expressed per unit area of the reference boundary $\partial_\sigma \mathcal{B}$.

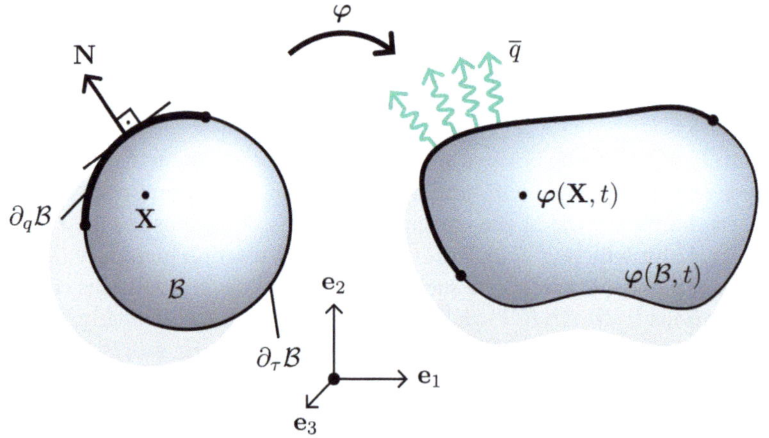

Figure 5.3.: Thermal part of the IBVP. Note that $\bar{q} = \mathbf{Q} \cdot \mathbf{N}$ is the prescribed rate of heat transfer accross the current boundary expressed per unit area of the reference boundary $\partial_q \mathcal{B}$.

5.1.2. Frame-invariant local form of the field equations

We next deal with the frame-invariant local form of the field equations corresponding to the GENERIC-based weak form (5.4). In this context, we consider alternative frame-invariant formulations related to the specific choice of the thermodynamic state variable

$\tau \in \{\theta, \eta, u\}$. Using standard arguments from the calculus of variations, weak form (5.4) gives rise to the corresponding local form of the field equations

$$\begin{aligned} \dot{\varphi} &= \rho^{-1}\mathbf{p}\,, \\ \dot{\mathbf{p}} &= \mathbf{b} + \mathrm{Div}\,(\mathbf{FS})\;, \end{aligned} \tag{5.7}$$

together with

$$\dot{\tau} = -\frac{1}{\partial_\tau \eta'} 2\mathbf{F}\partial_{\mathbf{C}}\eta' : \nabla(\rho^{-1}\mathbf{p}) - \frac{1}{\partial_\tau u'}\mathrm{Div}\mathbf{Q}\,. \tag{5.8}$$

These equations have to hold for all $\mathbf{X} \in \mathcal{B}$ and $t > 0$. The last equation governs the time-evolution of the thermodynamic state variable $\tau \in \{\theta, \eta, u\}$. Note that the second Piola-Kirchhoff stress tensor $\mathbf{S}$ in $(5.7)_2$ assumes the form (5.5), and the material heat flux vector $\mathbf{Q}$ in (5.8) is given by (5.6).

Choosing $\tau = \theta$ leads to the *temperature-based formulation*, where the second Piola-Kirchhoff stress tensor follows from (5.5) and assumes the form

$$\overline{\mathbf{S}} = 2\left[\partial_{\mathbf{C}}\overline{u}(\mathbf{C},\theta) - \theta\partial_{\mathbf{C}}\overline{\eta}(\mathbf{C},\theta)\right]\,. \tag{5.9}$$

The temperature typically enters the definition of the Helmholtz free energy density

$$\psi = \overline{\psi}(\mathbf{C},\theta)\,. \tag{5.10}$$

Then the entropy density is defined by

$$\eta = \overline{\eta}(\mathbf{C},\theta) = -\partial_\theta\overline{\psi}(\mathbf{C},\theta)\,. \tag{5.11}$$

The internal energy density function then follows from a Legende transformation of the free energy density. Accordingly, the internal energy density is given by

$$u = \overline{u}(\mathbf{C},\theta) = \overline{\psi}(\mathbf{C},\theta) + \theta\,\overline{\eta}(\mathbf{C},\theta)\,. \tag{5.12}$$

Taking into account the temperature formula (5.2) along with $\overline{\psi} = \overline{u} - \theta\overline{\eta}$, it can be seen that expression (5.9) for the second Piola-Kirchhoff stress tensor is equivalent to

$$\mathbf{S} = \overline{\mathbf{S}}(\mathbf{C},\theta) = 2\partial_{\mathbf{C}}\overline{\psi}(\mathbf{C},\theta)\,. \tag{5.13}$$

Equations (5.11) and (5.13) are sometimes called thermal equations of state (see, for example, [147]).

Alternatively, we may use the entropy density $\eta : \mathcal{B} \times \mathcal{I} \mapsto \mathbb{R}$ as thermodynamic state variable leading to the *entropy-based formulation*. Accordingly, choosing $\tau = \eta$, the second Piola-Kirchhoff stress tensor follows from (5.5) and assumes the form

$$\mathbf{S} = \widetilde{\mathbf{S}}(\mathbf{C},\eta) = 2\partial_{\mathbf{C}}\widetilde{u}(\mathbf{C},\eta)\,. \tag{5.14}$$

	free energy	internal energy	entropy	temperature
τ	$\psi'(\mathbf{C},\tau)$	$u'(\mathbf{C},\tau)$	$\eta'(\mathbf{C},\tau)$	$\Theta'(\mathbf{C},\tau)$
θ	$\overline{\psi}(\mathbf{C},\theta)$	$\overline{u}(\mathbf{C},\theta)$	$\overline{\eta}(\mathbf{C},\theta)$	θ
η	$\widetilde{\psi}(\mathbf{C},\eta)$	$\widetilde{u}(\mathbf{C},\eta)$	η	$\widetilde{\Theta}(\mathbf{C},\eta)$
u	$\widehat{\psi}(\mathbf{C},u)$	u	$\widehat{\eta}(\mathbf{C},u)$	$\widehat{\Theta}(\mathbf{C},u)$

Table 5.1.: Notation used in the present Chapter, depending on the choice of the thermodynamic state variable $\tau \in \{\theta, \eta, u\}$.

Temperature formula (5.2) leads to

$$\widetilde{\Theta}(\mathbf{C},\eta) = \partial_\eta \widetilde{u}(\mathbf{C},\eta)\ . \tag{5.15}$$

The last two equations belong to the caloric equations of state (see, for example, [136]).

Yet another option is to use the internal energy density $u : \mathcal{B} \times \mathcal{I} \mapsto \mathbb{R}$ as thermodynamic state variable leading to the *internal-energy-based formulation.* Thus, choosing $\tau = u$, the second Piola-Kirchhoff stress tensor follows from (5.5) and assumes the form

$$\mathbf{S} = \widehat{\mathbf{S}}(\mathbf{C},u) = -2\widehat{\Theta}\partial_{\mathbf{C}}\widehat{\eta}(\mathbf{C},u)\ . \tag{5.16}$$

Furthermore, temperature formula (5.2) yields

$$\widehat{\Theta}(\mathbf{C},u) = \frac{1}{\partial_u \widehat{\eta}(\mathbf{C},u)}\ . \tag{5.17}$$

The notation used in the present work is summarized in Table 5.1. A more detailed treatment of the alternative choice for the thermodynamic state variables can be found in [100, Sec. 2.3] (see also Section 4.2).

5.1.3. Balance laws

We next consider important balance laws of the underlying continuous formulation of finite-strain thermo-elasticity. The main goal of the present work is to preserve these balance laws under discretization. In this section we focus on the pure Neumann problem (i.e. $\partial_\sigma \mathcal{B} = \partial_q \mathcal{B} = \partial \mathcal{B}$).

The linear momentum of the continuum body is defined by $\mathbf{L} = \int_{\mathcal{B}} \mathbf{p}\,\mathrm{d}V$. Inserting $\mathbf{w}_\mathrm{p} = \boldsymbol{\xi}$ into $(5.4)_2$, where $\boldsymbol{\xi} \in \mathbb{R}^3$ is arbitrary but constant, leads to

$$\boldsymbol{\xi} \cdot \frac{\mathrm{d}\mathbf{L}}{\mathrm{d}t} = \boldsymbol{\xi} \cdot \left(\int_{\mathcal{B}} \mathbf{b}\,\mathrm{d}V + \int_{\partial\mathcal{B}} \mathbf{FSN}\,\mathrm{d}A \right)\ . \tag{5.18}$$

Due to the arbitrariness of $\boldsymbol{\xi} \in \mathbb{R}^3$, the last equation recovers the balance law for linear momentum. The parentheses on the right-hand side of (5.18) contain the resultant external load applied on the continuum body (see also Fig. 5.1).

The total angular momentum relative to the origin of the inertial frame is defined by $\mathbf{J} = \int_{\mathcal{B}} \boldsymbol{\varphi} \times \mathbf{p} \, \mathrm{d}V$. Substituting $\mathbf{w}_{\boldsymbol{\varphi}} = \mathbf{p} \times \boldsymbol{\xi}$, $\mathbf{w}_{\mathbf{p}} = \boldsymbol{\xi} \times \boldsymbol{\varphi}$, into (5.4) and subsequently adding the equations yields

$$\boldsymbol{\xi} \cdot \frac{\mathrm{d}\mathbf{J}}{\mathrm{d}t} = \boldsymbol{\xi} \cdot \left(\int_{\mathcal{B}} \boldsymbol{\varphi} \times \mathbf{b} \, \mathrm{d}V + \int_{\partial\mathcal{B}} \boldsymbol{\varphi} \times \mathbf{FSN} \, \mathrm{d}A \right) . \tag{5.19}$$

Due to the arbitrariness of $\boldsymbol{\xi} \in \mathbb{R}^3$, the last equation recovers the balance of angular momentum. Note that the parentheses on the right-hand side of (5.19) contain the resultant external torque about the origin (see also Fig. 5.1).

For verification of the balance of energy we choose the following test functions in (5.4)

$$\begin{aligned} \mathbf{w}_{\boldsymbol{\varphi}} &= \partial_{\boldsymbol{\varphi}} e' \,, \\ \mathbf{w}_{\mathbf{p}} &= \partial_{\mathbf{p}} e' \,, \\ \mathrm{w}_{\mathcal{T}} &= \partial_{\mathcal{T}} e' \,, \end{aligned} \qquad \text{or} \qquad \begin{aligned} \mathbf{w}_{\boldsymbol{\varphi}} &= -\mathbf{b} \,, \\ \mathbf{w}_{\mathbf{p}} &= \rho^{-1} \mathbf{p} \,, \\ \mathrm{w}_{\mathcal{T}} &= \partial_{\mathcal{T}} u' \,, \end{aligned}$$

where the total energy density e' is given by

$$e'(\boldsymbol{\varphi}, \mathbf{F}, \mathbf{p}, \mathcal{T}) = \frac{1}{2} \rho^{-1} \mathbf{p} \cdot \mathbf{p} + u'(\mathbf{C}, \mathcal{T}) - \mathbf{b} \cdot \boldsymbol{\varphi} \,. \tag{5.20}$$

Substituting these quantities into (5.4) and subsequently adding the three resulting equations, a straightforward calculation recovers the balance law for energy

$$\frac{\mathrm{d}\mathcal{E}'}{\mathrm{d}t} = \int_{\partial\mathcal{B}} \left(\rho^{-1} \mathbf{p} \cdot \mathbf{FSN} - \mathbf{N} \cdot \mathbf{Q} \right) \mathrm{d}A \,. \tag{5.21}$$

Concerning the balance of entropy, we insert $\mathrm{w}_{\mathcal{T}} = \partial_{\mathcal{T}} \eta'$ into $(5.4)_3$, to obtain

$$\begin{aligned} 0 &= \int_{\mathcal{B}} \left(\partial_{\mathcal{T}} \eta' \dot{\mathcal{T}} + \nabla(\rho^{-1}\mathbf{p}) : (2\mathbf{F} \partial_{\mathbf{C}} \eta') - \nabla \left(\frac{\partial_{\mathcal{T}} \eta'}{\partial_{\mathcal{T}} u'} \right) \cdot \mathbf{Q} \right) \mathrm{d}V + \int_{\partial\mathcal{B}} \frac{\partial_{\mathcal{T}} \eta'}{\partial_{\mathcal{T}} u'} \mathbf{Q} \cdot \mathbf{N} \, \mathrm{d}A \\ &= \int_{\mathcal{B}} \left(\frac{\mathrm{d}\eta'}{\mathrm{d}t} - \nabla \left(\frac{1}{\Theta'} \right) \cdot (\Theta')^2 \mathbf{K} \nabla \left(\frac{1}{\Theta'} \right) \right) \mathrm{d}V + \int_{\partial\mathcal{B}} \frac{1}{\Theta'} \mathbf{Q} \cdot \mathbf{N} \, \mathrm{d}A \,. \end{aligned}$$

Here, use has been made of formula (5.2) for the temperature along with expression (5.6) for the material heat flux vector. Moreover, the identity $\dot{\boldsymbol{\varphi}} = \rho^{-1}\mathbf{p}$ has again been taken into account. The above equation can be rewritten as

$$\frac{\mathrm{d}\mathcal{S}'}{\mathrm{d}t} = \underbrace{\int_{\mathcal{B}} \nabla \left(\frac{1}{\Theta'} \right) \cdot \Theta'^2 \mathbf{K} \nabla \left(\frac{1}{\Theta'} \right) \mathrm{d}V}_{\geq 0} - \int_{\partial\mathcal{B}} \frac{1}{\Theta} \mathbf{Q} \cdot \mathbf{N} \, \mathrm{d}A \,. \tag{5.22}$$

The last equation corresponds to the Clausius-Duhem form of the second law of thermodynamics (see, for example, [136, Sec. 5]).

5.2. Discretization in space

Next, we perform the discretization in space of the GENERIC-based weak form (5.4). To this end we apply the isoparametric finite element approach (see, for example, [145]), based on finite-dimensional approximations of the following quantities

$$\varphi^{\mathrm{h}}(\mathbf{X},t) = N^a(\mathbf{X})\,\mathbf{q}_a(t)\,, \qquad \mathbf{v}^{\mathrm{h}}(\mathbf{X},t) = N^a(\mathbf{X})\,\mathbf{v}_a(t)\,, \tag{5.23}$$

and

$$\tau^{\mathrm{h}}(\mathbf{X},t) = N^a(\mathbf{X})\,\tau_a(t)\,. \tag{5.24}$$

Here, the summation convention applies, where $a = 1,\dots,\mathrm{N}$, and N denotes the total number of nodes in the finite element mesh. Moreover, $N^a : \mathcal{B} \rightarrow \mathbb{R}$ are the nodal shape functions and $\mathbf{q}_a(t), \mathbf{v}_a(t) \in \mathbb{R}^3$, $\tau_a(t) \in \mathbb{R}$ are the respective nodal values at time t. Analogous approximations are used for the test functions $\mathbf{w}_\varphi$, $\mathbf{w}_\mathbf{p}$ and w_τ, denoted by $\mathbf{w}_\varphi^{\mathrm{h}}$, $\mathbf{w}_\mathbf{p}^{\mathrm{h}}$ and w_τ^{h}. Then weak form (5.4) leads to the following semi-discrete equations:

$$\begin{aligned}
0 &= \int_\mathcal{B} \mathbf{w}_\varphi^{\mathrm{h}} \cdot \left(\dot{\varphi}^{\mathrm{h}} - \mathbf{v}^{\mathrm{h}}\right) \mathrm{d}V\,,\\
0 &= \int_\mathcal{B} \left(\mathbf{w}_\mathbf{p}^{\mathrm{h}} \cdot \left(\rho\dot{\mathbf{v}}^{\mathrm{h}} - \mathbf{b}\right) + \nabla\mathbf{w}_\mathbf{p}^{\mathrm{h}} : \mathbf{F}^{\mathrm{h}}\mathbf{S}^{\mathrm{h}}\right) \mathrm{d}V - \int_{\partial_\sigma\mathcal{B}} \mathbf{w}_\mathbf{p}^{\mathrm{h}} \cdot \bar{\mathbf{t}}\,\mathrm{d}A\,,\\
0 &= \int_\mathcal{B} {\mathrm{w}_\tau}^{\mathrm{h}} \left(\dot{\tau}^{\mathrm{h}} + \nabla\mathbf{v}^{\mathrm{h}} : \left(\frac{2}{\Pi_h(\partial_\tau\eta^{\mathrm{h}})}\mathbf{F}^{\mathrm{h}}\partial_\mathbf{C}\eta^{\mathrm{h}}\right)\right) \mathrm{d}V\\
&\quad - \int_\mathcal{B} \nabla\left(\frac{{\mathrm{w}_\tau}^{\mathrm{h}}}{\Pi_h(\partial_\tau u^{\mathrm{h}})}\right) \cdot \mathbf{Q}^{\mathrm{h}}\,\mathrm{d}V + \int_{\partial_q\mathcal{B}} \frac{{\mathrm{w}_\tau}^{\mathrm{h}}}{\Pi_h(\partial_\tau u^{\mathrm{h}})}\bar{q}\,\mathrm{d}A\,,
\end{aligned} \tag{5.25}$$

where

$$\begin{aligned}
\mathbf{S}^{\mathrm{h}} &= 2\left(\partial_\mathbf{C} u^{\mathrm{h}} - \Theta^{\mathrm{h}}\partial_\mathbf{C}\eta^{\mathrm{h}}\right)\,,\\
\mathbf{Q}^{\mathrm{h}} &= (\Theta^{\mathrm{h}})^2\mathbf{K}^{\mathrm{h}}\nabla\left(\frac{1}{\Theta^{\mathrm{h}}}\right)\,,\\
\Theta^{\mathrm{h}} &= \frac{\Pi_h(\partial_\tau u^{\mathrm{h}})}{\Pi_h(\partial_\tau\eta^{\mathrm{h}})}\,.
\end{aligned} \tag{5.26}$$

In this connection, $u^{\mathrm{h}} = u'(\mathbf{C}^{\mathrm{h}}, \tau^{\mathrm{h}})$, $\eta^{\mathrm{h}} = \eta'(\mathbf{C}^{\mathrm{h}}, \tau^{\mathrm{h}})$, and $\mathbf{K}^{\mathrm{h}} = \mathbf{K}'(\mathbf{C}^{\mathrm{h}}, \tau^{\mathrm{h}})$. The interpolation formulas in (5.23) give rise to

$$\mathbf{F}^{\mathrm{h}} = \mathbf{q}_a \otimes \nabla N^a \quad \text{and} \quad \mathbf{C}^{\mathrm{h}} = \mathbf{q}_a \cdot \mathbf{q}_b \nabla N^a \otimes \nabla N^b\,. \tag{5.27}$$

Moreover, $\Pi_h(\partial_\tau u^{\mathrm{h}})$ denotes the L_2 projection of function $\partial_\tau u^{\mathrm{h}}$ into the finite-dimensional space spanned by the shape functions N^a, $a = 1,\dots,\mathrm{N}$. That is,

$$\Pi_h(\partial_\tau u^{\mathrm{h}}) = N^a(\partial_\tau u)_a\,, \tag{5.28}$$

where the nodal values $(\partial_\tau u)_a$ are determined by

$$\int_{\mathcal{B}} N^a \left(\partial_\tau u^{\mathrm{h}} - \Pi_h(\partial_\tau u^{\mathrm{h}})\right) \,\mathrm{d}V = 0\,, \tag{5.29}$$

for $a = 1, \ldots, \mathrm{N}$. In particular, (5.29) together with (5.28) constitute a linear system of algebraic equations given by

$$H^{ab}(\partial_\tau u)_b = \int_{\mathcal{B}} N^a \partial_\tau u^{\mathrm{h}} \,\mathrm{d}V\,, \tag{5.30}$$

where H^{ab} denote the components of the positive definite Gram matrix $[H^{ab}]$ defined by

$$H^{ab} = \int_{\mathcal{B}} N^a N^b \,\mathrm{d}V\,. \tag{5.31}$$

We note for later use that (5.28) in conjunction with (5.30) lead to the relationship

$$\Pi_h(\partial_\tau u^{\mathrm{h}}) = N^a H_{ab} \int_{\mathcal{B}} N^b \partial_\tau u^{\mathrm{h}} \,\mathrm{d}V\,, \tag{5.32}$$

where H_{ab} denote the components of the inverse of the Gram matrix, i.e. $[H_{ab}] = [H^{ab}]^{-1}$.

Analogous relationships hold for the projection of function $\partial_\tau \eta^{\mathrm{h}}$, $\Pi_h(\partial_\tau \eta^{\mathrm{h}})$. It is worth mentioning that the projections $\Pi_h(\partial_\tau u^{\mathrm{h}})$ and $\Pi_h(\partial_\tau \eta^{\mathrm{h}})$ are only required if the functions $\partial_\tau u^{\mathrm{h}}$ and $\partial_\tau \eta^{\mathrm{h}}$ do not belong to the finite element space spanned by the shape functions N^a. For example, in the temperature-based formulation (i.e. for $\tau = \theta$), $\partial_\theta u^{\mathrm{h}} = \partial_\theta \overline{u}(\mathbf{C}^{\mathrm{h}}, \theta^{\mathrm{h}})$ corresponds to the specific heat at constant deformation. Thus, if this quantity is prescribed to be constant, the projection $\Pi_h(\partial_\theta \overline{u})$ need not be performed. However, in general the temperature-based formulation requires both projections (i.e. $\Pi_h(\partial_\theta \overline{u})$ and $\Pi_h(\partial_\theta \overline{\eta})$).

In contrast to that, the two alternative formulations based on the choice $\tau \in \{\eta, u\}$ in general require only one projection. In particular, the formulation in terms of the internal energy density relies on $\Pi_h(\partial_u \widehat{\eta})$, whereas the entropy-based formulation relies on $\Pi_h(\partial_\eta \widetilde{u})$. Originally, the projection has been introduced in the framework of the entropy-based formulation in [102] (see also [106]).

In the next Section we will show that the proposed discretization in space of the GENERIC-based weak form (5.4) yields semi-discrete equations which can be brought into GENERIC form for finite dimensional systems (2.1). Thus, we speak of a *GENERIC-consistent space discretization*, see Section 2.1.2. In this connection the introduction of the above projections is essential for retaining consistency of the semi-discrete formulation.

5.2.1. GENERIC-consistent space discretization

In the following we will show that the proposed discretization in space of the GENERIC-based weak form (5.4) yields semi-discrete equations which can be brought into GENERIC form (2.1), thus we show that the above proposed space discretization is GENERIC-consistent in the sense of Section 2.1.2.

The present interpolation of the velocity field gives rise to nodal velocity vectors $\mathbf{v}_a$, cf. (5.23). We first introduce the conjugate nodal momentum vectors via a standard Legendre transformation. Accordingly, we express the kinetic energy $\mathcal{E}_{\text{kin}} = \mathcal{E}'_{\text{kin}}(\mathbf{p}) = \int_{\mathcal{B}} \frac{1}{2}\rho^{-1}\mathbf{p} \cdot \mathbf{p} \, \mathrm{d}V$ in terms of the velocity field by using the identity $\mathbf{p} = \rho\mathbf{v}$, leading to $\mathcal{E}_{\text{kin}} = \mathcal{T}(\mathbf{v}) = \int_{\mathcal{B}} \frac{1}{2}\rho\mathbf{v} \cdot \mathbf{v} \, \mathrm{d}V$. Using the discrete velocity field $\mathbf{v}^{\mathrm{h}}$ yields the kinetic energy of the discrete system given by

$$T = \mathcal{T}(\mathbf{v}^{\mathrm{h}}) = \frac{1}{2} M^{ab} \mathbf{v}_a \cdot \mathbf{v}_b \,, \tag{5.33}$$

where the components of the standard consistent mass matrix $[M^{ab}]$ are given by

$$M^{ab} = \int_{\mathcal{B}} \rho N^a N^b \, \mathrm{d}V \,. \tag{5.34}$$

Note that the mass matrix $[M^{ab}]$ is symmetric and positive definite. Now the conjugate nodal momentum vectors are defined by

$$\mathbf{p}^a = \partial_{\mathbf{v}_a} T = M^{ab} \mathbf{v}_b \,. \tag{5.35}$$

The last equation implies that the nodal velocity vectors can be written in terms of the nodal momentum vectors via

$$\mathbf{v}_a = M_{ab} \mathbf{p}^b \,, \tag{5.36}$$

where M_{ab} are the components of the inverse mass matrix, i.e. $[M_{ab}] = [M^{ab}]^{-1}$. Now we introduce the state vector of the discrete system at hand which is comprised of the nodal quantities $\mathbf{q}_a$, $\mathbf{p}^a$, τ_a, $a = 1, \ldots, \mathrm{N}$. We arrange these quantities in the nodal state vector of the system given by

$$\mathbf{z} = \begin{pmatrix} \mathbf{q}_1, & \ldots, & \mathbf{q}_N, & \mathbf{p}^1, & \ldots, & \mathbf{p}^N, & \tau_1, & \ldots, & \tau_N \end{pmatrix} . \tag{5.37}$$

Hereby the state vector has to be viewed as a column vector. Next we aim at the total energy and the total entropy of the discrete system which play the role of generators in the GENERIC formulation (2.1). The two required functions assume the form

$$\begin{aligned} \mathcal{E}(\mathbf{z}) &= \int_{\mathcal{B}} u^{\mathrm{h}} \, \mathrm{d}V + \frac{1}{2} M_{ab} \mathbf{p}^a \cdot \mathbf{p}^b - \mathbf{q}_a \cdot \int_{\mathcal{B}} N^a \mathbf{b} \, \mathrm{d}V \,, \\ \mathcal{S}(\mathbf{z}) &= \int_{\mathcal{B}} \eta^{\mathrm{h}} \, \mathrm{d}V \,. \end{aligned} \tag{5.38}$$

Concerning the derivatives $\partial_{\mathbf{z}}\mathcal{E}$ and $\partial_{\mathbf{z}}\mathcal{S}$, we obtain the individual contributions

$$\begin{aligned}\partial_{\mathbf{q}_a}\mathcal{E} &= \int_{\mathcal{B}} \partial_{\mathbf{F}} u^{\mathrm{h}} \nabla N^a \,\mathrm{d}V - \int_{\mathcal{B}} N^a \mathbf{b} \,\mathrm{d}V \,, \\ \partial_{\mathbf{p}^a}\mathcal{E} &= M_{ab}\mathbf{p}^b \,, \\ \partial_{\tau_a}\mathcal{E} &= \int_{\mathcal{B}} N^a \partial_\tau u^{\mathrm{h}} \,\mathrm{d}V \,,\end{aligned} \tag{5.39}$$

and

$$\begin{aligned}\partial_{\mathbf{q}_a}\mathcal{S} &= \int_{\mathcal{B}} \partial_{\mathbf{F}} \eta^{\mathrm{h}} \nabla N^a \,\mathrm{d}V \,, \\ \partial_{\mathbf{p}^a}\mathcal{S} &= \mathbf{0} \,, \\ \partial_{\tau_a}\mathcal{S} &= \int_{\mathcal{B}} N^a \partial_\tau \eta^{\mathrm{h}} \,\mathrm{d}V \,.\end{aligned} \tag{5.40}$$

Next, we take a closer look at the semi-discrete formulation emanating from the space-discrete weak form (5.25). To this end we focus on the pure Neumann problem and neglect the boundary integrals in (5.25). This is in accordance with the fact that the basic form (2.1) of the GENERIC formulation is valid for closed systems. The kinematic equation $(5.25)_1$ directly leads to $H^{ab}(\dot{\mathbf{q}}_a - \mathbf{v}_a) = \mathbf{0}$, where the components of the Gram matrix $[H^{ab}]$ have been introduced in (5.31). Due to the positive definiteness of the Gram matrix one obtains

$$\dot{\mathbf{q}}_a = M_{ab}\mathbf{p}^b = \partial_{\mathbf{p}^a}\mathcal{E} \,. \tag{5.41}$$

The space-discrete weak form of the balance law for linear momentum, $(5.25)_2$, yields

$$M^{ab}\dot{\mathbf{v}}_b = \int_{\mathcal{B}} N^a \mathbf{b} \,\mathrm{d}V - \int_{\mathcal{B}} \mathbf{F}^{\mathrm{h}} \mathbf{S}^{\mathrm{h}} \nabla N^a \,\mathrm{d}V \,. \tag{5.42}$$

Taking into account (5.35), $(5.26)_{1,3}$ and

$$\begin{aligned}\partial_{\mathbf{F}} u^{\mathrm{h}} &= 2\mathbf{F}^{\mathrm{h}} \partial_{\mathbf{C}} u^{\mathrm{h}}, \\ \partial_{\mathbf{F}} \eta^{\mathrm{h}} &= 2\mathbf{F}^{\mathrm{h}} \partial_{\mathbf{C}} \eta^{\mathrm{h}} \,,\end{aligned} \tag{5.43}$$

into account eq. (5.42) can be recast in the form

$$\dot{\mathbf{p}}^a = \int_{\mathcal{B}} N^a \mathbf{b} \,\mathrm{d}V - \int_{\mathcal{B}} \partial_{\mathbf{F}} u^{\mathrm{h}} \nabla N^a \,\mathrm{d}V + \int_{\mathcal{B}} \frac{\Pi_h(\partial_\tau u^{\mathrm{h}})}{\Pi_h(\partial_\tau \eta^{\mathrm{h}})} \partial_{\mathbf{F}} \eta^{\mathrm{h}} \nabla N^a \,\mathrm{d}V \,. \tag{5.44}$$

With regard to (5.32) and $(5.39)_3$, we get

$$\Pi_h(\partial_\tau u^{\mathrm{h}}) = N^a H_{ab} \partial_{\tau_b}\mathcal{E} \,. \tag{5.45}$$

Accordingly, taking into account the last equation and $(5.39)_1$, (5.44) can be rewritten as

$$\dot{\mathbf{p}}^a = -\partial_{\mathbf{q}_a}\mathcal{E} + \int_{\mathcal{B}} \frac{N^d}{\Pi_h(\partial_\tau \eta^{\mathrm{h}})} \partial_{\mathbf{F}}\eta^{\mathrm{h}} \nabla N^a \,\mathrm{d}V H_{db} \partial_{\tau_b}\mathcal{E} \,. \tag{5.46}$$

The space-discrete weak form of the energy equation, $(5.25)_3$, yields

$$\begin{aligned} H^{ab}\dot{\tau}_b = & -\mathbf{v}_d \cdot \int_{\mathcal{B}} \frac{N^a}{\Pi_h(\partial_\tau \eta^{\mathrm{h}})} \partial_{\mathbf{F}}\eta^{\mathrm{h}} \nabla N^d \,\mathrm{d}V \\ & + \nabla\left(\frac{N^a}{\Pi_h(\partial_\tau u^{\mathrm{h}})}\right) \cdot (\Theta^{\mathrm{h}})^2 \mathbf{K}^{\mathrm{h}} \nabla\left(\frac{\Pi_h(\partial_\tau \eta^{\mathrm{h}})}{\Pi_h(\partial_\tau u^{\mathrm{h}})}\right) \mathrm{d}V \,. \end{aligned} \tag{5.47}$$

In analogy to (5.45), we have

$$\Pi_h(\partial_\tau \eta^{\mathrm{h}}) = N^a H_{ab} \partial_{\tau_b}\mathcal{S} \,.$$

Using the last equation together with (5.36) and $(5.39)_2$, (5.47) can be recast in the form

$$\begin{aligned} \dot{\tau}_a = & - H_{ab}\partial_{\mathbf{p}^d}\mathcal{E} \cdot \int_{\mathcal{B}} \frac{N^b}{\Pi_h(\partial_\tau \eta^{\mathrm{h}})} \partial_{\mathbf{F}}\eta^{\mathrm{h}} \nabla N^d \,\mathrm{d}V \\ & + H_{ab} \int_{\mathcal{B}} \nabla\left(\frac{N^b}{\Pi_h(\partial_\tau u^{\mathrm{h}})}\right) \cdot (\Theta^{\mathrm{h}})^2 \mathbf{K}^{\mathrm{h}} \nabla\left(\frac{N^c}{\Pi_h(\partial_\tau u^{\mathrm{h}})}\right) \mathrm{d}V H_{cd} \partial_{\tau_d}\mathcal{S} \,. \end{aligned} \tag{5.48}$$

To summarize, the semi-discrete evolution equations resulting from the discretization in space of the underlying GENERIC-based weak form can be written as

$$\begin{aligned} \dot{\mathbf{q}}_a &= \partial_{\mathbf{p}^a}\mathcal{E} \,, \\ \dot{\mathbf{p}}^a &= -\partial_{\mathbf{q}_a}\mathcal{E} + \mathbf{l}^{a\cdot}_{\cdot b}\partial_{\tau_b}\mathcal{E} \,, \\ \dot{\tau}_a &= -\left(\mathbf{l}^{b\cdot}_{\cdot a}\right)^{\mathrm{T}} \partial_{\mathbf{p}^b}\mathcal{E} + m_{ab}\partial_{\tau_b}\mathcal{S} \,, \end{aligned} \tag{5.49}$$

where

$$\begin{aligned} \mathbf{l}^{a\cdot}_{\cdot b} &= \int_{\mathcal{B}} \frac{N^d}{\Pi_h(\partial_\tau \eta^{\mathrm{h}})} \partial_{\mathbf{F}}\eta^{\mathrm{h}} \nabla N^a \,\mathrm{d}V H_{db} \,, \\ m_{ab} &= H_{ac} \int_{\mathcal{B}} \nabla\left(\frac{N^c}{\Pi_h(\partial_\tau u^{\mathrm{h}})}\right) \cdot (\Theta^{\mathrm{h}})^2 \mathbf{K}^{\mathrm{h}} \nabla\left(\frac{N^d}{\Pi_h(\partial_\tau u^{\mathrm{h}})}\right) \mathrm{d}V H_{db} \,. \end{aligned} \tag{5.50}$$

The evolution equations (5.49) fit into the GENERIC framework for finite-dimensional systems provided by (2.1). In particular, (5.49) can be written in the form

$$\dot{\mathbf{z}} = \mathbf{L}\partial_{\mathbf{z}}\mathcal{E} + \mathbf{M}\partial_{\mathbf{z}}\mathcal{S} \,,$$

where the state vector is given by (5.37). Moreover, the Poisson matrix reads

$$\mathbf{L} = \left[\begin{array}{c|c|c} 0 & \mathrm{I} & 0 \\ \hline -\mathrm{I} & 0 & \left[\mathbf{l}^{a\,\cdot}_{\cdot\, b}\right] \\ \hline 0 & \left[-\mathbf{l}^{b\,\cdot}_{\cdot\, a}\right]^{\mathrm{T}} & 0 \end{array}\right], \tag{5.51}$$

where I is the identity matrix (with appropriate dimension corresponding to the partitioning of the state vector (5.37)), and matrix $\left[\mathbf{l}^{a\,\cdot}_{\cdot\, b}\right]$ consists of vectors $\mathbf{l}^{a\,\cdot}_{\cdot\, b}$ defined in $(5.50)_1$. Specifically, we have

$$\left[\mathbf{l}^{a\,\cdot}_{\cdot\, b}\right] = \begin{bmatrix} \mathbf{l}^{1\,\cdot}_{\cdot\, 1} & \cdots & \mathbf{l}^{1\,\cdot}_{\cdot\, \mathrm{N}} \\ \vdots & \ddots & \vdots \\ \mathbf{l}^{\mathrm{N}\,\cdot}_{\cdot\, 1} & \cdots & \mathbf{l}^{\mathrm{N}\,\cdot}_{\cdot\, \mathrm{N}} \end{bmatrix}. \tag{5.52}$$

The friction matrix is given by

$$\mathbf{M} = \left[\begin{array}{c|c|c} 0 & 0 & 0 \\ \hline 0 & 0 & 0 \\ \hline 0 & 0 & \left[m_{ab}\right] \end{array}\right], \tag{5.53}$$

where the coefficients m_{ab} have been introduced in $(5.50)_2$. It can be easily verified by a straightforward calculation that the two degeneration conditions (2.2) and (2.3) are satisfied. Thus the validity of (2.4) and (2.5) ensures thermodynamic consistency of the semi-discrete formulation.

Remark 6. Instead of using the interpolation of the velocity field (cf. (5.23)), one could as well interpolate the linear momentum density field. This would lead, for example, to minor changes in the partial derivative $(5.39)_2$ and the Poisson matrix (5.51). However, it would not affect the thermodynamic consistency of the resulting semi-discrete formulation. These observations are valid for constant mass density ρ. We prefer the interpolation of the velocity field, because (i) it does not put any restrictions on the mass density and, (ii) it leads to nodal velocities as degrees of freedom on which initial and boundary conditions for the velocity field can be directly imposed.

5.2.2. Balance laws

We next give a short summary of the balance laws pertaining to the space-discrete formulation described above. We restate the semi-discrete evolution equations derived in Section 5.2.1 in the form

$$\begin{aligned}
\dot{\mathbf{q}}_a &= \mathbf{v}_a \,, \\
M^{ab}\dot{\mathbf{v}}_b &= \int_{\mathcal{B}} N^a \mathbf{b}\,\mathrm{d}V - \int_{\mathcal{B}} \partial_{\mathbf{F}} u^{\mathrm{h}} \nabla N^a \,\mathrm{d}V + \int_{\mathcal{B}} \Theta^{\mathrm{h}} \partial_{\mathbf{F}} \eta^{\mathrm{h}} \nabla N^a \,\mathrm{d}V \,, \\
H^{ab}\dot{\tau}_b &= -\mathbf{v}_d \cdot \int_{\mathcal{B}} \frac{N^a}{\Pi_h(\partial_\tau \eta^{\mathrm{h}})} \partial_{\mathbf{F}} \eta^{\mathrm{h}} \nabla N^d \,\mathrm{d}V - \int_{\mathcal{B}} \nabla\left(\frac{N^a}{\Pi_h(\partial_\tau u^{\mathrm{h}})}\right) \cdot \mathbf{K}^{\mathrm{h}} \nabla \Theta^{\mathrm{h}} \,\mathrm{d}V \\
&\quad - \int_{\partial_q \mathcal{B}} \frac{N^a}{\Pi_h(\partial_\tau u^{\mathrm{h}})} \bar{q} \,\mathrm{d}A \,.
\end{aligned} \tag{5.54}$$

For simplicity of exposition we have neglected in $(5.54)_2$ the standard contribution of the external boundary tractions. The total linear momentum of the discrete system at hand is given by

$$\begin{aligned}
\mathbf{L}^{\mathrm{h}} &= \int_{\mathcal{B}} \rho \mathbf{v}^{\mathrm{h}} \,\mathrm{d}V \\
&= \int_{\mathcal{B}} \rho N^b \,\mathrm{d}V \mathbf{v}_b \\
&= \sum_{a=1}^{\mathrm{N}} \int_{\mathcal{B}} \rho N^a N^b \,\mathrm{d}V \mathbf{v}_b \\
&= \sum_{a=1}^{\mathrm{N}} M^{ab} \mathbf{v}_b \,.
\end{aligned} \tag{5.55}$$

Note that use has been made of the completeness condition $\sum_{a=1}^{\mathrm{N}} N^a = 1$ for the nodal shape functions. Now $(5.54)_2$ directly leads to the balance law for linear momentum

$$\frac{\mathrm{d}}{\mathrm{d}t}\mathbf{L}^{\mathrm{h}} = \int_{\mathcal{B}} \mathbf{b}\,\mathrm{d}V \,, \tag{5.56}$$

which can be regarded as the semi-discrete counterpart of (5.18). The total angular momentum relative to the origin of the inertial frame is given by

$$\begin{aligned}
\mathbf{J}^{\mathrm{h}} &= \int_{\mathcal{B}} \rho \boldsymbol{\varphi}^{\mathrm{h}} \times \mathbf{v}^{\mathrm{h}} \,\mathrm{d}V \\
&= M^{ab} \mathbf{q}_a \times \mathbf{v}_b \,.
\end{aligned} \tag{5.57}$$

A straightforward calculation based on $(5.54)_{1,2}$ yields

$$\boldsymbol{\xi} \cdot \frac{\mathrm{d}}{\mathrm{d}t}\mathbf{J}^{\mathrm{h}} = \boldsymbol{\xi} \cdot \int_{\mathcal{B}} \boldsymbol{\varphi}^{\mathrm{h}} \times \mathbf{b}\,\mathrm{d}V \,, \tag{5.58}$$

for arbitrary but constant $\boldsymbol{\xi} \in \mathbb{R}^3$. In the last equation the following relationship has been taken into account:

$$\begin{aligned}\boldsymbol{\xi} \cdot \left(\mathbf{q}_a \times (\partial_{\mathbf{F}} u^{\mathrm{h}} - \Theta^{\mathrm{h}} \partial_{\mathbf{F}} \eta^{\mathrm{h}}) \nabla N^a\right) &= \widehat{\boldsymbol{\xi}} : \left((\partial_{\mathbf{F}} u^{\mathrm{h}} - \Theta^{\mathrm{h}} \partial_{\mathbf{F}} \eta^{\mathrm{h}}) \nabla N^a \otimes \mathbf{q}_a\right) \\ &= \widehat{\boldsymbol{\xi}} : \left((\partial_{\mathbf{F}} u^{\mathrm{h}} - \Theta^{\mathrm{h}} \partial_{\mathbf{F}} \eta^{\mathrm{h}}) \mathbf{F}^{\mathrm{h,T}}\right) \\ &= \widehat{\boldsymbol{\xi}} : \left(\mathbf{F}^{\mathrm{h}} \mathbf{S}^{\mathrm{h}} \mathbf{F}^{\mathrm{h,T}}\right) \\ &= 0\,.\end{aligned} \tag{5.59}$$

Here, use has been made of (5.43) and $(5.26)_1$. Moreover, $\widehat{\boldsymbol{\xi}}$ is skew-symmetric such that $\widehat{\boldsymbol{\xi}}\mathbf{a} = \boldsymbol{\xi} \times \mathbf{a}$ for any $\mathbf{a} \in \mathbb{R}^3$. Relationship (5.58) can be viewed as semi-discrete counterpart of the balance law for angular momentum (5.19).

For completeness of exposition we eventually verify the balance laws for energy and entropy, although the GENERIC-consistent space discretization at hand guarantees thermodynamic consistency as has already been shown in Section 5.2.1. Multiplying $(5.54)_3$ by $(\partial_\tau u)_a$, and taking into account (5.28) together with $(5.26)_3$ yields

$$(\partial_\tau u)_a H^{ab} \dot{\tau}_b = -\mathbf{v}_d \cdot \int_{\mathcal{B}} \Theta^{\mathrm{h}} \partial_{\mathbf{F}} \eta^{\mathrm{h}} \nabla N^d \,\mathrm{d}V - \int_{\partial_q \mathcal{B}} \bar{q} \,\mathrm{d}A\,.$$

With regard to (5.30), the last equation can be rewritten as

$$\int_{\mathcal{B}} \partial_\tau u^{\mathrm{h}} \dot{\tau}^{\mathrm{h}} \,\mathrm{d}V + \int_{\mathcal{B}} \Theta^{\mathrm{h}} \partial_{\mathbf{F}} \eta^{\mathrm{h}} : \dot{\mathbf{F}}^{\mathrm{h}} \,\mathrm{d}V = -\int_{\partial_q \mathcal{B}} \bar{q} \,\mathrm{d}A\,. \tag{5.60}$$

Scalar multiplication of $(5.54)_2$ by $\mathbf{v}_a$ yields

$$\mathbf{v}_a \cdot M^{ab} \dot{\mathbf{v}}_b = \mathbf{v}_a \cdot \left(\int_{\mathcal{B}} N^a \mathbf{b} \,\mathrm{d}V - \int_{\mathcal{B}} \partial_{\mathbf{F}} u^{\mathrm{h}} \nabla N^a \,\mathrm{d}V + \int_{\mathcal{B}} \Theta^{\mathrm{h}} \partial_{\mathbf{F}} \eta^{\mathrm{h}} \nabla N^a \,\mathrm{d}V\right),$$

or

$$\frac{1}{2} \frac{\mathrm{d}}{\mathrm{d}t} \left(\mathbf{v}_a \cdot M^{ab} \mathbf{v}_b\right) = \int_{\mathcal{B}} \mathbf{v}^{\mathrm{h}} \cdot \mathbf{b} \,\mathrm{d}V - \int_{\mathcal{B}} \partial_{\mathbf{F}} u^{\mathrm{h}} : \dot{\mathbf{F}}^{\mathrm{h}} \,\mathrm{d}V + \int_{\mathcal{B}} \Theta^{\mathrm{h}} \partial_{\mathbf{F}} \eta^{\mathrm{h}} : \dot{\mathbf{F}}^{\mathrm{h}} \,\mathrm{d}V\,.$$

Substituting from the last equation into (5.60) leads to

$$\int_{\mathcal{B}} \left(\partial_\tau u^{\mathrm{h}} \dot{\tau}^{\mathrm{h}} + \partial_{\mathbf{F}} u^{\mathrm{h}} : \dot{\mathbf{F}}^{\mathrm{h}}\right) \mathrm{d}V + \frac{1}{2} \frac{\mathrm{d}}{\mathrm{d}t} \left(M^{ab} \mathbf{v}_a \cdot \mathbf{v}_b\right) - \int_{\mathcal{B}} \mathbf{v}^{\mathrm{h}} \cdot \mathbf{b} \,\mathrm{d}V = -\int_{\partial_q \mathcal{B}} \bar{q} \,\mathrm{d}A\,.$$

The last equation can be rewritten as

$$\frac{\mathrm{d}}{\mathrm{d}t} \mathcal{E}^{\mathrm{h}} = -\int_{\partial_q \mathcal{B}} \bar{q} \,\mathrm{d}A\,, \tag{5.61}$$

where the total energy of the discrete system assumes the form

$$\mathcal{E}^{\mathrm{h}} = \int_{\mathcal{B}} u^{\mathrm{h}} \,\mathrm{d}V + \frac{1}{2} M^{ab} \mathbf{v}_a \cdot \mathbf{v}_b - \int_{\mathcal{B}} \varphi^{\mathrm{h}} \cdot \mathbf{b} \,\mathrm{d}V \,. \tag{5.62}$$

Note that (5.61) can be regarded as semi-discrete version of the balance law for energy (5.21).

To recover the balance of entropy, multiply (5.54)3 by $(\partial_\tau \eta)_a$, and take into account (5.26)3, to obtain

$$(\partial_\tau \eta)_a H^{ab} \dot{\tau}_b = -\mathbf{v}_d \cdot \int_{\mathcal{B}} \partial_{\mathbf{F}} \eta^{\mathrm{h}} \nabla N^d \,\mathrm{d}V - \int_{\mathcal{B}} \nabla \left(\frac{1}{\Theta^{\mathrm{h}}} \right) \cdot \mathbf{K}^{\mathrm{h}} \nabla \Theta^{\mathrm{h}} \,\mathrm{d}V - \int_{\partial_q \mathcal{B}} \frac{1}{\Theta^{\mathrm{h}}} \bar{q} \,\mathrm{d}A \,.$$

The last equation can be rewritten as

$$\underbrace{\int_{\mathcal{B}} \left(\partial_\tau \eta^{\mathrm{h}} \dot{\tau}^{\mathrm{h}} + \partial_{\mathbf{F}} \eta^{\mathrm{h}} : \dot{\mathbf{F}}^{\mathrm{h}} \right) \mathrm{d}V}_{= \frac{\mathrm{d}}{\mathrm{d}t} \mathcal{S}^{\mathrm{h}}} = \underbrace{\int_{\mathcal{B}} \nabla \left(\frac{1}{\Theta^{\mathrm{h}}} \right) \cdot (\Theta^{\mathrm{h}})^2 \mathbf{K}^{\mathrm{h}} \nabla \left(\frac{1}{\Theta^{\mathrm{h}}} \right) \mathrm{d}V}_{\geq 0} - \int_{\partial_q \mathcal{B}} \frac{1}{\Theta^{\mathrm{h}}} \bar{q} \,\mathrm{d}A \,, \tag{5.63}$$

where the total entropy of the discrete system is defined by

$$\mathcal{S}^{\mathrm{h}} = \int_{\mathcal{B}} \eta^{\mathrm{h}} \,\mathrm{d}V = \int_{\mathcal{B}} \eta'(\mathbf{C}^{\mathrm{h}}, \tau^{\mathrm{h}}) \,\mathrm{d}V \,. \tag{5.64}$$

Note that (5.63) can be viewed as semi-discrete version of the balance of entropy (5.22).

5.3. Discretization in time

We aim at a second-order accurate, implicit time-stepping scheme which is capable of correctly reproducing the main balance laws outlined above. Due to its structure-preserving properties this type of integrator is called energy-momentum-entropy scheme. To devise such a scheme, we essentially apply the mid-point rule in which the derivatives of the internal energy density and the entropy density, respectively, are replaced by appropriate discrete derivatives.

We focus on a representative time interval $[t_{\mathrm{n}}, t_{\mathrm{n+1}}]$ with corresponding time-step size $\Delta t = t_{\mathrm{n+1}} - t_{\mathrm{n}}$. The discrete approximations at times t_{n} and $t_{\mathrm{n+1}}$ of a function $f(t)$ will be denoted by f_{n} and $f_{\mathrm{n+1}}$, respectively. Assume that the nodal state variables at time t_{n}, $\mathbf{q}_{a_{\mathrm{n}}}$, $\mathbf{v}_{a_{\mathrm{n}}}$, and $\tau_{a_{\mathrm{n}}}$ are given. The associated fields result from the nodal interpolation formulas (5.23) and are denoted by $\varphi^{\mathrm{h}}_{\mathrm{n}}, \mathbf{v}^{\mathrm{h}}_{\mathrm{n}} : \mathcal{B} \mapsto \mathbb{R}^3$ and $\tau^{\mathrm{h}}_{\mathrm{n}} : \mathcal{B} \mapsto \mathbb{R}$, $\tau \in \{\theta, \eta, u\}$.

We aim at the determination of the corresponding quantities at time t_{n+1} through the mid-point type discretization of the semi-discrete formulation (5.25) given by

$$
\begin{aligned}
0 &= \int_{\mathcal{B}} \mathbf{w}^{\mathrm{h}}_{\varphi} \cdot \left(\frac{\varphi^{\mathrm{h}}_{\mathrm{n}+1} - \varphi^{\mathrm{h}}_{\mathrm{n}}}{\Delta t} - \mathbf{v}^{\mathrm{h}}_{\mathrm{n}+\frac{1}{2}} \right) \mathrm{d}V \,, \\
0 &= \int_{\mathcal{B}} \left(\mathbf{w}^{\mathrm{h}}_{\mathbf{p}} \cdot \left(\rho \frac{\mathbf{v}^{\mathrm{h}}_{\mathrm{n}+1} - \mathbf{v}^{\mathrm{h}}_{\mathrm{n}}}{\Delta t} - \mathbf{b} \right) + \nabla \mathbf{w}^{\mathrm{h}}_{\mathbf{p}} : \mathbf{F}^{\mathrm{h}}_{\mathrm{n}+\frac{1}{2}} \mathbf{S}^{\mathrm{h}}_{\mathrm{algo}} \right) \mathrm{d}V - \int_{\partial_\sigma \mathcal{B}} \mathbf{w}^{\mathrm{h}}_{\mathbf{p}} \cdot \bar{\mathbf{t}}_{\mathrm{n}+\frac{1}{2}} \, \mathrm{d}A \,, \\
0 &= \int_{\mathcal{B}} \mathrm{w}^{\mathrm{h}}_{\tau} \left(\frac{\tau^{\mathrm{h}}_{\mathrm{n}+1} - \tau^{\mathrm{h}}_{\mathrm{n}}}{\Delta t} + \nabla \mathbf{v}^{\mathrm{h}}_{\mathrm{n}+\frac{1}{2}} : \left(\frac{2}{\Pi_h(\mathsf{D}_\tau \eta^{\mathrm{h}})} \mathbf{F}^{\mathrm{h}}_{\mathrm{n}+\frac{1}{2}} \mathsf{D}_{\mathbf{C}} \eta^{\mathrm{h}} \right) \right) \mathrm{d}V \\
&\quad - \int_{\mathcal{B}} \nabla \left(\frac{\mathrm{w}^{\mathrm{h}}_{\tau}}{\Pi_h(\mathsf{D}_\tau u^{\mathrm{h}})} \right) \cdot \mathbf{Q}^{\mathrm{h}}_{\mathrm{algo}} \, \mathrm{d}V + \int_{\partial_q \mathcal{B}} \frac{\mathrm{w}^{\mathrm{h}}_{\tau}}{\Pi_h(\mathsf{D}_\tau u^{\mathrm{h}})} \bar{q}_{\mathrm{n}+\frac{1}{2}} \, \mathrm{d}A \,,
\end{aligned}
\tag{5.65}
$$

where

$$
\begin{aligned}
\mathbf{S}^{\mathrm{h}}_{\mathrm{algo}} &= 2 \left(\mathsf{D}_{\mathbf{C}} u^{\mathrm{h}} - \Theta^{\mathrm{h}}_{\mathrm{algo}} \mathsf{D}_{\mathbf{C}} \eta^{\mathrm{h}} \right) \,, \\
\mathbf{Q}^{\mathrm{h}}_{\mathrm{algo}} &= (\Theta^{\mathrm{h}}_{\mathrm{algo}})^2 \mathbf{K}^{\mathrm{h}}_{\mathrm{algo}} \nabla \left(\frac{1}{\Theta^{\mathrm{h}}_{\mathrm{algo}}} \right) \,, \\
\Theta^{\mathrm{h}}_{\mathrm{algo}} &= \frac{\Pi_h(\mathsf{D}_\tau u^{\mathrm{h}})}{\Pi_h(\mathsf{D}_\tau \eta^{\mathrm{h}})} \,,
\end{aligned}
\tag{5.66}
$$

and $\mathbf{K}^{\mathrm{h}}_{\mathrm{algo}} = \mathbf{K}'(\mathbf{C}^{\mathrm{h}}_{\mathrm{n}+\frac{1}{2}}, \tau^{\mathrm{h}}_{\mathrm{n}+\frac{1}{2}})$.

The above discretization in time relies on the use of discrete derivatives in the sense of Gonzalez [46]. In particular, the discrete derivatives are applied to the internal energy density and the entropy density, respectively. For example, in the case of the internal energy density, the discrete derivatives are denoted by $\mathsf{D}_\tau u^{\mathrm{h}}$ and $\mathsf{D}_{\mathbf{C}} u^{\mathrm{h}}$, respectively. In particular, $\mathsf{D}_\tau u^{\mathrm{h}}$ is defined by

$$
\mathsf{D}_\tau u^{\mathrm{h}} = \frac{1}{2} \left(\mathrm{d}u_{\mathbf{C}^{\mathrm{h}}_{\mathrm{n}}}(\tau^{\mathrm{h}}_{\mathrm{n}}, \tau^{\mathrm{h}}_{\mathrm{n}+1}) + \mathrm{d}u_{\mathbf{C}^{\mathrm{h}}_{\mathrm{n}+1}}(\tau^{\mathrm{h}}_{\mathrm{n}}, \tau^{\mathrm{h}}_{\mathrm{n}+1}) \right) \,, \tag{5.67}
$$

where

$$
\mathrm{d}u_{\mathbf{C}}(\tau_{\mathrm{n}}, \tau_{\mathrm{n}+1}) = \partial_\tau u'(\mathbf{C}, \tau_{\mathrm{n}+\frac{1}{2}}) + \frac{u'(\mathbf{C}, \tau_{\mathrm{n}+1}) - u'(\mathbf{C}, \tau_{\mathrm{n}}) - \partial_\tau u'(\mathbf{C}, \tau_{\mathrm{n}+\frac{1}{2}}) \Delta\tau}{(\Delta\tau)^2} \Delta\tau \,,
$$

and $\Delta\tau = \tau_{\mathrm{n}+1} - \tau_{\mathrm{n}}$. Furthermore, $\mathsf{D}_{\mathbf{C}} u^{\mathrm{h}}$ is defined by

$$
\mathsf{D}_{\mathbf{C}} u^{\mathrm{h}} = \frac{1}{2} \left(\mathrm{d}u_{\tau^{\mathrm{h}}_{\mathrm{n}}}(\mathbf{C}^{\mathrm{h}}_{\mathrm{n}}, \mathbf{C}^{\mathrm{h}}_{\mathrm{n}+1}) + \mathrm{d}u_{\tau^{\mathrm{h}}_{\mathrm{n}+1}}(\mathbf{C}^{\mathrm{h}}_{\mathrm{n}}, \mathbf{C}^{\mathrm{h}}_{\mathrm{n}+1}) \right) \,, \tag{5.68}
$$

where

$$
\begin{aligned}
\mathrm{d}u_\tau(\mathbf{C}_{\mathrm{n}}, \mathbf{C}_{\mathrm{n}+1}) &= \partial_{\mathbf{C}} u'(\mathbf{C}_{\mathrm{n}+\frac{1}{2}}, \tau) \\
&\quad + \frac{u'(\mathbf{C}_{\mathrm{n}+1}, \tau) - u'(\mathbf{C}_{\mathrm{n}}, \tau) - \partial_{\mathbf{C}} u'(\mathbf{C}_{\mathrm{n}+\frac{1}{2}}, \tau) : \Delta\mathbf{C}}{\Delta\mathbf{C} : \Delta\mathbf{C}} \Delta\mathbf{C} \,,
\end{aligned}
$$

and $\Delta\mathbf{C} = \mathbf{C}_{\mathrm{n+1}} - \mathbf{C}_{\mathrm{n}}$. It can be verified by a straightforward calculation that the discrete derivatives (5.67) and (5.68) satisfy the directionality condition

$$\mathsf{D}_{\mathbf{C}} u^{\mathrm{h}} : \left(\mathbf{C}^{\mathrm{h}}_{\mathrm{n+1}} - \mathbf{C}^{\mathrm{h}}_{\mathrm{n}}\right) + \mathsf{D}_{\tau} u^{\mathrm{h}} \left(\tau^{\mathrm{h}}_{\mathrm{n+1}} - \tau^{\mathrm{h}}_{\mathrm{n}}\right) = u'(\mathbf{C}^{\mathrm{h}}_{\mathrm{n+1}}, \tau^{\mathrm{h}}_{\mathrm{n+1}}) - u'(\mathbf{C}^{\mathrm{h}}_{\mathrm{n}}, \tau^{\mathrm{h}}_{\mathrm{n}}) \,. \quad (5.69)$$

Analogous considerations apply to the discrete derivatives of the internal entropy density, $\mathsf{D}_{\tau}\eta^{\mathrm{h}}$ and $\mathsf{D}_{\mathbf{C}}\eta^{\mathrm{h}}$, respectively. Moreover, the time-average of any quantity $(\bullet)$ is given by $\frac{1}{2}\big((\bullet)_{\mathrm{n}} + (\bullet)_{\mathrm{n+1}}\big)$. In particular, this implies

$$\mathbf{C}_{\mathrm{n+\frac{1}{2}}} = \frac{1}{2}\left(\mathbf{C}_{\mathrm{n}} + \mathbf{C}_{\mathrm{n+1}}\right) \,. \quad (5.70)$$

Note that in general $\mathbf{C}_{\mathrm{n+\frac{1}{2}}} \neq \mathbf{F}^{\mathrm{T}}_{\mathrm{n+\frac{1}{2}}} \mathbf{F}_{\mathrm{n+\frac{1}{2}}}$. In (5.65) and (5.66), $\Pi_h(\mathsf{D}_{\tau} u^{\mathrm{h}})$ denotes the L_2 projection of $\mathsf{D}_{\tau} u^{\mathrm{h}}$ into the finite dimensional space spanned by the shape functions N^a, $a = 1, \dots, \mathrm{N}$. In analogy to (5.30),

$$H^{ab}(\mathsf{D}_{\tau} u)_b = \int_{\mathcal{B}} N^a \mathsf{D}_{\tau} u^{\mathrm{h}} \, \mathrm{d}V \,, \quad (5.71)$$

constitutes a linear system of algebraic equations for the determination of the nodal values $(\mathsf{D}_{\tau} u)_b$, leading to

$$\Pi_h(\mathsf{D}_{\tau} u^{\mathrm{h}}) = N^a (\mathsf{D}_{\tau} u)_a \,. \quad (5.72)$$

Analogous relations apply for $\Pi_h(\mathsf{D}_{\tau}\eta^{\mathrm{h}})$.

Remark 7. The standard mid-point rule can be recovered from (5.65) by simply replacing the discrete derivatives with the mid-point evaluation of the corresponding standard derivatives. That is, in the mid-point rule, instead of (5.67) and (5.68) one has to choose

$$\mathsf{D}_{\tau} u^{\mathrm{h}} = \partial_{\tau} u'(\mathbf{F}^{\mathrm{h,T}}_{\mathrm{n+\frac{1}{2}}} \mathbf{F}^{\mathrm{h}}_{\mathrm{n+\frac{1}{2}}}, \tau^{\mathrm{h}}_{\mathrm{n+\frac{1}{2}}}) \quad \text{and} \quad \mathsf{D}_{\mathbf{C}} u^{\mathrm{h}} = \partial_{\mathbf{C}} u'(\mathbf{F}^{\mathrm{h,T}}_{\mathrm{n+\frac{1}{2}}} \mathbf{F}^{\mathrm{h}}_{\mathrm{n+\frac{1}{2}}}, \tau^{\mathrm{h}}_{\mathrm{n+\frac{1}{2}}}) \,.$$

5.3.1. Discrete balance laws

We next show that the present discretization in space and time does indeed yield an EME scheme. In particular, the discrete fulfillment of the main balance laws can be shown along the lines of the semi-discrete formulation dealt with in Section 5.2.2. To this end we recast the scheme (5.65) in the form

$$\begin{aligned}
\frac{\mathbf{q}_{a_{n+1}} - \mathbf{q}_{a_n}}{\Delta t} &= \mathbf{v}_{a_{n+\frac{1}{2}}} \,, \\
M^{ab}\frac{\mathbf{v}_{b_{n+1}} - \mathbf{v}_{b_n}}{\Delta t} &= \int_{\mathcal{B}} N^a \mathbf{b} \, \mathrm{d}V - \int_{\mathcal{B}} \mathsf{D}_{\mathbf{F}} u^{\mathrm{h}} \nabla N^a \, \mathrm{d}V + \int_{\mathcal{B}} \Theta^{\mathrm{h}}_{\mathrm{algo}} \mathsf{D}_{\mathbf{F}} \eta^{\mathrm{h}} \nabla N^a \, \mathrm{d}V \,, \\
H^{ab}\frac{\tau_{b_{n+1}} - \tau_{b_n}}{\Delta t} &= -\mathbf{v}_{d_{n+\frac{1}{2}}} \cdot \int_{\mathcal{B}} \frac{N^a}{\Pi_h(\mathsf{D}_\tau \eta^{\mathrm{h}})} \mathsf{D}_{\mathbf{F}} \eta^{\mathrm{h}} \nabla N^d \, \mathrm{d}V \\
&\quad - \int_{\mathcal{B}} \nabla \left(\frac{N^a}{\Pi_h(\mathsf{D}_\tau u^{\mathrm{h}})} \right) \cdot \mathbf{K}^{\mathrm{h}}_{\mathrm{algo}} \nabla \Theta^{\mathrm{h}}_{\mathrm{algo}} \, \mathrm{d}V - \int_{\partial_q \mathcal{B}} \frac{N^a}{\Pi_h(\mathsf{D}_\tau u^{\mathrm{h}})} \bar{q}_{n+\frac{1}{2}} \, \mathrm{d}A \,,
\end{aligned} \tag{5.73}$$

where $\Theta^{\mathrm{h}}_{\mathrm{algo}}$ has been introduced in $(5.66)_3$. Moreover, similar to (5.43),

$$\begin{aligned}
\mathsf{D}_{\mathbf{F}} u^{\mathrm{h}} &= 2\mathbf{F}^{\mathrm{h}}_{n+\frac{1}{2}} \mathsf{D}_{\mathbf{C}} u^{\mathrm{h}} \,, \\
\mathsf{D}_{\mathbf{F}} \eta^{\mathrm{h}} &= 2\mathbf{F}^{\mathrm{h}}_{n+\frac{1}{2}} \mathsf{D}_{\mathbf{C}} \eta^{\mathrm{h}} \,.
\end{aligned} \tag{5.74}$$

Note that again the standard contribution of external tractions has been neglected in (5.73) for simplicity. Since the total linear momentum is given by $\mathbf{L}^{\mathrm{h}} = \sum_{a=1}^{\mathrm{N}} M^{ab}\mathbf{v}_b$, it immediately follows from $(5.73)_2$, that

$$\mathbf{L}^{\mathrm{h}}_{n+1} - \mathbf{L}^{\mathrm{h}}_{n} = \Delta t \int_{\mathcal{B}} \mathbf{b} \, \mathrm{d}V \,. \tag{5.75}$$

The last equation reflects the consistent approximation of the balance law for linear momentum. With regard to (5.57), the total angular momentum relative to the origin of the inertial frame assumes the form $\mathbf{J}^{\mathrm{h}} = M^{ab}\mathbf{q}_a \times \mathbf{v}_b$. Then we have

$$\mathbf{J}^{\mathrm{h}}_{n+1} - \mathbf{J}^{\mathrm{h}}_{n} = M^{ab} \left(\mathbf{q}_{a_{n+\frac{1}{2}}} \times (\mathbf{v}_{b_{n+1}} - \mathbf{v}_{b_n}) + (\mathbf{q}_{a_{n+1}} - \mathbf{q}_{a_n}) \times \mathbf{q}_{b_{n+\frac{1}{2}}} \right) . \tag{5.76}$$

Substituting from $(5.73)_{1,2}$ into the last equation leads to

$$\begin{aligned}
\boldsymbol{\xi} \cdot (\mathbf{J}^{\mathrm{h}}_{n+1} - \mathbf{J}^{\mathrm{h}}_{n}) &= \boldsymbol{\xi} \cdot \left(\mathbf{q}_{a_{n+\frac{1}{2}}} \times M^{ab}(\mathbf{v}_{b_{n+1}} - \mathbf{v}_{b_n}) \right) \\
&= \boldsymbol{\xi} \cdot \Delta t \int_{\mathcal{B}} \varphi^{\mathrm{h}}_{n+\frac{1}{2}} \times \mathbf{b} \, \mathrm{d}V \,,
\end{aligned} \tag{5.77}$$

for arbitrary but constant $\boldsymbol{\xi} \in \mathbb{R}^3$. In the last equation it has been taken into account that

$$\begin{aligned}
\boldsymbol{\xi} \cdot \left(\mathbf{q}_{a_{n+\frac{1}{2}}} \times (\mathsf{D}_{\mathbf{F}} u^{\mathrm{h}} - \Theta^{\mathrm{h}}_{\mathrm{algo}} \mathsf{D}_{\mathbf{F}} \eta^{\mathrm{h}}) \nabla N^a \right) &= \widehat{\boldsymbol{\xi}} : \left((\mathsf{D}_{\mathbf{F}} u^{\mathrm{h}} - \Theta^{\mathrm{h}}_{\mathrm{algo}} \mathsf{D}_{\mathbf{F}} \eta^{\mathrm{h}}) \mathbf{F}^{\mathrm{h,T}}_{n+\frac{1}{2}} \right) \\
&= \widehat{\boldsymbol{\xi}} : \left(\mathbf{F}^{\mathrm{h}}_{n+\frac{1}{2}} \mathbf{S}^{\mathrm{h}}_{\mathrm{algo}} \mathbf{F}^{\mathrm{h,T}}_{n+\frac{1}{2}} \right) \\
&= 0 \,.
\end{aligned} \tag{5.78}$$

This result follows from the symmetry of $\mathbf{S}^{\mathrm{h}}_{\mathrm{algo}}$ along with the skew-symmetry of $\widehat{\boldsymbol{\xi}}$. Note that (5.77) confirms the consistent approximation of the balance law for angular momentum.

To recover the discrete version of the balance of energy, multiply $(5.73)_3$ by $(\mathsf{D}_\tau u)_a$ and take into account (5.72) together with $(5.66)_3$ to obtain

$$(\mathsf{D}_\tau u)_a H^{ab}(\tau_{b_{\mathrm{n}+1}} - \tau_{b_{\mathrm{n}}}) = -\Delta t \mathbf{v}_{d_{\mathrm{n}+\frac{1}{2}}} \cdot \int_{\mathcal{B}} \Theta^{\mathrm{h}}_{\mathrm{algo}} \mathsf{D}_{\mathbf{F}} \eta^{\mathrm{h}} \nabla N^d \, \mathrm{d}V - \Delta t \int_{\partial_q \mathcal{B}} \overline{q}_{\mathrm{n}+\frac{1}{2}} \, \mathrm{d}A \, .$$

Employing (5.71) and $(5.73)_1$, the last equation can be rewritten in the form

$$\int_{\mathcal{B}} \mathsf{D}_\tau u^{\mathrm{h}} (\tau^{\mathrm{h}}_{\mathrm{n}+1} - \tau^{\mathrm{h}}_{\mathrm{n}}) \, \mathrm{d}V + \int_{\mathcal{B}} \Theta^{\mathrm{h}}_{\mathrm{algo}} \mathsf{D}_{\mathbf{F}} \eta^{\mathrm{h}} : (\mathbf{F}^{\mathrm{h}}_{\mathrm{n}+1} - \mathbf{F}^{\mathrm{h}}_{\mathrm{n}}) \, \mathrm{d}V = -\Delta t \int_{\partial_q \mathcal{B}} \overline{q}_{\mathrm{n}+\frac{1}{2}} \, \mathrm{d}A \, . \tag{5.79}$$

Scalar multiplication of $(5.73)_2$ by $\mathbf{v}_{a_{\mathrm{n}+\frac{1}{2}}}$ yields

$$\begin{aligned} \mathbf{v}_{a_{\mathrm{n}+\frac{1}{2}}} \cdot M^{ab} \frac{\mathbf{v}_{b_{\mathrm{n}+1}} - \mathbf{v}_{b_{\mathrm{n}}}}{\Delta t} &= \mathbf{v}_{a_{\mathrm{n}+\frac{1}{2}}} \cdot \int_{\mathcal{B}} N^a \mathbf{b} - \mathsf{D}_{\mathbf{F}} u^{\mathrm{h}} \nabla N^a \, \mathrm{d}V \\ &\quad + \mathbf{v}_{a_{\mathrm{n}+\frac{1}{2}}} \cdot \int_{\mathcal{B}} \Theta^{\mathrm{h}}_{\mathrm{algo}} \mathsf{D}_{\mathbf{F}} \eta^{\mathrm{h}} \nabla N^a \, \mathrm{d}V \, , \end{aligned}$$

or

$$\begin{aligned} \frac{M^{ab}}{2} \left(\mathbf{v}_{a_{\mathrm{n}+1}} \cdot \mathbf{v}_{b_{\mathrm{n}+1}} - \mathbf{v}_{a_{\mathrm{n}}} \cdot \mathbf{v}_{b_{\mathrm{n}}} \right) =& \Delta t \int_{\mathcal{B}} \mathbf{v}^{\mathrm{h}}_{\mathrm{n}+\frac{1}{2}} \cdot \mathbf{b} \, \mathrm{d}V - \int_{\mathcal{B}} \mathsf{D}_{\mathbf{F}} u^{\mathrm{h}} : (\mathbf{F}^{\mathrm{h}}_{\mathrm{n}+1} - \mathbf{F}^{\mathrm{h}}_{\mathrm{n}}) \, \mathrm{d}V \\ &+ \int_{\mathcal{B}} \Theta^{\mathrm{h}} \mathsf{D}_{\mathbf{F}} \eta^{\mathrm{h}} : (\mathbf{F}^{\mathrm{h}}_{\mathrm{n}+1} - \mathbf{F}^{\mathrm{h}}_{\mathrm{n}}) \, \mathrm{d}V \, . \end{aligned}$$

Substituting from the last equation into (5.79) and taking into account the relationship

$$\begin{aligned} \mathsf{D}_{\mathbf{F}} u^{\mathrm{h}} : (\mathbf{F}^{\mathrm{h}}_{\mathrm{n}+1} - \mathbf{F}^{\mathrm{h}}_{\mathrm{n}}) &= 2\mathbf{F}^{\mathrm{h}}_{\mathrm{n}+\frac{1}{2}} \mathsf{D}_{\mathbf{C}} u^{\mathrm{h}} : (\mathbf{F}^{\mathrm{h}}_{\mathrm{n}+1} - \mathbf{F}^{\mathrm{h}}_{\mathrm{n}}) \\ &= \mathsf{D}_{\mathbf{C}} u^{\mathrm{h}} : (\mathbf{C}^{\mathrm{h}}_{\mathrm{n}+1} - \mathbf{C}^{\mathrm{h}}_{\mathrm{n}}) \, , \end{aligned}$$

yields

$$\begin{aligned} &\int_{\mathcal{B}} \left(\mathsf{D}_\tau u^{\mathrm{h}} (\tau^{\mathrm{h}}_{\mathrm{n}+1} - \tau^{\mathrm{h}}_{\mathrm{n}}) + \mathsf{D}_{\mathbf{C}} u^{\mathrm{h}} : (\mathbf{C}^{\mathrm{h}}_{\mathrm{n}+1} - \mathbf{C}^{\mathrm{h}}_{\mathrm{n}}) \right) \mathrm{d}V + \frac{M^{ab}}{2} \left(\mathbf{v}_{a_{\mathrm{n}+1}} \cdot \mathbf{v}_{b_{\mathrm{n}+1}} - \mathbf{v}_{a_{\mathrm{n}}} \cdot \mathbf{v}_{b_{\mathrm{n}}} \right) \\ &= \int_{\mathcal{B}} (\boldsymbol{\varphi}^{\mathrm{h}}_{\mathrm{n}+1} - \boldsymbol{\varphi}^{\mathrm{h}}_{\mathrm{n}+1}) \cdot \mathbf{b} \, \mathrm{d}V - \Delta t \int_{\partial_q \mathcal{B}} \overline{q}_{\mathrm{n}+\frac{1}{2}} \, \mathrm{d}A \, . \end{aligned}$$

With regard to the total energy E^{h} introduced in (5.62), the last equation can be recast in the form

$$\mathcal{E}^{\mathrm{h}}_{\mathrm{n}+1} - \mathcal{E}^{\mathrm{h}}_{\mathrm{n}} = -\Delta t \int_{\partial_q \mathcal{B}} \overline{q}_{\mathrm{n}+\frac{1}{2}} \, \mathrm{d}A \, . \tag{5.80}$$

Comparison with (5.61) confirms that the present scheme correctly reproduces the balance of energy. Note that the fulfillment of the directionality condition (5.69) is of paramount importance for obtaining an energy consistent scheme.

Eventually, we aim at the discrete balance of entropy. To this end, multiply $(5.73)_3$ by $(\mathsf{D}_\tau\eta)_a$ and take into account $(5.66)_3$, to obtain

$$
\begin{aligned}
(\mathsf{D}_\tau\eta)_a H^{ab} \frac{\tau_{b_{\mathrm{n}+1}} - \tau_{b_{\mathrm{n}}}}{\Delta t} &= -\mathbf{v}_{d_{\mathrm{n}+\frac{1}{2}}} \cdot \int_{\mathcal{B}} \mathsf{D}_{\mathbf{F}}\eta^{\mathrm{h}} \nabla N^d \,\mathrm{d}V \\
&\quad - \int_{\mathcal{B}} \nabla \left(\frac{1}{\Theta^{\mathrm{h}}_{\mathrm{algo}}}\right) \cdot \mathbf{K}^{\mathrm{h}}_{\mathrm{algo}} \nabla \Theta^{\mathrm{h}}_{\mathrm{algo}} \,\mathrm{d}V - \int_{\partial_q \mathcal{B}} \frac{\overline{q}_{\mathrm{n}+\frac{1}{2}}}{\Theta^{\mathrm{h}}_{\mathrm{algo}}} \,\mathrm{d}A\,.
\end{aligned}
$$

The last equation can be rewritten as

$$
\begin{aligned}
&\int_{\mathcal{B}} \left(\mathsf{D}_\tau\eta^{\mathrm{h}}(\tau^{\mathrm{h}}_{\mathrm{n}+1} - \tau^{\mathrm{h}}_{\mathrm{n}}) + \mathsf{D}_{\mathbf{C}}\eta^{\mathrm{h}} : (\mathbf{C}^{\mathrm{h}}_{\mathrm{n}+1} - \mathbf{C}^{\mathrm{h}}_{\mathrm{n}})\right) \mathrm{d}V \\
&\quad = \underbrace{\Delta t \int_{\mathcal{B}} \nabla \left(\frac{1}{\Theta^{\mathrm{h}}_{\mathrm{algo}}}\right) \cdot (\Theta^{\mathrm{h}}_{\mathrm{algo}})^2 \mathbf{K}^{\mathrm{h}}_{\mathrm{algo}} \nabla \left(\frac{1}{\Theta^{\mathrm{h}}_{\mathrm{algo}}}\right) \mathrm{d}V}_{\geq 0} - \Delta t \int_{\partial_q \mathcal{B}} \frac{\overline{q}_{\mathrm{n}+\frac{1}{2}}}{\Theta^{\mathrm{h}}_{\mathrm{algo}}} \,\mathrm{d}A\,.
\end{aligned}
$$

The discrete derivatives of the entropy density, namely $\mathsf{D}_\tau\eta^{\mathrm{h}}$ and $\mathsf{D}_{\mathbf{C}}\eta^{\mathrm{h}}$, respectively, satisfy an associated directionality condition in analogy to (5.69). Accordingly, the left-hand side of the last equation equals $\mathcal{S}^{\mathrm{h}}_{\mathrm{n}+1} - \mathcal{S}^{\mathrm{h}}_{\mathrm{n}}$, where $\mathcal{S}^{\mathrm{h}}$ is the total entropy of the discrete system introduced in (5.64). Comparison with (5.63) confirms that the present scheme correctly reproduces the balance of entropy.

5.4. Numerical investigations

As has been shown in the last section the newly developed schemes are thermodynamically consistent and comply with the balance laws for linear and angular momentum, respectively. These structure-preserving properties hold for arbitrarily large time-steps. Depending on the specific choice of the thermodynamic state variable $\tau \in \{\theta, \eta, u\}$, we obtain three alternative EME schemes. The respective scheme is denoted as $(\mathrm{EME})_\tau$. For example, choosing the temperature as state variable leads to the temperature-based $(\mathrm{EME})_\theta$ scheme.

We apply the newly developed $(\mathrm{EME})_\tau$ schemes to representative numerical examples dealing with finite-strain thermo-elastodynamics. In particular, we treat the three examples which have been previously investigated in Section 4.5. In Chapter 4 the time discretization was confined to the mid-point rule which led to a genuine lack of numerical stability. In particular the lack of stability of the mid-point rule is especially pronounced in the case of large time steps. Therefore, in the numerical examples presented below we focus on large time steps to show that the new $(\mathrm{EME})_\tau$ schemes remain stable.

5.4.1. Flying L-shaped block

First we consider the example given in Section 4.5.1 and choose the largest time-step size for all three $(\mathrm{EME})_\tau$ schemes. The total linear momentum of the block is a conserved quantity since in the initial loading phase the external forces are equilibrated. As after the loading phase ($t > 5$s) no external torque is acting on the block the total angular momentum is a conserved quantity as well. All $(\mathrm{EME})_\tau$ integrators under consideration are capable to conserve both momentum maps (up to numerical round-off), independent of the time step size which can be observed from Fig. 5.4, where representative numerical results of the $(\mathrm{EME})_u$ integrator are shown.

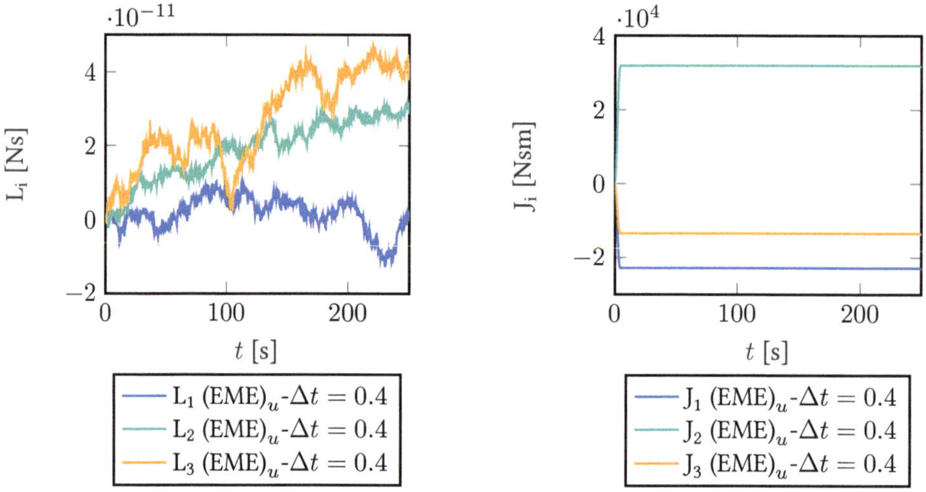

Figure 5.4.: L-shaped block: Algorithmic conservation of linear momentum $(\mathrm{EME})_u$ scheme (left), Total discrete angular momentum $(\mathrm{EME})_u$ scheme (right)

During the loading phase ($0 \leq t \leq 5$) the total energy of the block increases, whereas after the loading phase the total energy has to be a conserved quantity. All $(\mathrm{EME})_\tau$ integrators exactly reproduce this conservation law (up to numerical round-off), see Fig. 5.5. This is in sharp contrast to the mid-point based schemes. In particular, as has been shown in Section 4.5.1, all mid-point based schemes exhibit numerical instability even though the formulation in terms of the internal energy density is capable to conserve the energy.

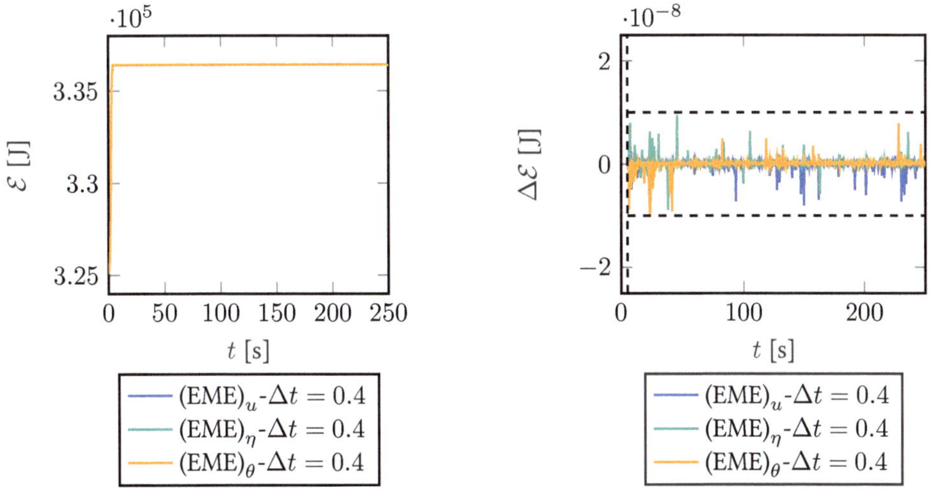

Figure 5.5.: L-shaped block: Total energy (EME)$_\tau$ schemes (left), Incremental change of total energy (EME)$_\tau$ schemes (right)

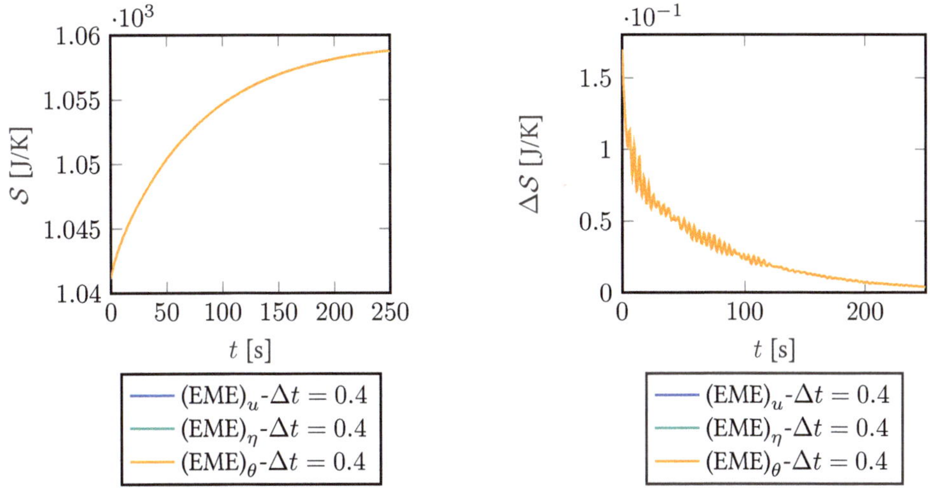

Figure 5.6.: L-shaped block: total entropy (EME)$_\tau$ schemes (left), Incremental change of total entropy (EME)$_\tau$ schemes (right)

Since the (EME)$_\tau$ integrators are consistent with the second law of thermodynamics (independent of the time-step size), the total entropy is a non-decreasing function over time, as can be observed from Fig. 5.6. The above investigations indicate that, in contrast to the mid-point based schemes (see Section 4.5.1) the (EME)$_\tau$ integrators are numerically stabile even for large time steps. After the loading phase, the environment of the present example can be classified as thermally perfect in the sense of [146]. Therefore, $\mathcal{L}$ introduced in (4.92) plays the role of a Lyapunov function. All of the $(\text{EME})_\tau$ integrators correctly deliver a decreasing value of $\mathcal{L}$ over time (Fig. 5.7) thus confirming numerical stability. This is in sharp contrast to the mid-point based schemes dealt with in Section 4.5.1.

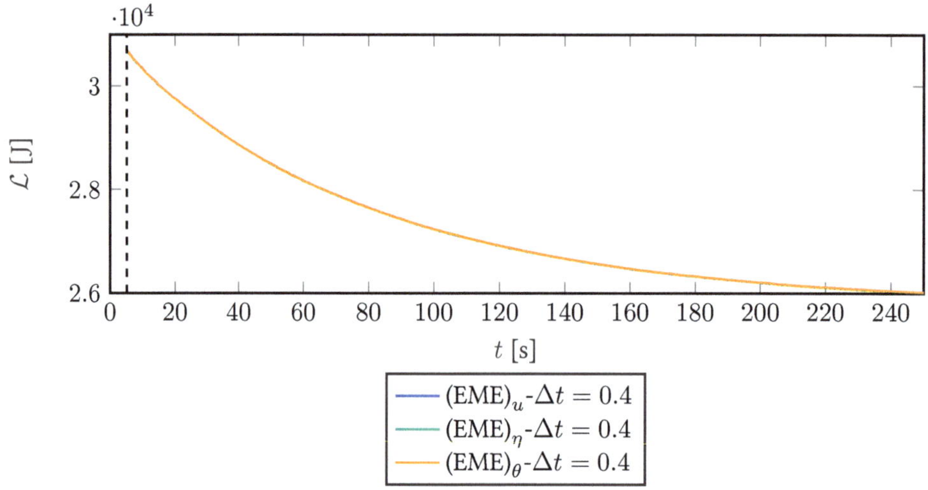

Figure 5.7.: L-shaped block: Lyapunov function computed with the $(\text{EME})_\tau$ schemes

Eventually, the motion of the L-shaped block is illustrated in Fig. 5.8 with snapshots at successive points in time. In addition to that, the distribution of the temperature over the block is shown.

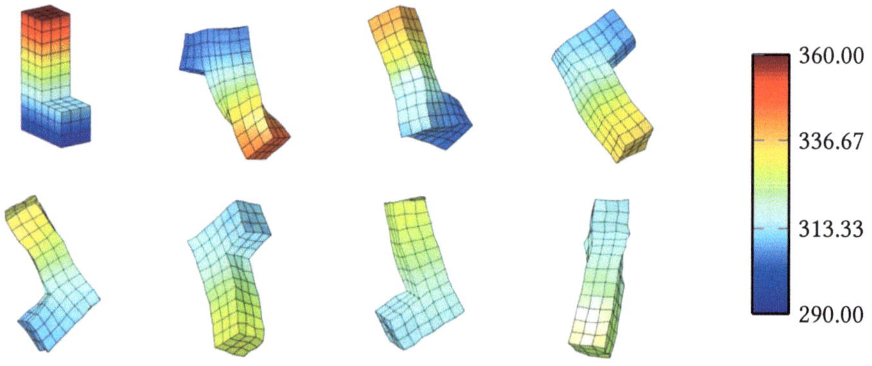

Figure 5.8.: L-shaped block: Snapshots of the motion along with the temperature distribution over the block at $t \in \{0, 32, 64, 96, 128, 160, 192, 224\}$, obtained with the $(\text{EME})_\theta$ scheme and time step $\Delta t = 0.4\text{s}$

5.4.2. Rotating disc

Next we consider the example given in Section 4.5.2 and choose the largest time-step size for all three $(\mathrm{EME})_\tau$ schemes. As neither external loads ($\partial_\sigma \mathcal{B} = \emptyset$) nor displacement boundary conditions ($\partial_\varphi \mathcal{B} = \emptyset$) are acting on the disc, the thermomechanical system at hand has translational and rotational symmetry. Therefore the corresponding momentum maps are first integrals of the motion. All $(\mathrm{EME})_\tau$ integrators at hand are capable to conserve the respective momentum map. Representative numerical results of the $(\mathrm{EME})_\eta$ integrator are shown in Fig. 5.9. During initial period $t \in [0, 4]$s the total energy of the system is expected to increase due to the prescribed heat flow into the system. After initial period the system is classified as closed and therefore the total energy of the system should stay constant. All $(\mathrm{EME})_\tau$ integrators are capable to correctly reproduce the first law of thermodynamics (see Fig. 5.10) in contrast to the mid-point based schemes, where only the formulation in the internal energy density was in accordance with the first law of thermodynamics, see Section 4.5.2. The total entropy of the system is expected to increase due to the prescribed heat flow into the system during the initial period $t \in [0, 4]$s. As the system is closed after the initial period, the total entropy of the system should be a non-decreasing function, whereby the irreversibility is caused by heat conduction.

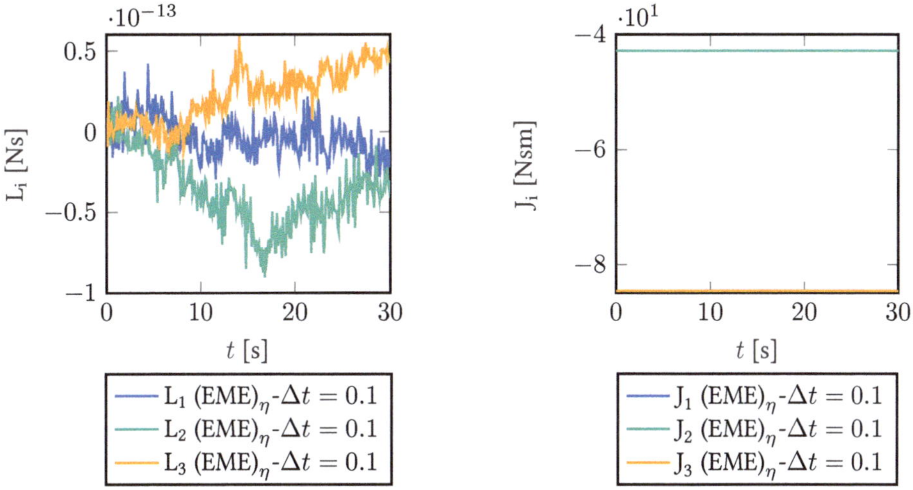

Figure 5.9.: Rotating disc: Total linear momentum (left) and total angular momentum (right)

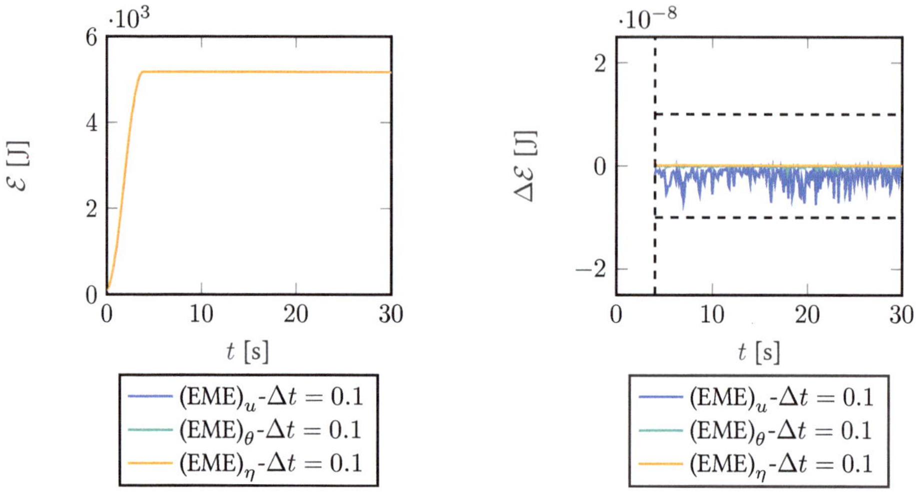

Figure 5.10.: Rotating disc: Total energy (left) and incremental change of total energy (right)

All $(\text{EME})_\tau$ integrators correctly reproduce the second law of thermodynamics as can be observed from Fig. 5.11. Again this is in contrast to the mid-point based schemes investigated in Section 4.5.2, where only the formulation in terms of the entropy density was shown to be consistent with the second law.

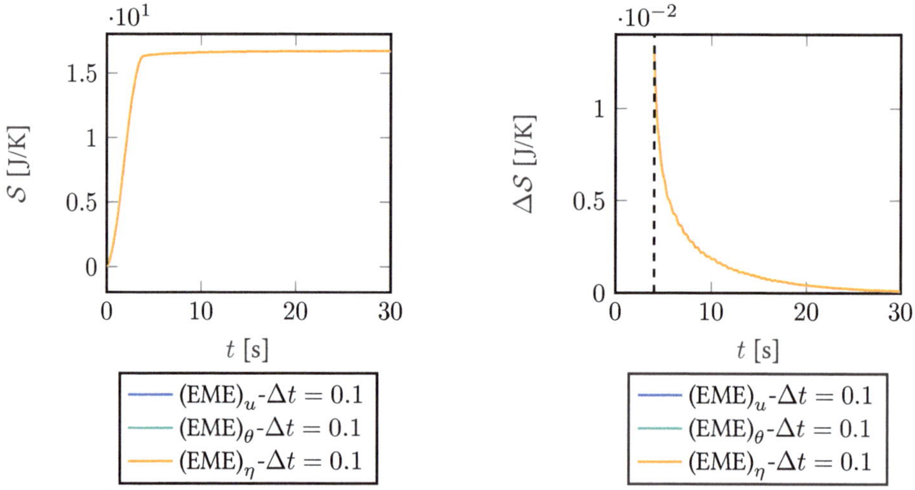

Figure 5.11.: Rotating disc: Total entropy (left) and incremental change of total entropy (right)

To investigate the numerical stability of the $(\text{EME})_\tau$ schemes further, we consider again the Lyapunov function $\mathcal{L}$ introduced in (4.92). After the initial period the system is closed and therefore $\mathcal{L}$ has to decrease with time.

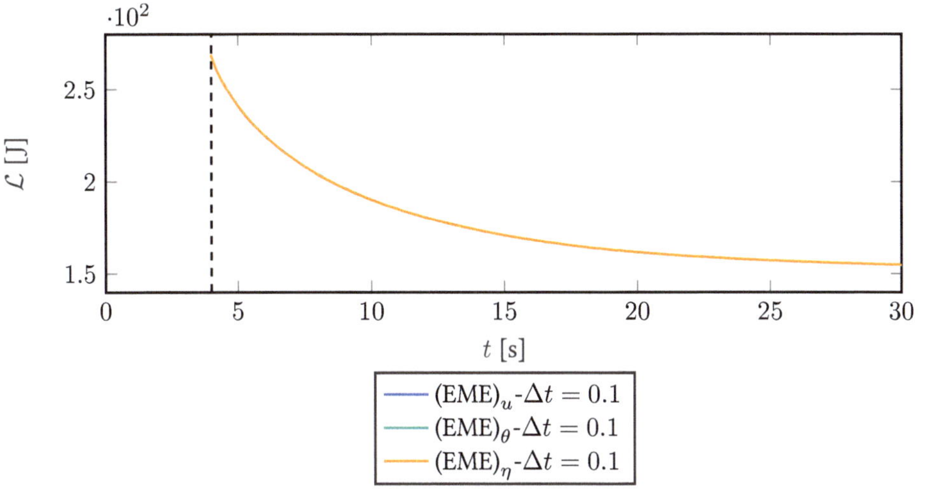

Figure 5.12.: Rotating disc: Lyapunov function $(\text{EME})_\tau$ schemes

All $(\text{EME})_\tau$ integrators correctly reproduce this behaviour as can be observed from Fig. 5.12. This again confirms numerical stability of the EME schemes at hand and is in contrast to the mid-point based schemes investigated in Section 4.5.2.

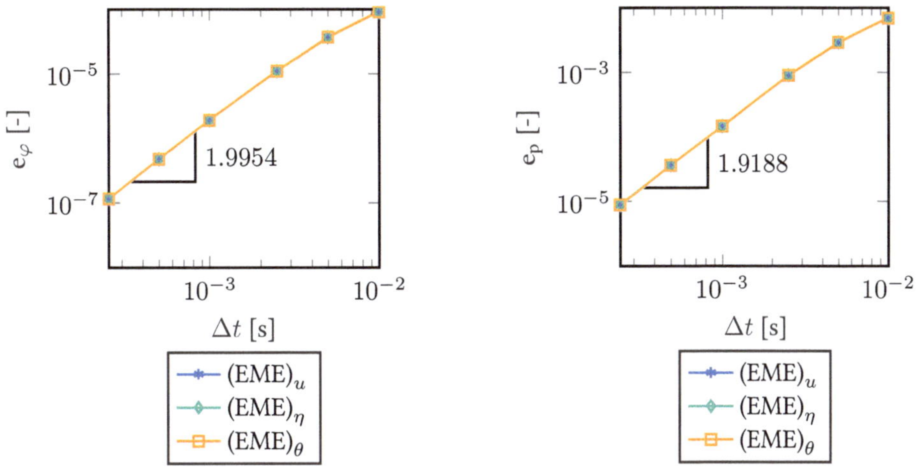

Figure 5.13.: Rotating disc: Error in the position (left) and error in the velocity (right)

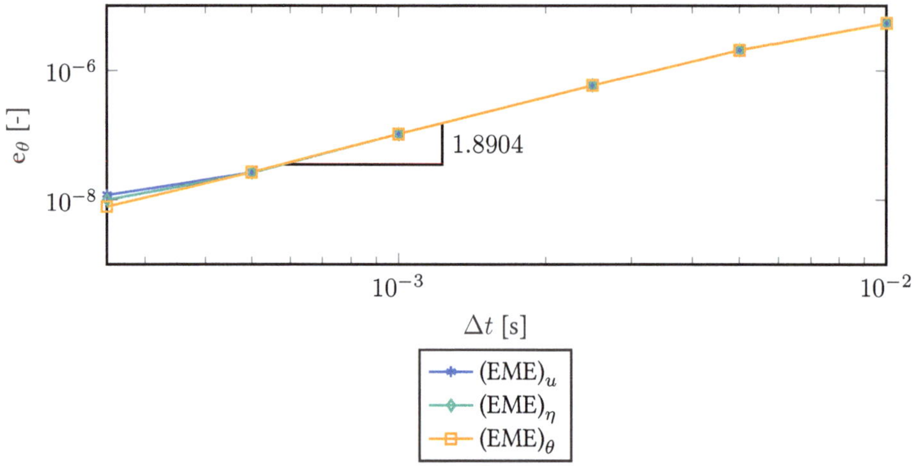

Figure 5.14.: Rotating disc: Error in the absolute temperature

Further we investigate the order of accuracy of the present structure-preserving methods. Therefore we introduce the relative L_2 error norm in the position, the velocity and the absolute temperature, respectively

$$\mathrm{e}_{\varphi} = \frac{||\varphi - \varphi_r||_{L_2}}{||\varphi_r||_{L_2}}, \quad \mathrm{e}_{\mathrm{p}} = \frac{||\mathbf{v} - \mathbf{v}_r||_{L_2}}{||\mathbf{v}_r||_{L_2}}, \quad \mathrm{e}_{\Theta} = \frac{||\Theta - \Theta_r||_{L_2}}{||\Theta_r||_{L_2}},$$
$$|| \bullet ||_{L_2} = \left[\int_{\mathcal{B}_I} \langle \bullet, \bullet \rangle \, \mathrm{d}V \right]^{\frac{1}{2}} .$$

In the above equations $\bullet_r$ with $\bullet \in \{\varphi, \mathbf{v}, \Theta\}$ is the reference solution at time $t = 4.02$s calculated with the smallest time step size. We consider the motion of the rotating disk in the time interval $[4, 4.02]$s. To this end, the simulation is started at time t= 4s by using the data obtained with the $(\mathrm{EME})_\theta$ scheme and Δt= 0.04s. As can be observed from Figs. 5.13 and 5.14 all schemes exhibit second order accuracy in φ, $\mathbf{v}$ and Θ. Eventually, the motion of the disc is illustrated in Fig. 5.15 with snapshots at successive points in time. In addition to that, the distribution of the temperature over the disc is shown.

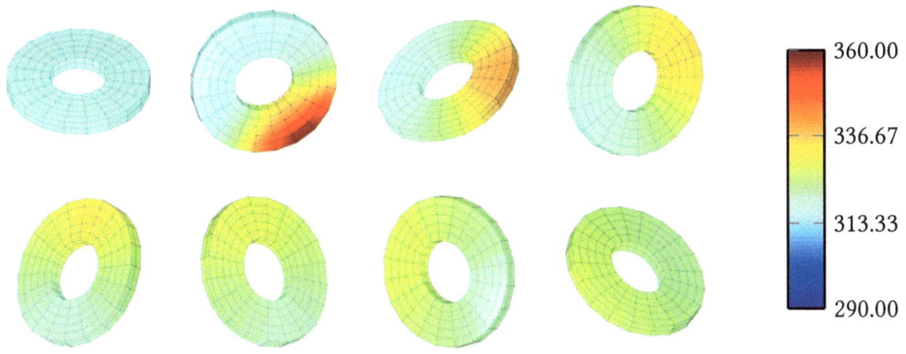

Figure 5.15.: Rotating disc: Snapshots of the motion at successive points in time $t \in \{0, 4, 8, 12, 16, 18, 24, 28\}$s, and corresponding temperature distribution, calculated with the (EME)$_\theta$ scheme and $\Delta t = 0.1$s

5.4.3. Rotating disc in a thermally perfect environment

Finally we examine the example given in Section 4.5.3 and consider only the largest time-step size. Again we focus on the temperature-based (EME)$_\theta$ integrator, which makes possible to impose temperature Dirichlet boundary conditions in a standard manner. As in the previous example the system at hand has both translational and rotational symmetry. Accordingly, the corresponding momentum maps are exactly conserved by the (EME)$_\theta$ integrator, see Fig. 5.16.

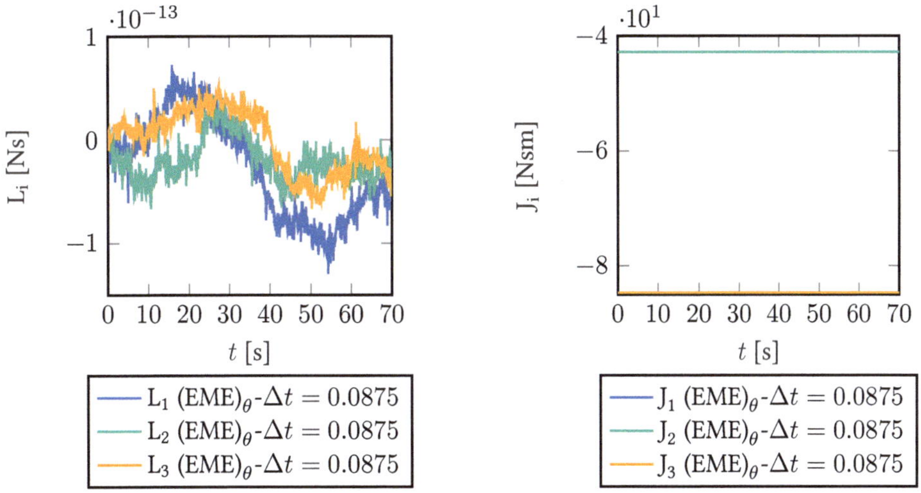

Figure 5.16.: Rotating disc in a thermally perfect environment: Total linear momentum (left), total angular momentum (right)

According to the temperature boundary conditions heat is flowing out of the disc. The corresponding time-evolution of the total energy is depicted in Fig. 5.17 (left). In addition to that, the total entropy is as well decreasing over time, see Fig. 5.17 (right).

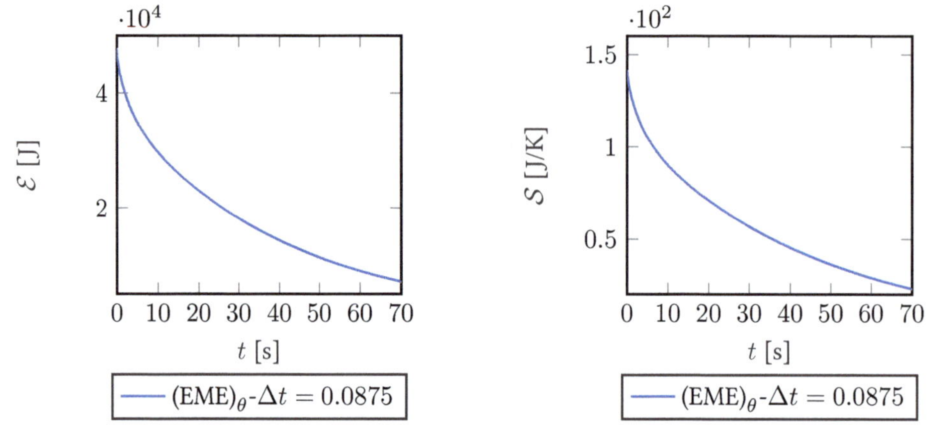

Figure 5.17.: Rotating disc in a thermally perfect environment: Total energy (EME)$_\theta$ scheme (left), and total entropy (right)

By design of the thermally perfect environment, functional $\mathcal{L}$ introduced in (4.92) is a Lyapunov function throughout the whole simulation time. Correspondingly, the (EME)$_\theta$ integrator yields a steadily decreasing value of $\mathcal{L}$ even for very large time steps (see Fig. 5.18). In particular, in contrast to the mid-point rule (cf. Section 4.5.3), the present scheme does stay numerically stable.

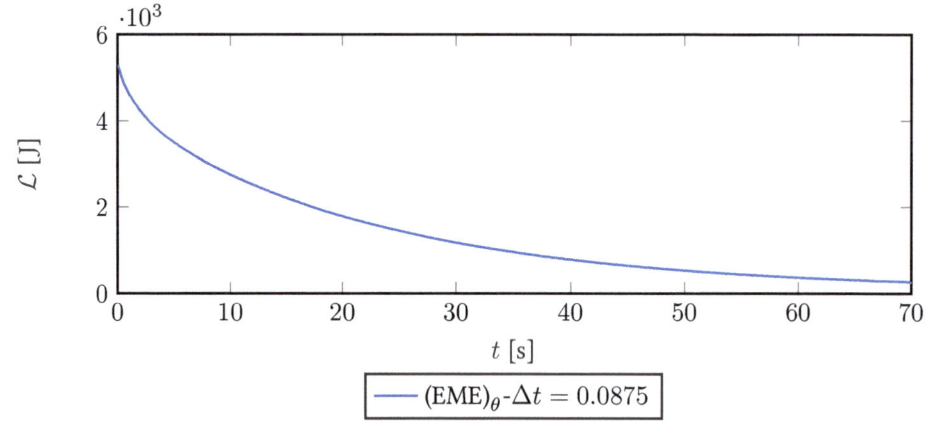

Figure 5.18.: Rotating disc in a thermally perfect environment: Lyapunov function

Eventually, both the motion and the evolution of the temperature distribution are shown in Fig. 5.19 with snapshots at successive points in time.

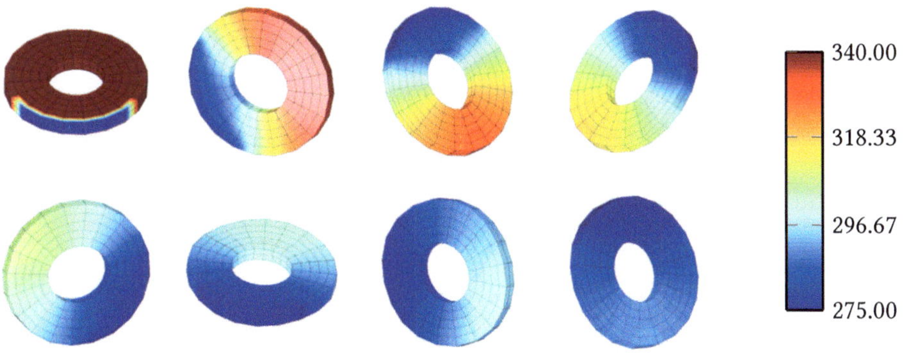

Figure 5.19.: Rotating disc in a thermally perfect environment: Snapshots of the motion at $t \in \{0, 10, 20, 30, 40, 50, 60, 70\}$s and corresponding temperature distribution, calculated with the $(\mathrm{EME})_\theta$ scheme and $\Delta t = 0.0875$s

6. GENERIC-based numerical methods for large-strain thermo-viscoelasticity[1]

Large-strain thermo-viscoelasticity is described in the framework of GENERIC. To this end, a new material representation of the inelastic part of the dissipative bracket is proposed. The bracket form of GENERIC generates the governing equations for large-strain thermo-viscoelasticity including the nonlinear evolution law for the internal variables associated with inelastic deformations. The GENERIC formalism facilitates the free choice of the thermodynamic variable. In particular, one may choose (i) the internal energy density, (ii) the entropy density, or (iii) the absolute temperature as the thermodynamic state variable. A mixed finite element method is proposed for the discretization in space which preserves the GENERIC form of the resulting semi-discrete evolution equations. The *GENERIC-consistent space discretization* makes the design of structure-preserving time-stepping schemes possible. The mid-point type discretization in time yields three alternative schemes. Depending on the specific choice of the thermodynamic variable, these schemes are shown to be partially structure-preserving. Numerical investigations are presented which confirm the theoretical findings and shed light on the numerical stability of the newly developed schemes.

6.1. GENERIC-based formulation of large strain thermo-viscoelasticity

The present Chapter relies on a material description of large-strain thermo-viscoelasticity within the framework of GENERIC. The developments presented herein extend the work in the previous Chapters 4 and 5 on thermo-elasticity to the realm of inelastic deformations. In the GENERIC framework the time-evolution of any functional $\mathcal{A}$ is governed by

$$\frac{\mathrm{d}\mathcal{A}}{\mathrm{d}t} = \{\mathcal{A}, \mathcal{E}\} + [\mathcal{A}, \mathcal{S}] \,. \tag{6.1}$$

[1] This Chapter is based on [4]

The evolution equation (6.1) is valid for closed systems in which the boundaries are disregarded. We consider a continuum body with material points $\mathbf{X} = X^A \mathbf{e}_A$ in the reference configuration $\mathcal{B} \subset \mathbb{R}^3$ (Fig. 6.1). The material coordinates X^A refer to canonical base vectors $\mathbf{e}_A \in \mathbb{R}^3$. Here and in the sequel, the summation convention applies to repeated indices.

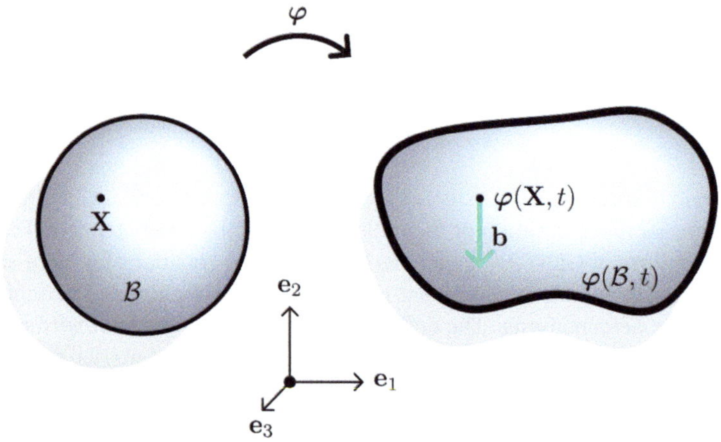

Figure 6.1.: Reference configuration $\mathcal{B}$ and deformed configuration $\varphi(\mathcal{B}, t)$ at time t. For now the focus is on the motion of isolated thermomechanically coupled solids. That is, external tractions acting on the boundary as well heat fluxes across the boundary are disregarded.

Within the Lagrangian description of continuum mechanics the deformed configuration of the body at time t is characterized by the deformation map $\varphi : \mathcal{B} \times \mathcal{I} \mapsto \mathbb{R}^3$, where $\mathcal{I} = [0, T]$ is the time interval of interest. Accordingly, the placement at time t of the material point $\mathbf{X} \in \mathcal{B}$ is given by $\mathbf{x} = \varphi(\mathbf{X}, t)$. The corresponding velocity field $\mathbf{v} : \mathcal{B} \times \mathcal{I} \mapsto \mathbb{R}^3$ is defined by $\mathbf{v} = \dot{\varphi}$, where a superposed dot denotes the material time derivative. The deformation gradient is given by

$$\mathbf{F} = \partial_{\mathbf{X}} \varphi \,, \tag{6.2}$$

or, in components, $(\mathbf{F})^{i}{}_{\cdot A} = \partial x^i / \partial X^A$. Furthermore, the right Cauchy-Green tensor reads

$$\mathbf{C} = \mathbf{F}^{\mathrm{T}} \mathbf{F} \,. \tag{6.3}$$

In the GENERIC (6.1), we consider functionals of the form

$$\mathcal{A} = \mathcal{A}(\varphi, \mathbf{p}, \tau, \mathbf{C}_p^{-1}) = \int_{\mathcal{B}} a(\varphi, \mathbf{F}, \mathbf{p}, \tau, \mathbf{C}_p^{-1}) \, \mathrm{d}V \,, \tag{6.4}$$

where $\mathbf{p} = \rho \mathbf{v}$ is the linear momentum density and $\rho : \mathcal{B} \mapsto \mathbb{R}_+$ is the mass density in the reference configuration. Moreover, $\tau : \mathcal{B} \times \mathcal{I} \mapsto \mathbb{R}$ is a generalized thermodynamic field which may be chosen from among three alternatives, $\tau \in \{\theta, \eta, u\}$. The three alternative fields are the absolute temperature θ, the entropy density η, and the internal energy density u. To describe inelastic deformations, we make use of the internal variable $\mathbf{C}_p^{-1} : \mathcal{B}\times\mathcal{I} \mapsto \mathbb{R}^{3\times 3}$. Since we focus on isotropic inelastic deformations, $\mathbf{C}_p^{-1}$ is assumed to be symmetric, i.e. $\mathbf{C}_p^{-\mathrm{T}} = \mathbf{C}_p^{-1}$. Therefore the density function a in (6.4) is subject to the same isotropic restrictions as the Helmholtz free energy function, to be introduced in Section 6.2.

GENERIC (6.1) decouples reversible and irreversible processes. In particular, the Poisson bracket $\{\cdot,\cdot\}$ is responsible for reversible processes, while the dissipative (or friction) bracket $[\cdot,\cdot]$ embodies irreversible processes. Accordingly, these two brackets constitute fundamental entities of the GENERIC framework and shall be specified next. Since inelastic deformations are purely irreversible, the Poisson bracket retains its thermo-elastic form (see eq. (4.34))

$$\{\mathcal{A},\mathcal{B}\} = \int_{\mathcal{B}} \left(\delta_{\boldsymbol{\varphi}} a + \mathrm{Div}\left(\frac{\delta_\tau a}{\partial_\tau \eta}\partial_{\mathbf{F}}\eta\right)\right)\cdot\delta_{\mathbf{p}} b - \left(\delta_{\boldsymbol{\varphi}} b + \mathrm{Div}\left(\frac{\delta_\tau b}{\partial_\tau \eta}\partial_{\mathbf{F}}\eta\right)\right)\cdot\delta_{\mathbf{p}} a \,\mathrm{d}V\,, \tag{6.5}$$

where $\mathcal{A}$ and $\mathcal{B}$ are arbitrary functionals of the form (6.4). Furthermore, the functional derivatives in (6.5) are given by

$$\begin{aligned} \delta_{\boldsymbol{\varphi}} a &= \partial_{\boldsymbol{\varphi}} a - \mathrm{Div}\partial_{\mathbf{F}} a\,,\\ \delta_{\mathbf{p}} a &= \partial_{\mathbf{p}} a\,,\\ \delta_\tau a &= \partial_\tau a\,,\\ \delta_{\mathbf{C}_p^{-1}} a &= \partial_{\mathbf{C}_p^{-1}} a\,. \end{aligned} \tag{6.6}$$

In (6.5), η denotes the entropy density giving rise to the total entropy of the system

$$\mathcal{S}(\boldsymbol{\varphi},\tau,\mathbf{C}_p^{-1}) = \int_{\mathcal{B}} \eta(\mathbf{F},\tau,\mathbf{C}_p^{-1})\,\mathrm{d}V\,. \tag{6.7}$$

The dissipative bracket for the thermo-viscoelastic problem at hand can be additively decomposed into a part due to heat conduction and a part taking into account inelastic deformations:

$$[\mathcal{A},\mathcal{B}] = [\mathcal{A},\mathcal{B}]_{\mathrm{cond}} + [\mathcal{A},\mathcal{B}]_{\mathrm{inel}}\,. \tag{6.8}$$

In analogy to thermo-elasticity with heat conduction, the dissipative bracket accounting for heat conduction is given by (see eq. (4.35))

$$[\mathcal{A},\mathcal{B}]_{\mathrm{cond}} = \int_{\mathcal{B}} \nabla\left(\frac{\delta_\tau a}{\partial_\tau u}\right)\cdot\Theta^2\mathbf{K}\nabla\left(\frac{\delta_\tau b}{\partial_\tau u}\right)\mathrm{d}V\,. \tag{6.9}$$

Here, $\mathbf{K} = \mathbf{K}(\mathbf{C}, \tau)$ is the material conductivity tensor, for which we assume that $\mathbf{K}^{\mathrm{T}} = \mathbf{K}$ and $\mathbf{a} : \mathbf{K} : \mathbf{a} \geq 0$ holds for all $\mathbf{a} \in \mathbb{R}^3$. An important element of the GENERIC framework is that the absolute temperature takes the form

$$\Theta(\mathbf{F}, \tau, \mathbf{C}_p^{-1}) = \frac{\partial_\tau u(\mathbf{F}, \tau, \mathbf{C}_p^{-1})}{\partial_\tau \eta(\mathbf{F}, \tau, \mathbf{C}_p^{-1})} \,. \tag{6.10}$$

In (6.9) and (6.10), u denotes the internal energy density which, together with the kinetic energy and the potential of dead loads, constitutes the total energy of the system

$$\mathcal{E}(\boldsymbol{\varphi}, \mathbf{p}, \tau, \mathbf{C}_p^{-1}) = \int_{\mathcal{B}} \left(\frac{1}{2} \rho^{-1} \mathbf{p} \cdot \mathbf{p} + u(\mathbf{F}, \tau, \mathbf{C}_p^{-1}) - \mathbf{b} \cdot \boldsymbol{\varphi} \right) \mathrm{d}V \,. \tag{6.11}$$

Here, $\mathbf{b} : \mathcal{B} \mapsto \mathbb{R}^3$ represent prescribed body forces which, for simplicity, are assumed to be dead loads.

Concerning the contribution of inelastic deformations, we introduce the following form of the dissipative bracket (see Section 6.2 for further details):

$$\begin{aligned} &[\mathcal{A}, \mathcal{B}]_{\text{inel}} \\ &= \int_{\mathcal{B}} 2 \left(\frac{\delta_\tau a}{\partial_\tau u} \partial_{\mathbf{C}_p^{-1}} u \mathbf{C}_p^{-1} - \delta_{\mathbf{C}_p^{-1}} a \mathbf{C}_p^{-1} \right) : \Theta \boldsymbol{\mathcal{N}} : 2 \left(\frac{\delta_\tau b}{\partial_\tau u} \partial_{\mathbf{C}_p^{-1}} u \mathbf{C}_p^{-1} - \delta_{\mathbf{C}_p^{-1}} b \mathbf{C}_p^{-1} \right) \mathrm{d}V \,. \end{aligned} \tag{6.12}$$

The fourth-order inelastic material flow tensor $\boldsymbol{\mathcal{N}}$ has the properties $\boldsymbol{\mathcal{N}}^{\mathrm{T}} = \boldsymbol{\mathcal{N}}$ (major symmetry) and $\mathbf{M} : \boldsymbol{\mathcal{N}} : \mathbf{M} \geq 0$ (positive semi-definiteness) for all $\mathbf{M}$ given in (6.17).

It can be easily verified that the Poisson bracket (6.5) is skew-symmetric ($\{\mathcal{A}, \mathcal{B}\} = -\{\mathcal{B}, \mathcal{A}\}$), the dissipative bracket (6.12) is symmetric ($[\mathcal{A}, \mathcal{B}] = [\mathcal{B}, \mathcal{A}]$) and further satisfies the non-negativity condition $[\mathcal{A}, \mathcal{A}] \geq 0$. Moreover, the two brackets satisfy the fundamental degeneracy (or non-interaction) conditions $\{\mathcal{A}, \mathcal{S}\} = 0$ and $[\mathcal{A}, \mathcal{E}] = 0$. In conjunction with these properties of the two brackets, GENERIC (6.1) ensures for closed systems (i) the conservation of total energy ($\mathrm{d}\mathcal{E}/\mathrm{d}t = 0$), and (ii) a non-decreasing total entropy ($\mathrm{d}\mathcal{S}/\mathrm{d}t \geq 0$). Due to these structural properties, the GENERIC-based formulation offers an ideal starting point for the design of structure-preserving numerical methods.

6.1.1. Local evolution equations

We provide a short outline of the local evolution equations emanating from GENERIC (6.1). With regard to (6.1) the total energy $\mathcal{E}$ generates the reversible contribution through the Poisson bracket $\{\mathcal{A}, \mathcal{E}\}$. Using expression (6.11) for the total energy along with formulas (6.6) for the variational derivatives, Poisson bracket (6.5) yields

$$\{\mathcal{A}, \mathcal{E}\} = \int_{\mathcal{B}} \left(\delta_{\boldsymbol{\varphi}} a + \operatorname{Div} \left(\frac{\delta_\tau a}{\partial_\tau \eta} \partial_{\mathbf{F}} \eta \right) \right) \cdot \rho^{-1} \mathbf{p} + (\mathbf{b} + \operatorname{Div} \mathbf{P}) \cdot \delta_{\mathbf{p}} a \, \mathrm{d}V \,, \tag{6.13}$$

where the first Piola-Kirchhoff stress tensor $\mathbf{P} = \mathbf{P}(\mathbf{F}, \tau, \mathbf{C}_p^{-1})$ takes the form

$$\mathbf{P} = \partial_{\mathbf{F}} u - \Theta \partial_{\mathbf{F}} \eta \,. \tag{6.14}$$

The irreversible contribution to GENERIC (6.1) is generated by the total entropy $\mathcal{S}$ through the dissipative bracket $[\mathcal{A}, \mathcal{S}]$. Inserting expression (6.7) for the total entropy into the dissipative bracket (6.8) leads to

$$[\mathcal{A}, \mathcal{S}] = \int_{\mathcal{B}} \nabla \left(\frac{\delta_\tau a}{\partial_\tau u} \right) \cdot \mathbf{Q} \, \mathrm{d}V + \int_{\mathcal{B}} 2 \left(\frac{\delta_\tau a}{\partial_\tau u} \partial_{\mathbf{C}_p^{-1}} u \mathbf{C}_p^{-1} - \delta_{\mathbf{C}_p^{-1}} a \mathbf{C}_p^{-1} \right) : \boldsymbol{\mathcal{N}} : \mathbf{M} \, \mathrm{d}V \,, \tag{6.15}$$

where the material heat flux vector $\mathbf{Q} = \mathbf{Q}(\mathbf{F}, \tau, \mathbf{C}_p^{-1})$ is given by

$$\mathbf{Q} = \Theta^2 \mathbf{K} \nabla \left(\frac{1}{\Theta} \right) = -\mathbf{K} \nabla \Theta \,, \tag{6.16}$$

and $\mathbf{M} = \mathbf{M}(\mathbf{F}, \tau, \mathbf{C}_p^{-1})$ denotes the material representation of the Mandel stress tensor, which takes the form

$$\mathbf{M} = 2 \left(\partial_{\mathbf{C}_p^{-1}} u - \Theta \partial_{\mathbf{C}_p^{-1}} \eta \right) \mathbf{C}_p^{-1} \,. \tag{6.17}$$

It is important to realize that in (6.14), (6.16) and (6.17), Θ is the temperature field which has been introduced in (6.10). These relationships are representative for the GENERIC-based formulation.

With regard to the left-hand side of GENERIC (6.1) and expression (6.4) for functional $\mathcal{A}$, the time derivative of $\mathcal{A}$ can be written as

$$\frac{\mathrm{d}}{\mathrm{d}t} \mathcal{A} = \int_{\mathcal{B}} \left(\delta_{\boldsymbol{\varphi}} a \cdot \dot{\boldsymbol{\varphi}} + \delta_{\mathbf{p}} a \cdot \dot{\mathbf{p}} + \delta_\tau a \dot{\tau} + \delta_{\mathbf{C}_p^{-1}} a : \dot{\overline{\mathbf{C}_p^{-1}}} \right) \mathrm{d}V \,. \tag{6.18}$$

Substituting (6.18), (6.13) and (6.15) into GENERIC (6.1), we arrive at

$$\begin{aligned} 0 = & \int_{\mathcal{B}} \delta_{\boldsymbol{\varphi}} a \cdot \left(\dot{\boldsymbol{\varphi}} - \rho^{-1} \mathbf{p} \right) \mathrm{d}V \\ & + \int_{\mathcal{B}} \delta_{\mathbf{p}} a \cdot \left(\dot{\mathbf{p}} - (\mathbf{b} + \mathrm{Div} \mathbf{P}) \right) \mathrm{d}V \\ & + \int_{\mathcal{B}} \delta_\tau a \left(\dot{\tau} - \frac{2}{\partial_\tau u} \left(\partial_{\mathbf{C}_p^{-1}} u \mathbf{C}_p^{-1} \right) : \boldsymbol{\mathcal{N}} : \mathbf{M} \right) - \mathrm{Div} \left(\frac{\delta_\tau a}{\partial_\tau \eta} \partial_{\mathbf{F}} \eta \right) \cdot \rho^{-1} \mathbf{p} \, \mathrm{d}V \\ & - \int_{\mathcal{B}} \nabla \left(\frac{\delta_\tau a}{\partial_\tau u} \right) \cdot \mathbf{Q} + \delta_{\mathbf{C}_p^{-1}} a : \left(\dot{\overline{\mathbf{C}_p^{-1}}} + 2 \left[\boldsymbol{\mathcal{N}} : \mathbf{M} \right] \mathbf{C}_p^{-1} \right) \mathrm{d}V \,. \end{aligned} \tag{6.19}$$

This equation has to hold for arbitrary functionals $\mathcal{A}$ of the form (6.4). Standard arguments now lead to the local evolution equations emanating from GENERIC (6.1):

$$\begin{aligned}
\dot{\varphi} &= \rho^{-1}\mathbf{p}\,,\\
\dot{\mathbf{p}} &= \mathbf{b} + \mathrm{Div}\mathbf{P}\,,\\
\dot{\tau} &= \frac{2}{\partial_\tau u}\left(\partial_{\mathbf{C}_p^{-1}} u \mathbf{C}_p^{-1}\right) : \boldsymbol{\mathcal{N}} : \mathbf{M} - \frac{1}{\partial_\tau \eta}\partial_{\mathbf{F}}\eta : \nabla(\rho^{-1}\mathbf{p}) - \frac{1}{\partial_\tau u}\mathrm{Div}\mathbf{Q}\,,\\
\dot{\overline{\mathbf{C}_p^{-1}}} &= -2\,(\boldsymbol{\mathcal{N}} : \mathbf{M})\,\mathbf{C}_p^{-1}\,.
\end{aligned} \tag{6.20}$$

We stress again that in the above formulas the first Piola-Kirchhoff stress tensor $\mathbf{P}$, the Mandel stress tensor $\mathbf{M}$ and the heat flux vector $\mathbf{Q}$ are given by formulas (6.14), (6.17) and (6.16), respectively.

6.1.2. GENERIC-based weak form of the IBVP

Starting from the GENERIC-based local evolution equations (6.20), we deduce the weak form of the initial boundary value problem (IBVP) pertaining to finite-strain thermo-viscoelasticity. To this end, we decompose the boundary $\partial\mathcal{B}$ of the continuum body into a displacement boundary $\partial_\varphi\mathcal{B}$, on which $\varphi = \overline{\varphi}$, and a traction boundary $\partial_\sigma\mathcal{B}$, on which $\mathbf{PN} = \overline{\mathbf{t}}$, where $\overline{\varphi}$ and $\overline{\mathbf{t}}$ are prescribed functions for $t \geq 0$ (Fig. 6.2). Moreover, $\partial_\varphi\mathcal{B} \cup \partial_\sigma\mathcal{B} = \partial\mathcal{B}$ and $\partial_\varphi\mathcal{B} \cap \partial_\sigma\mathcal{B} = \emptyset$.

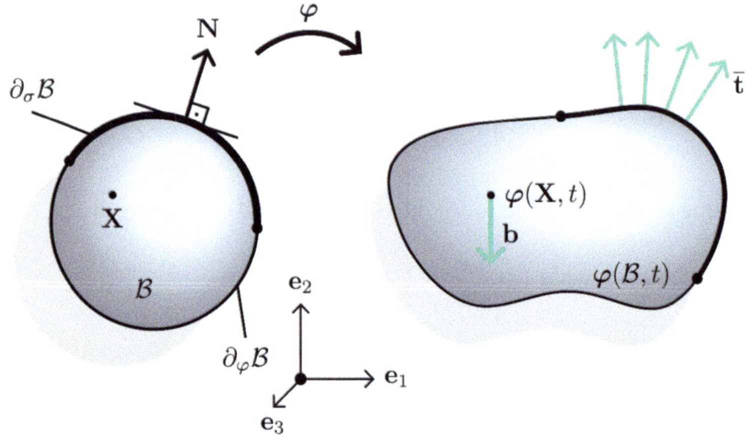

Figure 6.2.: Mechanical part of the IBVP. Note that $\overline{\mathbf{t}} = \mathbf{PN}$ denotes prescribed external Piola tractions acting on the current boundary expressed per unit area of the reference boundary $\partial_\sigma\mathcal{B}$.

Similarly, for the thermal part we consider the subsets of the boundary $\partial_\tau\mathcal{B}$ and $\partial_q\mathcal{B}$, with the properties $\partial_\tau\mathcal{B} \cup \partial_q\mathcal{B} = \partial\mathcal{B}$ and $\partial_\tau\mathcal{B} \cap \partial_q\mathcal{B} = \emptyset$ (Fig. 6.3). The thermodynamic state variable is prescribed on $\partial_\tau \mathcal{B}$, i.e. $\tau = \overline{\tau}$, whereas the heat flux is prescribed on $\partial_q \mathcal{B}$, i.e. $\mathbf{Q} \cdot \mathbf{N} = \overline{q}$. Both $\overline{\tau}$ and $\overline{q}$ are assumed to be given for $t \geq 0$.

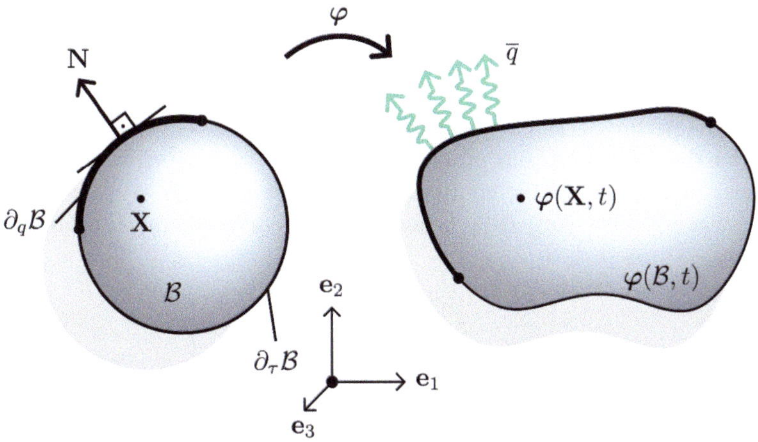

Figure 6.3.: Thermal part of the IBVP. Note that $\overline{q} = \mathbf{Q} \cdot \mathbf{N}$ is the prescribed rate of heat transfer across the current boundary expressed per unit area of the reference boundary $\partial_q \mathcal{B}$.

We aim at the determination of the motion $\varphi : \mathcal{B} \times \mathcal{I} \mapsto \mathbb{R}^3$, the linear momentum $\mathbf{p} : \mathcal{B} \times \mathcal{I} \mapsto \mathbb{R}^3$, the thermodynamic state variable $\tau : \mathcal{B} \times \mathcal{I} \mapsto \mathbb{R}$, and the inelastic deformation $\mathbf{C}_p^{-1} : \mathcal{B} \times \mathcal{I} \mapsto \mathbb{R}^{3\times 3}$. The unknown fields are subject to initial conditions of the form $\varphi(\cdot, 0) = \mathbf{X}$, $\mathbf{p}(\cdot, 0) = \rho \mathbf{V}_0$, $\tau(\cdot, 0) = \tau^{\text{ini}}$ and $\mathbf{C}_p^{-1}(\cdot, 0) = \mathbf{I}$ in $\mathcal{B}$. Here, $\mathbf{V}_0$ is a prescribed material velocity field, and τ^{ini} is a prescribed field of the thermodynamic state variable $\tau \in \{\theta, \eta, u\}$. The unknown fields are determined by applying a space-time discretization to the weak form of the IBVP at hand.

The weak form of the IBVP can be obtained by scalar multiplying the local evolution equations (6.20) by suitable test functions and subsequently integrating over domain $\mathcal{B}$. The standard procedure yields

$$
\begin{aligned}
0 = &\int_{\mathcal{B}} \mathbf{w}_\varphi \cdot (\dot{\varphi} - \mathbf{v}) \, \mathrm{d}V \\
&+ \int_{\mathcal{B}} (\mathbf{w}_\mathbf{p} \cdot (\rho \dot{\mathbf{v}} - \mathbf{b}) + \mathbf{P} : \nabla \mathbf{w}_\mathbf{p}) \, \mathrm{d}V - \int_{\partial_\sigma \mathcal{B}} \mathbf{w}_\mathbf{p} \cdot \bar{\mathbf{t}} \, \mathrm{d}A \\
&+ \int_{\mathcal{B}} \mathrm{w}_\tau \left(\dot{\tau} - \frac{2}{\partial_\tau u} \left(\partial_{\mathbf{C}_p^{-1}} u \mathbf{C}_p^{-1} \right) : \boldsymbol{\mathcal{N}} : \mathbf{M} + \frac{1}{\partial_\tau \eta} \partial_\mathbf{F} \eta : \nabla \mathbf{v} \right) \mathrm{d}V \\
&- \int_{\mathcal{B}} \nabla \left(\frac{\mathrm{w}_\tau}{\partial_\tau u} \right) \cdot \mathbf{Q} \, \mathrm{d}V + \int_{\partial_q \mathcal{B}} \frac{\mathrm{w}_\tau}{\partial_\tau u} \overline{q} \, \mathrm{d}A \\
&+ \int_{\mathcal{B}} \mathbf{w}_{\mathbf{C}_p^{-1}} : \left(\dot{\overline{\mathbf{C}_p^{-1}}} + 2 \left(\boldsymbol{\mathcal{N}} : \mathbf{M} \right) \mathbf{C}_p^{-1} \right) \mathrm{d}V \, ,
\end{aligned}
\tag{6.21}
$$

where the velocity field $\mathbf{v} = \rho^{-1}\mathbf{p}$ has been introduced. In weak form (6.21), $\mathbf{w}_\varphi, \mathbf{w}_\mathbf{p} : \mathcal{B} \mapsto \mathbb{R}^3$, $\mathrm{w}_\tau : \mathcal{B} \mapsto \mathbb{R}$ and $\mathbf{w}_{\mathbf{C}_p^{-1}} : \mathcal{B} \mapsto \mathbb{R}^{3\times 3}$ are test functions that have to satisfy the boundary conditions $\mathbf{w}_\varphi = \mathbf{0}$ and $\mathbf{w}_\mathbf{p} = \mathbf{0}$ on $\partial_\varphi \mathcal{B}$, and $\mathrm{w}_\tau = 0$ on $\partial_\tau \mathcal{B}$.

6.1.3. Balance laws

We verify the pertinent balance laws of the IBVP at hand. For that purpose it suffices to consider the pure Neumann problem (i.e. $\partial_\sigma \mathcal{B} = \partial_q \mathcal{B} = \partial \mathcal{B}$). We start with the total linear momentum of the continuum body defined by $\mathbf{L} = \int_\mathcal{B} \mathbf{p}\, \mathrm{d}V$. Choosing $(\mathbf{w}_\varphi, \mathbf{w}_\mathbf{p}, \mathrm{w}_\tau, \mathbf{w}_{\mathbf{C}_p^{-1}}) = (\mathbf{0}, \boldsymbol{\xi}, 0, \mathbf{0})$, where $\boldsymbol{\xi} \in \mathbb{R}^3$ is arbitrary but constant, weak form (6.21) leads to

$$\boldsymbol{\xi} \cdot \frac{\mathrm{d}\mathbf{L}}{\mathrm{d}t} = \boldsymbol{\xi} \cdot \left(\int_\mathcal{B} \mathbf{b}\, \mathrm{d}V + \int_{\partial\mathcal{B}} \bar{\mathbf{t}}\, \mathrm{d}A \right) . \tag{6.22}$$

Due to the arbitrariness of $\boldsymbol{\xi} \in \mathbb{R}^3$, (6.22) coincides with the balance law for linear momentum. Note that the parentheses on the right-hand side of (6.22) contain the resultant external force exerted on the continuum body (see also Fig. 6.2).

The total angular momentum relative to the origin of the inertial frame is defined by $\mathbf{J} = \int_\mathcal{B} \boldsymbol{\varphi} \times \mathbf{p}\, \mathrm{d}V$. Choosing $(\mathbf{w}_\varphi, \mathbf{w}_\mathbf{p}, \mathrm{w}_\tau, \mathbf{w}_{\mathbf{C}_p^{-1}}) = (\mathbf{p} \times \boldsymbol{\xi}, \boldsymbol{\xi} \times \boldsymbol{\varphi}, 0, \mathbf{0})$, weak form (6.21) yields

$$\boldsymbol{\xi} \cdot \frac{\mathrm{d}\mathbf{J}}{\mathrm{d}t} = \boldsymbol{\xi} \cdot \left(\int_\mathcal{B} \boldsymbol{\varphi} \times \mathbf{b}\, \mathrm{d}V + \int_{\partial\mathcal{B}} \boldsymbol{\varphi} \times \bar{\mathbf{t}}\, \mathrm{d}A \right) . \tag{6.23}$$

Note that the symmetry condition $\mathbf{F}\mathbf{P}^\mathrm{T} = \mathbf{P}\mathbf{F}^\mathrm{T}$ has been employed. Since $\boldsymbol{\xi} \in \mathbb{R}^3$ is arbitrary, (6.23) corresponds to the balance of angular momentum. The parentheses on the right-hand side of (6.23) contain the resultant external torque about the origin (see also Fig. 6.2).

Next, we substitute $(\mathbf{w}_\varphi, \mathbf{w}_\mathbf{p}, \mathrm{w}_\tau, \mathbf{w}_{\mathbf{C}_p^{-1}}) = (-\mathbf{b}, \rho^{-1}\mathbf{p}, \partial_\tau u, \partial_{\mathbf{C}_p^{-1}} u)$ into weak form (6.21). A straightforward calculation leads to the balance law for total energy

$$\frac{\mathrm{d}}{\mathrm{d}t} \int_\mathcal{B} \left(\frac{1}{2} \rho^{-1} \mathbf{p} \cdot \mathbf{p} + u \right) \mathrm{d}V = \int_\mathcal{B} \mathbf{b} \cdot \dot{\boldsymbol{\varphi}}\, \mathrm{d}V + \int_{\partial\mathcal{B}} \left(\rho^{-1} \mathbf{p} \cdot \bar{\mathbf{t}} - \bar{q} \right) \mathrm{d}A . \tag{6.24}$$

For the balance of entropy we choose $(\mathbf{w}_\varphi, \mathbf{w}_\mathbf{p}, \mathrm{w}_\tau, \mathbf{w}_{\mathbf{C}_p^{-1}}) = (\mathbf{0}, \rho^{-1}\mathbf{p}, \partial_\tau \eta, \partial_{\mathbf{C}_p^{-1}} \eta)$ in the weak form (6.21) to obtain

$$\begin{aligned} 0 = &\int_\mathcal{B} \left(\partial_\mathbf{F} \eta : \dot{\mathbf{F}} + \partial_\tau \eta \dot{\tau} + 2 \partial_{\mathbf{C}_p^{-1}} \eta : \left(\left(\boldsymbol{\mathcal{N}} : \mathbf{M} \right) \mathbf{C}_p^{-1} \right) - 2 \left(\partial_{\mathbf{C}_p^{-1}} u \mathbf{C}_p^{-1} \right) : \boldsymbol{\mathcal{N}} : \mathbf{M} \right) \mathrm{d}V \\ &+ \int_\mathcal{B} \partial_{\mathbf{C}_p^{-1}} \eta : \dot{\overline{\mathbf{C}_p^{-1}}} - \nabla \left(\frac{\partial_\tau \eta}{\partial_\tau u} \right) \cdot \mathbf{Q}\, \mathrm{d}V + \int_{\partial\mathcal{B}} \frac{\partial_\tau \eta}{\partial_\tau u} \bar{q}\, \mathrm{d}A \\ = &\int_\mathcal{B} \left(\frac{\mathrm{d}\eta}{\mathrm{d}t} - \frac{1}{\Theta} \mathbf{M} : \boldsymbol{\mathcal{N}} : \mathbf{M} - \nabla \left(\frac{1}{\Theta} \right) \cdot \Theta^2 \mathbf{K} \nabla \left(\frac{1}{\Theta} \right) \right) \mathrm{d}V + \int_{\partial\mathcal{B}} \frac{1}{\Theta} \bar{q}\, \mathrm{d}A . \end{aligned}$$

Here, use has been made of formula (6.10) for the temperature along with expressions (6.16) for the material heat flux vector and (6.17) for the Mandel stress tensor, respectively. The above equation can be rewritten as

$$\frac{\mathrm{d}\mathcal{S}}{\mathrm{d}t} + \int_{\partial\mathcal{B}} \frac{1}{\Theta}\bar{q}\,\mathrm{d}A = \int_{\mathcal{B}} \underbrace{\nabla\left(\frac{1}{\Theta}\right)\cdot\Theta^2\mathbf{K}\nabla\left(\frac{1}{\Theta}\right)}_{=\mathcal{D}_{\mathrm{cond}}\geq 0} + \underbrace{\frac{1}{\Theta}\mathbf{M}:\boldsymbol{\mathcal{N}}:\mathbf{M}}_{=\mathcal{D}_{\mathrm{inel}}\geq 0}\,\mathrm{d}V \geq 0\,, \tag{6.25}$$

where $\mathcal{D}_{\mathrm{inel}}$ is the local production of entropy due to inelastic deformations and $\mathcal{D}_{\mathrm{cond}}$ is the local production of entropy due to heat conduction. The last equation corresponds to the Clausius-Duhem form of the second law of thermodynamics (see, for example, [136, Sec. 5]).

6.2. Inelastic part of the dissipative bracket

The present model for large-strain thermo-viscoelasticity relies on the introduction of the internal variable $\mathbf{C}_p^{-1} : \mathcal{B}\times\mathcal{I} \mapsto \mathbb{R}^{3\times 3}$, whose time-evolution accounts for local inelastic deformations. Following [148], $\mathbf{C}_p^{-1}$ can be associated with the multiplicative decomposition of the deformation gradient [149]

$$\mathbf{F} = \mathbf{F}_e\mathbf{F}_p\,, \tag{6.26}$$

into an elastic part $\mathbf{F}_e$ and an inelastic (or viscous) part $\mathbf{F}_p$. Decomposition (6.26) gives rise to the relationship

$$\mathbf{C}_p = \mathbf{F}_p^{\mathrm{T}}\mathbf{F}_p\,. \tag{6.27}$$

The restriction to the isotropic case implies that the free energy takes the separable form (see, for example, [150])

$$\Psi = \Psi(\mathbf{F},\tau,\mathbf{C}_p^{-1}) = \Psi^{\mathrm{eq}}(\mathbf{F},\tau) + \Psi^{\mathrm{neq}}(\mathbf{F},\tau,\mathbf{C}_p^{-1})\,, \tag{6.28}$$

where Ψ^{eq} is the equilibrium part and Ψ^{neq} is the non-equilibrium part.

Since inelastic deformations are purely irreversible in nature, they lead to a contribution to the dissipative bracket in GENERIC (6.1), cf. (6.8). To derive the inelastic dissipative bracket (6.12), we resort to the dissipative bracket derived in [99] within a temperature-based framework for GENERIC. In particular, [99] consider functionals of the form

$$\overline{\mathrm{A}}(\boldsymbol{\varphi},\mathbf{p},\theta,\mathbf{F}_p) = \int_{\mathcal{B}} \bar{\mathrm{a}}(\boldsymbol{\varphi},\mathbf{F},\mathbf{p},\theta,\mathbf{F}_p)\,\mathrm{d}V\,, \tag{6.29}$$

with associated density function $\bar{\mathrm{a}}(\boldsymbol{\varphi},\mathbf{F},\mathbf{p},\theta,\mathbf{F}_p)$. The inelastic dissipative bracket from [99] is given by

$$\left[\overline{\mathrm{A}},\overline{\mathrm{B}}\right]_{\mathrm{inel}} = \int_{\mathcal{B}} \left(\frac{\delta_\theta\bar{\mathrm{a}}}{\partial_\theta\bar{\mathrm{u}}}\partial_{\mathbf{F}_p}\bar{\mathrm{u}} - \delta_{\mathbf{F}_p}\bar{\mathrm{a}}\right) : \Theta\mathbf{N} : \left(\frac{\delta_\theta\bar{\mathrm{b}}}{\partial_\theta\bar{\mathrm{u}}}\partial_{\mathbf{F}_p}\bar{\mathrm{u}} - \delta_{\mathbf{F}_p}\bar{\mathrm{b}}\right)\mathrm{d}V\,, \tag{6.30}$$

where $\overline{\mathrm{u}}(\mathbf{F},\theta,\mathbf{F}_p)$ is the internal energy density and $\mathbf{N}$ is a fourth-order inelastic flow tensor which has the properties $\mathbf{N}^{\mathrm{T}} = \mathbf{N}$ (major symmetry) and $\mathbf{A} : \mathbf{N} : \mathbf{A} \geq 0$ for all $\mathbf{A} \in \mathbb{R}^{3\times 3}$. In components, these properties read

$$(\mathbf{N})^{\alpha\cdot\beta\cdot}_{\cdot I\cdot J} = (\mathbf{N})^{\beta\cdot\alpha\cdot}_{\cdot J\cdot I} \qquad \text{and} \qquad (\mathbf{A})^{\cdot I}_{\alpha\cdot}\,(\mathbf{N})^{\alpha\cdot\beta\cdot}_{\cdot I\cdot J}\,(\mathbf{A})^{\cdot J}_{\beta\cdot} \geq 0\,. \tag{6.31}$$

As has been shown in [99], inelastic dissipative bracket (6.30) comes along with the flow rule

$$\dot{\mathbf{F}}_p = -\mathbf{N} : \mathbf{\Sigma}\,,$$

where

$$\mathbf{\Sigma} = \partial_{\mathbf{F}_p}\overline{\mathrm{u}} - \Theta\partial_{\mathbf{F}_p}\overline{\eta}\,.$$

In the last equation, $\overline{\eta}(\mathbf{F},\theta,\mathbf{F}_p)$ is the entropy density of the temperature-based formulation.

6.2.1. Change of variables

We perform a change of variables to transform the inelastic bracket (6.30) to the present setting which is based on functionals of the form (6.4). In particular, to link the current density functions $a(\boldsymbol{\varphi},\mathbf{F},\mathbf{p},\tau,\mathbf{C}_p^{-1})$ to those in (6.29), we express the generalized thermodynamic state variable $\tau \in \{\theta,\eta,u\}$ in terms of the temperature by inverting relation (6.10) to get

$$\tau = \overline{\tau}(\mathbf{F},\theta,\mathbf{C}_p^{-1})\,. \tag{6.32}$$

Moreover, relation (6.27) implies

$$\mathbf{C}_p^{-1} = \mathbf{F}_p^{-1}\mathbf{F}_p^{-\mathrm{T}}\,. \tag{6.33}$$

Now, the two density functions under consideration can be connected through

$$\overline{\mathrm{a}}(\boldsymbol{\varphi},\mathbf{F},\mathbf{p},\theta,\mathbf{F}_p) = a(\boldsymbol{\varphi},\mathbf{F},\mathbf{p},\overline{\tau}(\mathbf{F},\theta,\mathbf{C}_p^{-1}),\mathbf{C}_p^{-1})\,,$$

where relationships (6.32) and (6.33) are employed on the right-hand side of the last equation. A straightforward application of the chain rule to the last equation yields

$$\begin{aligned} \partial_\theta\overline{\mathrm{a}} &= \partial_\tau a\,\partial_\theta\overline{\tau}\,, \\ \partial_{\mathbf{F}_p}\overline{\mathrm{a}} &= \partial_{\mathbf{F}_p}a + \partial_\tau a\,\partial_{\mathbf{F}_p}\overline{\tau}\,. \end{aligned} \tag{6.34}$$

Note that $\delta_\theta\overline{\mathrm{a}} = \partial_\theta\overline{\mathrm{a}}$, and $\delta_{\mathbf{F}_p}\overline{\mathrm{a}} = \partial_{\mathbf{F}_p}\overline{\mathrm{a}}$. Furthermore, with regard to (6.10) and (6.32) we have $\theta = \Theta(\mathbf{F},\overline{\tau}(\mathbf{F},\theta,\mathbf{C}_p^{-1}),\mathbf{C}_p^{-1})$, from which follows that

$$\begin{aligned} 1 &= \partial_\theta\Theta = \partial_\tau\Theta\,\partial_\theta\overline{\tau}\,, \\ \mathbf{0} &= \partial_{\mathbf{F}_p}\Theta = \partial_\tau\Theta\,\partial_{\mathbf{F}_p}\overline{\tau} + \partial_{\mathbf{F}_p}\Theta\,. \end{aligned}$$

We thus obtain

$$\partial_\theta \overline{\mathcal{T}} = \frac{1}{\partial_\mathcal{T}\Theta} \,, \qquad \partial_{\mathbf{F}_p} \overline{\mathcal{T}} = -\frac{\partial_{\mathbf{F}_p}\Theta}{\partial_\mathcal{T}\Theta} \,. \tag{6.35}$$

Substituting from (6.35) into (6.34) yields

$$\partial_\theta \bar{\mathrm{a}} = \frac{\partial_\mathcal{T} a}{\partial_\mathcal{T}\Theta} \,, \qquad \partial_{\mathbf{F}_p} \bar{\mathrm{a}} = \partial_{\mathbf{F}_p} a - \frac{\partial_\mathcal{T} a}{\partial_\mathcal{T}\Theta} \partial_{\mathbf{F}_p}\Theta \,. \tag{6.36}$$

Note that $\delta_\mathcal{T} a = \partial_\mathcal{T} a$ and $\delta_{\mathbf{F}_p} a = \partial_{\mathbf{F}_p} a$. Making use of (6.36), the terms in the parenthesis of (6.30) can be rewritten as

$$\frac{\delta_\theta \bar{\mathrm{a}}}{\partial_\theta \bar{\mathrm{u}}} \partial_{\mathbf{F}_p} \bar{\mathrm{u}} - \delta_{\mathbf{F}_p} \bar{\mathrm{a}} = \frac{\delta_\mathcal{T} a}{\partial_\mathcal{T} u} \partial_{\mathbf{F}_p} u - \delta_{\mathbf{F}_p} a \,. \tag{6.37}$$

Taking into account the relationship

$$\partial_{\mathbf{F}_p} a = -2\mathbf{F}_p^{-\mathrm{T}} \partial_{\mathbf{C}_p^{-1}} a \mathbf{C}_p^{-1} \,, \tag{6.38}$$

the inelastic dissipative bracket (6.30) can be recast in the form

$$\begin{aligned} &[\mathcal{A}, \mathcal{B}]_{\text{inel}} \\ &= \int_\mathcal{B} 2\left(\frac{\delta_\mathcal{T} a}{\partial_\mathcal{T} u} \partial_{\mathbf{C}_p^{-1}} u \mathbf{C}_p^{-1} - \delta_{\mathbf{C}_p^{-1}} a \mathbf{C}_p^{-1}\right) : \Theta \boldsymbol{\mathcal{N}} : 2\left(\frac{\delta_\mathcal{T} b}{\partial_\mathcal{T} u} \partial_{\mathbf{C}_p^{-1}} u \mathbf{C}_p^{-1} - \delta_{\mathbf{C}_p^{-1}} b \mathbf{C}_p^{-1}\right) \mathrm{d}V \,. \end{aligned}$$

This bracket coincides with the one introduced in (6.12). Note that in the above formula, $\delta_{\mathbf{C}_p^{-1}} a = \partial_{\mathbf{C}_p^{-1}} a$. Moreover, $\boldsymbol{\mathcal{N}}$ is the material form of the fourth-order flow tensor $\mathbf{N}$ in (6.30). In components,

$$(\boldsymbol{\mathcal{N}})^{I\,\cdot\,K\,\cdot}_{\,\cdot\,J\,\cdot\,L} = (\mathbf{F}_p^{-1})^{I\,\cdot}_{\,\cdot\,\alpha} (\mathbf{F}_p^{-1})^{K\,\cdot}_{\,\cdot\,\gamma} (\mathbf{N})^{\alpha\,\cdot\,\gamma\,\cdot}_{\,\cdot\,J\,\cdot\,L} \,. \tag{6.39}$$

Accordingly, the material flow tensor inherits symmetry and positive semi-definiteness. That is, in components,

$$(\boldsymbol{\mathcal{N}})^{I\,\cdot\,K\,\cdot}_{\,\cdot\,J\,\cdot\,L} = (\boldsymbol{\mathcal{N}})^{K\,\cdot\,I\,\cdot}_{\,\cdot\,L\,\cdot\,J} \quad \text{and} \quad (\mathbf{M})_{I\,\cdot}^{\,\cdot\,J} (\boldsymbol{\mathcal{N}})^{I\,\cdot\,K\,\cdot}_{\,\cdot\,J\,\cdot\,L} (\mathbf{M})_{K\,\cdot}^{\,\cdot\,L} \geq 0 \tag{6.40}$$

for all $\mathbf{M}$ introduced in eq. (6.17). As has been shown in Section 6.1.1, the above inelastic dissipative bracket gives rise to flow rules of the form (cf. $(6.20)_4$)

$$\dot{\overline{\mathbf{C}_p^{-1}}} = -2\,(\boldsymbol{\mathcal{N}} : \mathbf{M})\,\mathbf{C}_p^{-1} \,. \tag{6.41}$$

It is worth noting that this flow rule can be viewed as material version of the viscoelastic evolution equation

$$\mathcal{L}_v \mathbf{b}_e = -2\left(\boldsymbol{\mathcal{V}}^{-1} : \boldsymbol{\tau}^{\mathrm{neq}}\right)\mathbf{b}_e\,, \tag{6.42}$$

derived in [148, 150] and being applied in, e.g., [151]. In the last equation, $\mathbf{b}_e = \mathbf{F}_e\mathbf{F}_e^{\mathrm{T}} = \mathbf{F}\mathbf{C}_p^{-1}\mathbf{F}^{\mathrm{T}}$, $\mathcal{L}_v\mathbf{b}_e = \mathbf{F}\dot{\overline{\mathbf{C}_p^{-1}}}\mathbf{F}^{\mathrm{T}}$, $\boldsymbol{\tau}^{\mathrm{neq}} = 2\mathbf{F}\partial_{\mathbf{C}}\Psi^{\mathrm{neq}}\mathbf{F}^{\mathrm{T}}$, and $\boldsymbol{\mathcal{V}}^{-1}$ is an isotropic, positive definite fourth-order tensor. Using these relationships, flow rule (6.42) can be recast in the form

$$\dot{\overline{\mathbf{C}_p^{-1}}} = -2\mathbf{F}^{-1}\left(\boldsymbol{\mathcal{V}}^{-1} : \left(\mathbf{F}^{-\mathrm{T}}\mathbf{M}\mathbf{F}^{\mathrm{T}}\right)\right)\mathbf{F}\mathbf{C}_p^{-1}\,, \tag{6.43}$$

where the relation

$$2\mathbf{C}\partial_{\mathbf{C}}\Psi^{\mathrm{neq}} = 2\partial_{\mathbf{C}_p^{-1}}\Psi^{\mathrm{neq}}\mathbf{C}_p^{-1} = \mathbf{M}\,,$$

has been employed. Note that in the last equation, definition (6.17) of the Mandel stress tensor has been taken into account. Comparing (6.43) with present flow rule (6.41), leads to the conclusion that the respective fourth-order flow tensors are related by

$$(\boldsymbol{\mathcal{N}})^{I\cdot K\cdot}_{\cdot J\cdot L} = (\mathbf{F}^{-1})^{I\cdot}_{\cdot a}(\mathbf{F})^{b\cdot}_{\cdot J}(\mathbf{F}^{-1})^{K\cdot}_{\cdot c}(\mathbf{F})^{d\cdot}_{\cdot L}(\boldsymbol{\mathcal{V}}^{-1})^{a\cdot c\cdot}_{\cdot b\cdot d}\,.$$

We further remark that the present viscoelastic evolution equation (6.41) can also be brought into the form

$$\dot{\mathbf{C}}_p = 2\mathbf{C}_p\boldsymbol{\mathcal{N}} : \mathbf{M}\,.$$

This version of the viscoelastic evolution equation has been used in [106], see also [76]. Thus, we conclude that the newly proposed inelastic dissipative bracket (6.12) gives rise to evolution equations for the internal variables that have been previously developed in the context of finite deformation thermo-viscoelasticity.

6.3. Discretization in space

Concerning the space discretization of the present GENERIC-based formulation, we essentially apply an isoparametric finite element approach (see, for example, [145]). Our main goal is to achieve a GENERIC-consistent space discretization in the sense of Section 2.1.2. In particular, a GENERIC-consistent space discretization ensures that the discrete formulation inherits the fundamental balance laws for both the energy and the entropy from the underlying continuous formulation (cf. Section 6.1.3).

We first restate the governing equations of the IBVP to be discretized in space and time. With regard to the GENERIC-based weak form (6.21), we consider the following set of equations:

$$
\begin{aligned}
0 &= \dot{\varphi} - \mathbf{v}\,, \\
0 &= \dot{\overline{\mathbf{C}_p^{-1}}} + 2\left(\boldsymbol{\mathcal{N}} : \mathbf{M}\right)\mathbf{C}_p^{-1}\,, \\
0 &= \int_{\mathcal{B}} \mathrm{w}_{u_\mathcal{T}} \left(\partial_\mathcal{T} u - u_\mathcal{T}\right)\,\mathrm{d}V\,, \\
0 &= \int_{\mathcal{B}} \mathrm{w}_{\eta_\mathcal{T}} \left(\partial_\mathcal{T} \eta - \eta_\mathcal{T}\right)\,\mathrm{d}V\,, \\
0 &= \int_{\mathcal{B}} \left(\mathbf{w}_\mathbf{p}\cdot\left(\rho\dot{\mathbf{v}} - \mathbf{b}\right) + \mathbf{P} : \nabla\mathbf{w}_\mathbf{p}\right)\mathrm{d}V - \int_{\partial_\sigma\mathcal{B}} \mathbf{w}_\mathbf{p}\cdot\bar{\mathbf{t}}\,\mathrm{d}A\,, \\
0 &= \int_{\mathcal{B}} \mathrm{w}_\mathcal{T}\left(\dot{\mathcal{T}} - \frac{2}{u_\mathcal{T}}\left(\partial_{\mathbf{C}_p^{-1}} u\,\mathbf{C}_p^{-1}\right) : \boldsymbol{\mathcal{N}} : \mathbf{M} + \frac{1}{\eta_\mathcal{T}}\partial_\mathbf{F}\eta : \nabla\mathbf{v}\right)\mathrm{d}V \\
&\quad - \int_{\mathcal{B}} \nabla\left(\frac{\mathrm{w}_\mathcal{T}}{u_\mathcal{T}}\right)\cdot\mathbf{Q}\,\mathrm{d}V + \int_{\partial_q\mathcal{B}} \frac{\mathrm{w}_\mathcal{T}}{u_\mathcal{T}}\bar{q}\,\mathrm{d}A\,\mathrm{d}V\,.
\end{aligned}
\tag{6.44}
$$

Here, the first equation represents the local form of the kinematic relationship $\dot{\varphi} = \mathbf{v}$, while the second equation is the viscoelastic evolution equation. The third and fourth equation serve the purpose to introduce the new fields $u_\mathcal{T}$ and $\eta_\mathcal{T}$. This procedure facilitates a mixed finite element approach which turns out to be crucial for a GENERIC-consistent discrete formulation. Due to the arbitrariness of the test functions $\mathrm{w}_{u_\mathcal{T}}$ and $\mathrm{w}_{\eta_\mathcal{T}}$, $(6.44)_3$ and $(6.44)_4$ impose the conditions $u_\mathcal{T} = \partial_\mathcal{T} u$ and $\eta_\mathcal{T} = \partial_\mathcal{T}\eta$, respectively. Moreover,

$$
\begin{aligned}
\mathbf{P} &= \partial_\mathbf{F} u - \Theta\partial_\mathbf{F}\eta\,, \\
\mathbf{M} &= 2\left(\partial_{\mathbf{C}_p^{-1}} u - \Theta\partial_{\mathbf{C}_p^{-1}}\eta\right)\mathbf{C}_p^{-1}
\end{aligned}
\tag{6.45}
$$

are the GENERIC-specific representations of the first Piola-Kirchhoff stress tensor and the material Mandel stress tensor previously introduced in (6.14) and (6.17), respectively. Similarly, the material heat flux vector $\mathbf{Q}$ has been introduced in (6.16). We emphasize again that the GENERIC-based formulation is based on expression (6.10) for the temperature field. In the present mixed formulation this implies

$$
\Theta = \frac{u_\mathcal{T}}{\eta_\mathcal{T}}\,. \tag{6.46}
$$

The finite element method is based on finite dimensional approximations of the following quantities

$$
\begin{aligned}
\varphi^\mathrm{h} &= N^a\mathbf{q}_a\,, \\
\mathbf{v}^\mathrm{h} &= N^a\mathbf{v}_a\,, \\
\mathcal{T}^\mathrm{h} &= N^a\mathcal{T}_a\,,
\end{aligned}
\tag{6.47}
$$

and

$$
\begin{aligned}
u_\tau^{\mathrm{h}} &= N^a (u_\tau)_a \;, \\
\eta_\tau^{\mathrm{h}} &= N^a (\eta_\tau)_a \;.
\end{aligned}
\tag{6.48}
$$

As before, the summation convention applies, where $a = 1, \ldots, \mathrm{N}$, and N denotes the total number of nodes in the finite element mesh. Moreover, $N^a : \mathcal{B} \rightarrow \mathbb{R}$ are the nodal shape functions with associated nodal values $\mathbf{q}_a, \mathbf{v}_a \in \mathbb{R}^3$ and $\tau_a, (u_\tau)_a, (\eta_\tau)_a \in \mathbb{R}$. Analogous approximations are used for the test functions $\mathrm{w}_{u_\tau}, \mathrm{w}_{\eta_\tau}, \mathbf{w}_{\mathrm{p}}$, and w_τ, denoted by $\mathrm{w}_{u_\tau}^{\mathrm{h}}, \mathrm{w}_{\eta_\tau}^{\mathrm{h}}, \mathbf{w}_{\mathrm{p}}^{\mathrm{h}}$, and w_τ^{h}.

In what follows, we summarize the space-discrete version of (6.44). Nodal collocation of kinematic equation $(6.44)_1$ yields

$$
\dot{\mathbf{q}}_a = \mathbf{v}_a \;, \tag{6.49}
$$

for $a = 1, \ldots, \mathrm{N}$. Viscoelastic evolution equation $(6.44)_2$ is collocated at the integration points $\mathbf{X}_g \in \mathcal{B}$ used for the numerical evaluation of the volume integrals in (6.44). Accordingly,

$$
\overline{(\mathbf{C}_p^{-1})}^{\boldsymbol{\cdot}}_g = -2 \left(\boldsymbol{\mathcal{N}}_g : \mathbf{M}_g \right) (\mathbf{C}_p^{-1})_g \;, \tag{6.50}
$$

for $g = 1, \ldots, \mathrm{G}$, where G denotes the total number of integration points. Here and in the sequel, index g indicates evaluation at the integration point $\mathbf{X}_g \in \mathcal{B}$. For example,

$$
\begin{aligned}
(\mathbf{C}_p^{-1})_g &= \mathbf{C}_p^{-1}(\mathbf{X}_g, t) \;, \\
\mathbf{M}_g &= 2 \left(\partial_{\mathbf{C}_p^{-1}} u_g - \Theta_g \partial_{\mathbf{C}_p^{-1}} \eta_g \right) (\mathbf{C}_p^{-1})_g \;,
\end{aligned}
\tag{6.51}
$$

where

$$
\begin{aligned}
u_g &= u \left(\mathbf{F}_g^{\mathrm{h}}, \tau_g^{\mathrm{h}}, (\mathbf{C}_p^{-1})_g \right) \;, \\
\eta_g &= \eta \left(\mathbf{F}_g^{\mathrm{h}}, \tau_g^{\mathrm{h}}, (\mathbf{C}_p^{-1})_g \right) \;,
\end{aligned}
\tag{6.52}
$$

are the internal energy and entropy densities at point $\mathbf{X}_g$. Furthermore, the discrete deformation gradient at $\mathbf{X}_g$, $\mathbf{F}_g^{\mathrm{h}}$, and the discrete generalized thermal variable τ_g^{h}, follow from $(6.47)_{1,3}$ and thus take the form

$$
\begin{aligned}
\mathbf{F}_g^{\mathrm{h}} &= \mathbf{q}_a(t) \otimes \nabla N^a(\mathbf{X}_g) \;, \\
\tau_g^{\mathrm{h}} &= N^a(\mathbf{X}_g) \tau_a(t) \;.
\end{aligned}
\tag{6.53}
$$

Similarly, the discrete temperature at $\mathbf{X}_g$ follows from (6.46) and is given by

$$
\Theta_g = \frac{(u_\tau^{\mathrm{h}})_g}{(\eta_\tau^{\mathrm{h}})_g} \;, \tag{6.54}
$$

where, in view of interpolations (6.48), $(u_\tau^{\mathrm{h}})_g = u_\tau^{\mathrm{h}}(\mathbf{X}_g, t)$ and $(\eta_\tau^{\mathrm{h}})_g = \eta_\tau^{\mathrm{h}}(\mathbf{X}_g, t)$. In the discrete setting, the fields u_τ^{h} and η_τ^{h} are determined through $(6.44)_{3,4}$. In particular, inserting the approximations (6.48) along with the corresponding formulas for $\mathrm{w}^{\mathrm{h}}_{u_\tau}$, $\mathrm{w}^{\mathrm{h}}_{\eta_\tau}$ into $(6.44)_{3,4}$, we obtain

$$
\begin{aligned}
0 &= \sum_{g=1}^{G} N^a(\mathbf{X}_g)\left(\partial_\tau u_g - N^b(\mathbf{X}_g)(u_\tau)_b\right) w_g \ , \\
0 &= \sum_{g=1}^{G} N^a(\mathbf{X}_g)\left(\partial_\tau \eta_g - N^b(\mathbf{X}_g)(u_\tau)_b\right) w_g \ ,
\end{aligned}
\tag{6.55}
$$

for $a = 1, \ldots, \mathrm{N}$. To calculate the spatial integrals, appropriate numerical quadrature formulas of the form

$$
\int_{\mathcal{B}} f(\mathbf{X})\, \mathrm{d}V \approx \sum_{g=1}^{G} f(\mathbf{X}_g) w_g \ , \tag{6.56}
$$

have been applied to obtain (6.55)[2]. Here, w_g play the role of generalized weighting coefficients resulting from the specific quadrature rule along with the isoparametric description of reference domain $\mathcal{B}$. Now, $(6.55)_1$ can be solved for the nodal quantities $(u_\tau)_a$, $a = 1, \ldots, \mathrm{N}$. To this end, we introduce the components H^{ab} of the positive definite Gram matrix $[H^{ab}]$,

$$
H^{ab} = \sum_{g=1}^{G} N^a(\mathbf{X}_g) N^b(\mathbf{X}_g) w_g \ , \tag{6.57}
$$

so that $(6.55)_1$ can be rewritten as

$$
H^{ab}(u_\tau)_b = \sum_{g=1}^{G} N^a(\mathbf{X}_g) \partial_\tau u_g w_g \ . \tag{6.58}
$$

The components H_{ab} of the inverse Gram matrix, $[H_{ab}] = [H^{ab}]^{-1}$, satisfy the relationship .

$$
H_{ab} H^{bc} = \delta_{a \cdot}^{\cdot c} \ , \tag{6.59}
$$

where $\delta_{a \cdot}^{\cdot c}$ denotes the Kronecker delta. Now, interpolation formula $(6.48)_1$ along with (6.58) lead to the result

$$
u_\tau^{\mathrm{h}} = N^a H_{ab} \sum_{g=1}^{G} N^b(\mathbf{X}_g) \partial_\tau u_g w_g \ , \tag{6.60}
$$

[2] The summation over g will always be stated explicitly, so that the summation convention does not apply to index g.

which will be utilized below. Similarly, interpolation formula $(6.48)_2$ in conjunction with $(6.55)_2$ yield the result

$$\eta_\tau^\mathrm{h} = N^a H_{ab} \sum_{g=1}^{G} N^b(\mathbf{X}_g)\partial_\tau \eta_g w_g \,. \tag{6.61}$$

Next, we turn to the discretization of $(6.44)_5$, which can be done in a straightforward way to obtain

$$M^{ab}\dot{\mathbf{v}}_b - \sum_{g=1}^{G} N^a(\mathbf{X}_g)\mathbf{b}_g w_g + \sum_{g=1}^{G} \mathbf{P}_g \nabla N^a(\mathbf{X}_g) w_g = 0 \,. \tag{6.62}$$

For simplicity, we have neglected the contribution of the external tractions which could be easily added to the above equation. In the above equation, first Piola-Kirchhoff stress tensor $\mathbf{P}_g$ at point $\mathbf{X}_g \in \mathcal{B}$ is given by

$$\mathbf{P}_g = \partial_\mathbf{F} u_g - \Theta_g \partial_\mathbf{F} \eta_g \,. \tag{6.63}$$

Moreover, in (6.62), the components M^{ab} of the mass matrix are given by

$$M^{ab} = \sum_{g=1}^{G} \rho(\mathbf{X}_g) N^a(\mathbf{X}_g) N^b(\mathbf{X}_g) w_g \,. \tag{6.64}$$

We further introduce nodal momentum vectors $\mathbf{p}^a$ conjugate to nodal position vectors $\mathbf{q}_a$ through the standard relation

$$\mathbf{p}^a = M^{ab}\mathbf{v}_b \,. \tag{6.65}$$

Eventually, we consider the space-discrete version of $(6.44)_6$. Straight-forward application of our approach yields

$$\begin{aligned} 0 =& H^{ab}\dot{\tau}_b - \sum_{g=1}^{G} \frac{2N^a(\mathbf{X}_g)}{(u_\tau^\mathrm{h})_g} \left(\partial_{\mathbf{C}_p^{-1}} u_g (\mathbf{C}_p^{-1})_g\right) : \boldsymbol{\mathcal{N}}_g : \mathbf{M}_g w_g \\ &+ \mathbf{v}_c \cdot \sum_{g=1}^{G} \frac{N^a(\mathbf{X}_g)}{(\eta_\tau^\mathrm{h})_g} \partial_\mathbf{F} \eta_g \nabla N^c(\mathbf{X}_g) w_g - \sum_{g=1}^{G} \nabla\left(\frac{N^a(\mathbf{X}_g)}{(u_\tau^\mathrm{h})_g}\right) \cdot \mathbf{Q}_g w_g \,, \end{aligned} \tag{6.66}$$

where the material heat flux vector $\mathbf{Q}_g$ at point $\mathbf{X}_g \in \mathcal{B}$ is given by

$$\mathbf{Q}_g = \Theta_g^2 \mathbf{K}_g \nabla\left(\frac{(\eta_\tau^\mathrm{h})_g}{(u_\tau^\mathrm{h})_g}\right) \,.$$

For simplicity, in the space-discrete evolution equation (6.66) for the generalized nodal thermal variable τ_b, the term accounting for heat transfer across the boundary has been neglected. To summarize, the resulting evolution equations for the space-discrete system at hand can be written in the form

$$
\begin{aligned}
\dot{\mathbf{q}}_a =& M_{ab}\mathbf{p}^b \,, \\
\dot{\mathbf{p}}^a =& \sum_{g=1}^{G} N^a(\mathbf{X}_g)\mathbf{b}_g w_g - \sum_{g=1}^{G} \mathbf{P}_g \nabla N^a(\mathbf{X}_g) w_g \,, \\
\dot{\tau}_a =& H_{ab} \sum_{g=1}^{G} \frac{2N^b(\mathbf{X}_g)}{(u_\tau^{\mathrm{h}})_g} \left(\partial_{\mathbf{C}_p^{-1}} u_g (\mathbf{C}_p^{-1})_g \right) : \boldsymbol{\mathcal{N}}_g : \mathbf{M}_g w_g \\
& - \mathbf{v}_c \cdot H_{ab} \sum_{g=1}^{G} \frac{N^b(\mathbf{X}_g)}{(\eta_\tau^{\mathrm{h}})_g} \partial_{\mathbf{F}} \eta_g \nabla N^c(\mathbf{X}_g) w_g + H_{ab} \sum_{g=1}^{G} \nabla \left(\frac{N^b(\mathbf{X}_g)}{(u_\tau^{\mathrm{h}})_g} \right) \mathbf{Q}_g w_g \,, \\
\dot{\overline{(\mathbf{C}_p^{-1})}}_g =& -2 \left(\boldsymbol{\mathcal{N}}_g : \mathbf{M}_g \right) (\mathbf{C}_p^{-1})_g \,,
\end{aligned}
\tag{6.67}
$$

for $1, \ldots, N$ and $1, \ldots, G$. In $(6.67)_1$, M_{ab} stands for the components of the inverse mass matrix satisfying $M_{ab}M^{bc} = \delta_{a\cdot}^{\cdot c}$. The set of equations (6.67) constitute nonlinear first-order ordinary differential equations for the determination of the unknowns which can be collected in the state vector

$$
\mathbf{z} = \left(\mathbf{q}_1, \ldots, \mathbf{q}_N, \mathbf{p}^1, \ldots, \mathbf{p}^N, \tau_1, \ldots \tau_N, (\mathbf{C}_p^{-1})_1^{AB}, \ldots, (\mathbf{C}_p^{-1})_G^{AB} \right) . \tag{6.68}
$$

In the sequel, state vector (6.68) will be viewed as column vector. In particular, this implies that the six independent components $(\mathbf{C}_p^{-1})_g^{AB}$, $g = 1, \ldots, G$, of the internal variable $(\mathbf{C}_p^{-1})_g$ (at quadrature point $\mathbf{X}_g$) are arranged in a column vector.

The set of evolution equations (6.67) fits into the GENERIC framework for discrete systems, as shown next.

6.3.1. GENERIC-consistent space discretization

Our discretization approach presented above is GENERIC-consistent in the sense of [2] and thus can be framed in the context of GENERIC for discrete systems. To see this, the set of evolution equations (6.67) needs to be put into the form

$$
\dot{\mathbf{z}} = \mathbf{L}\nabla\mathcal{E}(\mathbf{z}) + \mathbf{M}\nabla\mathcal{S}(\mathbf{z}) \,. \tag{6.69}
$$

Here, the focus is again on closed systems in which the boundary contributions are disregarded. In analogy to (6.1), the time-evolution of the state vector is decomposed additively into a reversible part generated by the total energy $\mathcal{E}$ and an irreversible part generated by the total entropy $\mathcal{S}$. Poisson matrix $\mathbf{L}$ needs be skew-symmetric, while friction matrix $\mathbf{M}$ needs be symmetric positive semi-definite.

In the above equation, the total energy of the discrete system under consideration is given by

$$\mathcal{E}(\mathbf{z}) = \frac{1}{2} M_{ab} \mathbf{p}^a \cdot \mathbf{p}^b + \sum_{g=1}^{G} \left[u \left(\mathbf{F}_g^{\mathrm{h}}, \tau_g^{\mathrm{h}}, (\mathbf{C}_p^{-1})_g \right) - \mathbf{q}_a N^a(\mathbf{X}_g) \mathbf{b}(\mathbf{X}_g) \right] w_g \,. \tag{6.70}$$

This is the space-discrete version of total energy (6.11). Similarly, the space-discrete version of total entropy (6.7) takes the form

$$\mathcal{S}(\mathbf{z}) = \sum_{g=1}^{G} \eta \left(\mathbf{F}_g^{\mathrm{h}}, \tau_g^{\mathrm{h}}, (\mathbf{C}_p^{-1})_g \right) w_g \,. \tag{6.71}$$

To get the gradient of the total energy, $\nabla \mathcal{E}(\mathbf{z})$, we consider the derivative of (6.70) with respect to time. Accordingly, the left-hand side of (6.70) yields

$$\frac{\mathrm{d}}{\mathrm{d}t} \mathcal{E}(\mathbf{z}) = \partial_{\mathbf{q}_a} \mathcal{E} \cdot \dot{\mathbf{q}}_a + \partial_{\mathbf{p}^a} \mathcal{E} \cdot \dot{\mathbf{p}}^a + \partial_{\tau_a} \mathcal{E} \cdot \dot{\tau}_a + \sum_{g=1}^{G} \partial_{(\mathbf{C}_p^{-1})_g^{AB}} \mathcal{E} \, \overline{(\mathbf{C}_p^{-1})}{}_g^{\!\!\cdot\, AB} \,, \tag{6.72}$$

while the right-hand side of (6.70) gives

$$\begin{aligned} \frac{\mathrm{d}}{\mathrm{d}t} \mathcal{E}(\mathbf{z}) = {} & \dot{\mathbf{p}}^a \cdot \mathbf{p}^b M_{ab} + \dot{\mathbf{q}}_a \cdot \sum_{g=1}^{G} \left[\partial_{\mathbf{F}} u_g \nabla N^a(\mathbf{X}_g) - N^a(\mathbf{X}_g) \mathbf{b}(\mathbf{X}_g) \right] w_g \\ & + \dot{\tau}_a \sum_{g=1}^{G} N^a(\mathbf{X}_g) \partial_\tau u_g w_g + \sum_{g=1}^{G} w_g \partial_{\mathbf{C}_p^{-1}} u_g : \overline{(\mathbf{C}_p^{-1})}{}^{\!\cdot}_g \,. \end{aligned} \tag{6.73}$$

Comparing (6.72) with (6.73) yields the following expressions for the respective derivatives of the total energy

$$\begin{aligned} \partial_{\mathbf{q}_a} \mathcal{E} &= \sum_{g=1}^{G} \left[\partial_{\mathbf{F}} u_g \nabla N^a(\mathbf{X}_g) - N^a(\mathbf{X}_g) \mathbf{b}(\mathbf{X}_g) \right] w_g \,, \\ \partial_{\mathbf{p}^a} \mathcal{E} &= M_{ab} \mathbf{p}^b \,, \\ \partial_{\tau_a} \mathcal{E} &= \sum_{g=1}^{G} N^a(\mathbf{X}_g) \partial_\tau u_g w_g \,, \\ \frac{\partial \mathcal{E}}{\partial (\mathbf{C}_p^{-1})_g^{AB}} &= w_g \frac{\partial u_g}{\partial (\mathbf{C}_p^{-1})^{AB}} \,. \end{aligned} \tag{6.74}$$

In particular, inserting from $(6.74)_3$ into (6.60), we obtain the important relationship

$$u_\tau^{\mathrm{h}} = N^a H_{ab} \partial_{\tau_b} \mathcal{E} \,. \tag{6.75}$$

Similarly, the derivatives of total entropy (6.71) take the form

$$
\begin{aligned}
\partial_{\mathbf{q}_a}\mathcal{S} &= \sum_{g=1}^{G} \partial_{\mathbf{F}}\eta_g \nabla N^a(\mathbf{X}_g) w_g \ , \\
\partial_{\mathbf{p}^a}\mathcal{S} &= \mathbf{0} \ , \\
\partial_{\tau_a}\mathcal{S} &= \sum_{g=1}^{G} N^a(\mathbf{X}_g)\partial_{\tau}\eta_g w_g \ , \\
\frac{\partial \mathcal{S}}{\partial (\mathbf{C}_p^{-1})_g^{AB}} &= w_g \frac{\partial \eta_g}{\partial (\mathbf{C}_p^{-1})^{AB}} \ .
\end{aligned}
\tag{6.76}
$$

Inserting from (6.76)3 into (6.61), we obtain

$$
\eta_\tau^{\mathrm{h}} = N^a H_{ab} \partial_{\tau_b}\mathcal{S} \ . \tag{6.77}
$$

Now, guided by discrete GENERIC (6.69), the evolution equations in (6.67) can be recast. In particular, taking into account (6.74)2, kinematic relation (6.67)1 can be rewritten as

$$
\dot{\mathbf{q}}_a = \partial_{\mathbf{p}^a}\mathcal{E} \ . \tag{6.78}
$$

Next, evolution equation (6.67)2 for the nodal momentum vectors together with (6.63) and (6.54) yields

$$
\dot{\mathbf{p}}^a = -\sum_{g=1}^{G} \left[\partial_{\mathbf{F}} u_g \nabla N^a(\mathbf{X}_g) - N^a(\mathbf{X}_g)\mathbf{b}(\mathbf{X}_g)\right] w_g + \sum_{g=1}^{G} \frac{(u_\tau^{\mathrm{h}})_g}{(\eta_\tau^{\mathrm{h}})_g} \partial_{\mathbf{F}}\eta_g \nabla N^a(\mathbf{X}_g) w_g \ .
$$

Taking into account (6.74)1 and (6.75), we obtain

$$
\dot{\mathbf{p}}^a = -\partial_{\mathbf{q}_a}\mathcal{E} + \partial_{\tau_b}\mathcal{E}\, H_{bc} \sum_{g=1}^{G} \frac{N^c(\mathbf{X}_g)}{(\eta_\tau^{\mathrm{h}})_g} \partial_{\mathbf{F}}\eta_g \nabla N^a(\mathbf{X}_g) w_g \ . \tag{6.79}
$$

Concerning the evolution of the nodal thermal variable τ_a, (6.67)3 along with (6.51)2, (6.54) and (6.77) lead to

$$
\begin{aligned}
\dot{\tau}_a =& H_{ab} \sum_{g=1}^{G} \frac{4N^b(\mathbf{X}_g)N^c(\mathbf{X}_g)}{(u_\tau^{\mathrm{h}})_g(\eta_\tau^{\mathrm{h}})_g} \left(\partial_{\mathbf{C}_p^{-1}} u_g (\mathbf{C}_p^{-1})_g\right) : \boldsymbol{\mathcal{N}}_g : \left(\partial_{\mathbf{C}_p^{-1}} u_g (\mathbf{C}_p^{-1})_g\right) w_g\, H_{cd}\partial_{\tau_d}\mathcal{S} \\
&- H_{ab} \sum_{g=1}^{G} \frac{4N^b(\mathbf{X}_g)}{(\eta_\tau^{\mathrm{h}})_g} \left(\partial_{\mathbf{C}_p^{-1}} \eta_g (\mathbf{C}_p^{-1})_g\right) : \boldsymbol{\mathcal{N}}_g : \left(\partial_{\mathbf{C}_p^{-1}} u_g (\mathbf{C}_p^{-1})_g\right) w_g \\
&- \partial_{\mathbf{p}^c}\mathcal{E} \cdot H_{ab} \sum_{g=1}^{G} \frac{N^b(\mathbf{X}_g)}{(\eta_\tau^{\mathrm{h}})_g} \partial_{\mathbf{F}}\eta_g \nabla N^c(\mathbf{X}_g) w_g \\
&+ H_{ab} \sum_{g=1}^{G} \Theta_g^2 \nabla \left(\frac{N^b(\mathbf{X}_g)}{(u_\tau^{\mathrm{h}})_g}\right) \cdot \mathbf{K}_g \nabla \left(\frac{N^c(\mathbf{X}_g)}{(u_\tau^{\mathrm{h}})_g}\right) w_g\, H_{cd}\partial_{\tau_d}\mathcal{S} \ .
\end{aligned}
\tag{6.80}
$$

Eventually, evolution equation $(6.67)_4$ for the internal variable, together with $(6.51)_2$, (6.54) and (6.77) result in

$$\begin{aligned}\dot{\overline{(\mathbf{C}_p^{-1})}}_g = &- \frac{4N^a(\mathbf{X}_g)}{(\eta_\tau^{\mathrm{h}})_g}\left(\boldsymbol{\mathcal{N}}_g : \left(\partial_{\mathbf{C}_p^{-1}} u_g(\mathbf{C}_p^{-1})_g\right)\right)(\mathbf{C}_p^{-1})_g\, H_{ab}\partial_{\tau_b}\mathcal{S} \\ &+ 4\Theta_g\left(\boldsymbol{\mathcal{N}}_g : \left(\partial_{\mathbf{C}_p^{-1}} \eta_g(\mathbf{C}_p^{-1})_g\right)\right)(\mathbf{C}_p^{-1})_g\,.\end{aligned} \tag{6.81}$$

Next, we aim at introducing relation $(6.76)_4$ into the last term of (6.81). To this end, we introduce the fourth-order tensor $\boldsymbol{\mathcal{M}}$ with components

$$(\boldsymbol{\mathcal{M}})^{ABIJ} = (\mathbf{C}_p^{-1})^{CB}(\boldsymbol{\mathcal{N}})^{A\cdot I\cdot}_{\cdot C\cdot K}(\mathbf{C}_p^{-1})^{JK}\,. \tag{6.82}$$

Note that $\boldsymbol{\mathcal{M}}$ enjoys major symmetry, $(\boldsymbol{\mathcal{M}})^{ABIJ} = (\boldsymbol{\mathcal{M}})^{IJAB}$, due to the major symmetry of the material flow tensor $\boldsymbol{\mathcal{N}}$ and the symmetry of $\mathbf{C}_p^{-1}$. Now, (6.81) can be recast in index form

$$\dot{\overline{(\mathbf{C}_p^{-1})}}{}_g^{AB} = -4(\boldsymbol{\mathcal{M}})_g^{ABIJ}\frac{\partial u_g}{\partial(\mathbf{C}_p^{-1})^{IJ}}\frac{N^a(\mathbf{X}_g)}{(\eta_\tau^{\mathrm{h}})_g}H_{ab}\,\partial_{\tau_b}\mathcal{S} + 4\frac{\Theta_g}{w_g}(\boldsymbol{\mathcal{M}})_g^{ABIJ}\frac{\partial\mathcal{S}}{\partial(\mathbf{C}_p^{-1})_g^{IJ}}\,. \tag{6.83}$$

Altogether, evolution equations (6.67) pertaining to the state variables of the discrete system at hand can be recast in the form

$$\begin{aligned}\dot{\mathbf{q}}_a &= \partial_{\mathbf{p}^a}\mathcal{E}\,, \\ \dot{\mathbf{p}}^a &= -\,\partial_{\mathbf{q}_a}\mathcal{E} + \mathbf{l}^{a\cdot}_{\cdot b}\,\partial_{\tau_b}\mathcal{E}\,, \\ \dot{\tau}_a &= -\,(\mathbf{l}^{b\cdot}_{\cdot a})^{\mathrm{T}}\,\partial_{\mathbf{p}^b}\mathcal{E} + m_{ab}\,\partial_{\tau_b}\mathcal{S} + \sum_{g=1}^{G} m_{g,a}^{IJ}\frac{\partial\mathcal{S}}{\partial(\mathbf{C}_p^{-1})_g^{IJ}}\,, \\ \dot{\overline{(\mathbf{C}_p^{-1})}}{}_g^{AB} &= m_{g,b}^{AB}\,\partial_{\tau_b}\mathcal{S} + m_g^{ABIJ}\frac{\partial\mathcal{S}}{\partial(\mathbf{C}_p^{-1})_g^{IJ}}\,,\end{aligned} \tag{6.84}$$

where

$$\begin{aligned}\mathbf{l}^{a\cdot}_{\cdot b} &= H_{bc}\sum_{g=1}^{G}\frac{N_g^c}{(\eta_\tau^{\mathrm{h}})_g}\partial_{\mathbf{F}}\eta_g\nabla N_g^a w_g\,, \\ m_{ab} &= H_{ac}\sum_{g=1}^{G}\Theta_g^2\nabla\left(\frac{N_g^c}{(u_\tau^{\mathrm{h}})_g}\right)\cdot\mathbf{K}_g\nabla\left(\frac{N_g^d}{(u_\tau^{\mathrm{h}})_g}\right)w_g\,H_{db} \\ &\quad + H_{ac}\sum_{g=1}^{G}\frac{4N_g^cN_g^d}{(u_\tau^{\mathrm{h}})_g(\eta_\tau^{\mathrm{h}})_g}\frac{\partial u_g}{\partial(\mathbf{C}_p^{-1})^{AB}}(\boldsymbol{\mathcal{M}})_g^{ABIJ}\frac{\partial u_g}{\partial(\mathbf{C}_p^{-1})^{IJ}}w_g\,H_{db}\,, \\ m_{g,b}^{AB} &= -\,4(\boldsymbol{\mathcal{M}})_g^{ABIJ}\frac{\partial u_g}{\partial(\mathbf{C}_p^{-1})^{IJ}}\frac{N_g^a}{(\eta_\tau^{\mathrm{h}})_g}H_{ab}\,, \\ m_g^{ABIJ} &= 4\frac{\Theta_g}{w_g}(\boldsymbol{\mathcal{M}})_g^{ABIJ}\,.\end{aligned} \tag{6.85}$$

For simplicity of exposition, in (6.85), $N_g^a = N^a(\mathbf{X}_g)$. Note that the properties $m_{ab} = m_{ba}$ and $m_g^{ABIJ} = m_g^{IJAB}$ hold.

Now, evolution equations (6.84) give rise to specific forms of Poisson matrix $\mathbf{L}$ and friction matrix $\mathbf{M}$ in GENERIC (6.69). In particular, the Poisson matrix takes the form

$$
\mathbf{L} = \left[\begin{array}{c|c|c|c} 0 & \mathrm{I} & 0 & 0 \\ \hline -\mathrm{I} & 0 & \left[\mathbf{1}^{a\,\cdot}_{\cdot\, b}\right] & 0 \\ \hline 0 & \left[-\mathbf{1}^{b\,\cdot}_{\cdot\, a}\right]^{\mathrm{T}} & 0 & 0 \\ \hline 0 & 0 & 0 & 0 \end{array}\right], \tag{6.86}
$$

where I is the identity matrix (with appropriate dimension corresponding to the partitioning of the state vector (6.68)), and matrix $\left[\mathbf{1}^{a\,\cdot}_{\cdot\, b}\right]$ consists of vectors $\mathbf{1}^{a\,\cdot}_{\cdot\, b}$ defined in $(6.85)_1$. Specifically, we have

$$
\left[\mathbf{1}^{a\,\cdot}_{\cdot\, b}\right] = \begin{bmatrix} \mathbf{1}^{1\,\cdot}_{\cdot\,1} & \cdots & \mathbf{1}^{1\,\cdot}_{\cdot\,N} \\ \vdots & \ddots & \vdots \\ \mathbf{1}^{N\,\cdot}_{\cdot\,1} & \cdots & \mathbf{1}^{N\,\cdot}_{\cdot\,N} \end{bmatrix}. \tag{6.87}
$$

It can be observed that Poisson matrix (6.86) is skew-symmetric. Furthermore, the friction matrix is given by

$$
\mathbf{M} = \left[\begin{array}{c|c|c|c} 0 & 0 & 0 & 0 \\ \hline 0 & 0 & 0 & 0 \\ \hline 0 & 0 & \left[m_{ab}\right] & \left[\mathbf{m}_{g,a}^{IJ}\right] \\ \hline 0 & 0 & \left[\mathbf{m}_{g,b}^{AB}\right]^{\mathrm{T}} & \left[\mathbf{m}_g^{ABIJ}\right] \end{array}\right], \tag{6.88}
$$

Here, the block matrices $\left[m_{ab}\right] \in \mathbb{R}^{N\times N}$, $\left[\mathbf{m}_{g,b}^{IJ}\right] \in \mathbb{R}^{N\times 6G}$, and $\left[\mathbf{m}_g^{ABIJ}\right]$ contain the components defined in $(6.85)_{2-4}$ and take the form

$$
\left[m_{ab}\right] = \begin{bmatrix} m_{11} & \cdots & m_{1N} \\ \vdots & \ddots & \vdots \\ m_{N1} & \cdots & m_{NN} \end{bmatrix}, \qquad \left[\mathbf{m}_{g,a}^{IJ}\right] = \begin{bmatrix} \left[m_{1,1}^{IJ}\right] & \cdots & \left[m_{G,1}^{IJ}\right] \\ \vdots & \ddots & \vdots \\ \left[m_{1,N}^{IJ}\right] & \cdots & \left[m_{G,N}^{IJ}\right] \end{bmatrix}
$$

and

$$\left[\mathbf{m}_g^{ABIJ}\right] = \begin{bmatrix} \left[m_1^{ABIJ}\right] & \cdots & 0 \\ \vdots & \ddots & \vdots \\ 0 & \cdots & \left[m_G^{ABIJ}\right] \end{bmatrix} .$$

Note that the last matrix is block-diagonal and symmetric. It can be easily observed that friction matrix (6.88) is symmetric and positive-semidefinite.

6.3.2. Conservation properties

We next verify the conservation properties of the semi-discrete formulation of the closed system dealt with in Section 6.3.1. Since the evolution equations pertaining to the semi-discrete formulation can be brought into GENERIC form (6.69), (i) conservation of total energy, and (ii) non-decreasing total entropy are automatically satisfied. To see this, we first consider the time-derivative of the total energy,

$$\begin{aligned} \frac{\mathrm{d}}{\mathrm{d}t}\mathcal{E}(\mathbf{z}) &= \nabla\mathcal{E}(\mathbf{z})\dot{\mathbf{z}} \\ &= \nabla\mathcal{E}(\mathbf{z})\mathbf{L}\nabla\mathcal{E}(\mathbf{z}) + \nabla\mathcal{E}(\mathbf{z})\mathbf{M}\nabla\mathcal{S}(\mathbf{z}) \, , \end{aligned} \tag{6.89}$$

where (6.69) has been used. The first term on the right-hand side of the last equation vanishes due to the skew-symmetry of Poisson matrix (6.86). The second term vanishes too, since the non-interaction condition

$$\mathbf{M}\nabla\mathcal{E}(\mathbf{z}) = \mathbf{0} \, . \tag{6.90}$$

holds. The last equation can be easily verified by a straight-forward calculation. Accordingly, (6.89) yields the conservation law $\mathrm{d}\mathcal{E}/\mathrm{d}t = 0$. Similarly, the time-derivative of the total entropy yields

$$\begin{aligned} \frac{\mathrm{d}}{\mathrm{d}t}\mathcal{S}(\mathbf{z}) &= \nabla\mathcal{S}(\mathbf{z})\dot{\mathbf{z}} \\ &= \nabla\mathcal{S}(\mathbf{z})\mathbf{L}\nabla\mathcal{E}(\mathbf{z}) + \nabla\mathcal{S}(\mathbf{z})\mathbf{M}\nabla\mathcal{S}(\mathbf{z}) \, . \end{aligned} \tag{6.91}$$

It can be verified by a straight-forward calculation that the second non-interaction condition

$$\mathbf{L}\nabla\mathcal{S}(\mathbf{z}) = \mathbf{0} \, . \tag{6.92}$$

is satisfied. In addition to that, the positive semi-definiteness of friction matrix (6.88) implies the result $\mathrm{d}\mathcal{S}/\mathrm{d}t \geq 0$. In particular, this result represents the semi-discrete version of (6.25) (apart from the boundary term in (6.25) originating from heat flux across the boundary). Thus, the total entropy of the closed system at hand is non-decreasing, due to the irreversible nature of heat conduction and visco-elastic deformations.

Since the material response is assumed to be frame-indifferent (or objective), specific symmetry properties are inherent to the discrete system under consideration. In particular, invariance under rigid rotations implies satisfaction of the following conditions (see Appendix C.1 for further details):

$$\begin{aligned} \mathbf{0} &= \mathbf{q}_a \times \partial_{\mathbf{q}_a} \mathcal{E} \, , \\ \mathbf{0} &= \mathbf{q}_a \times \mathbf{l}^{a\cdot}_{\cdot b} \, , \end{aligned} \tag{6.93}$$

for all $b = 1, \ldots, N$. We tacitly assume that no resultant external torque is acting on the system (i.e. the right-hand side of (6.23) is assumed to vanish). The semi-discrete version of the angular momentum relative to the origin is given by

$$\mathbf{J}^{\mathrm{h}} = \int_{\mathcal{B}} \rho \boldsymbol{\varphi}^{\mathrm{h}} \times \mathbf{v}^{\mathrm{h}} \, \mathrm{d}V = \mathbf{q}_a \times \mathbf{p}^a \, , \tag{6.94}$$

where formulas $(6.47)_{1,2}$ along with definition (6.65) of the nodal momentum vectors $\mathbf{p}^a$ have been used. The time-derivative of (6.94) reads

$$\begin{aligned} \frac{\mathrm{d}}{\mathrm{d}t} \mathbf{J}^{\mathrm{h}} &= \dot{\mathbf{q}}_a \times \mathbf{p}^a + \mathbf{q}_a \times \dot{\mathbf{p}}^a \\ &= M_{ab} \mathbf{p}^b \times \mathbf{p}^a + \mathbf{q}_a \times \left(-\partial_{\mathbf{q}_a} \mathcal{E} + \mathbf{l}^{a\cdot}_{\cdot b} \, \partial_{\tau_b} \mathcal{E} \right) \, , \end{aligned} \tag{6.95}$$

where $(6.67)_1$ and $(6.84)_2$ have been used. Now the right-hand side of the last equation vanishes due to (i) the symmetry of M_{ab} together with the skew-symmetry of the vector cross product, and (ii) symmetry conditions (6.93). Result $\mathrm{d}\mathbf{J}^{\mathrm{h}}/\mathrm{d}t = \mathbf{0}$ implies conservation of total angular momentum.

6.3.3. Choice of the thermodynamic state variable

We shortly outline the impact of the specific choice of the thermodynamic state variable, $\tau \in \{\theta, \eta, u\}$, on the structure of GENERIC (6.69). For simplicity, in this section we neglect body forces, that is, $\mathbf{b} = \mathbf{0}$ and still focus on closed systems.

Choosing the internal energy density as thermodynamic state variable, i.e. $\tau = u$, the total energy (6.70) takes a particularly simple form given by

$$\hat{\mathcal{E}}(\mathbf{z}) = \frac{1}{2} M_{ab} \mathbf{p}^a \cdot \mathbf{p}^b + \sum_{g=1}^{G} u_g^{\mathrm{h}} w_g \, . \tag{6.96}$$

As before, u_g^{h} stands for $u^{\mathrm{h}}(\mathbf{X}_g, t)$. In particular, interpolation formula (6.53)$_2$ gives rise to $u_g^{\mathrm{h}} = N^a(\mathbf{X}_g)u_a(t)$. The derivatives of the total energy in (6.74) simplify to

$$\begin{aligned}
\partial_{\mathbf{q}_a}\hat{\mathcal{E}} &= \mathbf{0} \,, \\
\partial_{\mathbf{p}^a}\hat{\mathcal{E}} &= M_{ab}\mathbf{p}^b \,, \\
\partial_{u_a}\hat{\mathcal{E}} &= \sum_{g=1}^{G} N^a(\mathbf{X}_g)w_g \,, \\
\frac{\partial\hat{\mathcal{E}}}{\partial(\mathbf{C}_p^{-1})_g^{AB}} &= 0 \,.
\end{aligned} \tag{6.97}$$

Consequently,

$$\begin{aligned}
\frac{\mathrm{d}}{\mathrm{d}t}\hat{\mathcal{E}}(\mathbf{z}) &= \nabla\hat{\mathcal{E}}(\mathbf{z})\dot{\mathbf{z}} \\
&= M_{ab}\mathbf{p}^a \cdot \dot{\mathbf{p}}^b + \sum_{g=1}^{G} N^a(\mathbf{X}_g)\dot{u}_a w_g \,.
\end{aligned} \tag{6.98}$$

Moreover, for the choice $\tau = u$ friction matrix (6.88) attains a particularly simple block-diagonal form, since $m_{g,b}^{AB} = 0$, and coefficients m_{ab} only contain contributions due to heat conduction (cf. (6.85)).

Choosing the internal entropy density as thermodynamic state variable, i.e. $\tau = \eta$, yields a particularly simple form of the total entropy (6.71) given by

$$\tilde{\mathcal{S}}(\mathbf{z}) = \sum_{g=1}^{G} \eta_g^{\mathrm{h}} w_g \,, \tag{6.99}$$

where interpolation formula (6.53)$_2$ gives rise to $\eta_g^{\mathrm{h}} = N^a(\mathbf{X}_g)\eta_a(t)$. Consequently, the derivatives of the total entropy in (6.76) simplify to

$$\begin{aligned}
\partial_{\mathbf{q}_a}\tilde{\mathcal{S}} &= \mathbf{0} \,, \\
\partial_{\mathbf{p}^a}\tilde{\mathcal{S}} &= \mathbf{0} \,, \\
\partial_{\eta_a}\tilde{\mathcal{S}} &= \sum_{g=1}^{G} N^a(\mathbf{X}_g)w_g \,, \\
\frac{\partial\tilde{\mathcal{S}}}{\partial(\mathbf{C}_p^{-1})_g^{AB}} &= 0 \,.
\end{aligned} \tag{6.100}$$

Thus

$$\frac{\mathrm{d}}{\mathrm{d}t}\tilde{\mathcal{S}}(\mathbf{z}) = \nabla\tilde{\mathcal{S}}(\mathbf{z})\dot{\mathbf{z}} \\ = \sum_{g=1}^{G} N^a(\mathbf{X}_g)\dot{\eta}_a w_g \,. \tag{6.101}$$

Moreover, the choice $\tau = \eta$ leads to $\mathbf{l}^{a\cdot}_{\cdot b} = \mathbf{0}$ (see (6.85)), so that Poisson matrix (6.86) yields a particularly simple form.

In contrast to the above considerations, selecting the total temperature as thermodynamic state variable, i.e. $\tau = \theta$, essentially does not lead to any simplifications. We eventually remark that these conclusions also affect the discretization in time, which will be treated next.

6.4. Discretization in time

We now turn to the discretization in time of the semi-discrete GENERIC-consistent evolution equations derived in Section 6.3.1. To this end, we focus on a representative time interval $[t_{\mathrm{n}}, t_{\mathrm{n}+1}]$ with corresponding time-step size $\Delta t = t_{\mathrm{n}+1} - t_{\mathrm{n}}$. We aim at second-order accurate, implicit time-stepping schemes based on the mid-point rule. Application of the mid-point rule to (6.69) yields

$$\mathbf{z}_{\mathrm{n}+1} - \mathbf{z}_{\mathrm{n}} = \Delta t\mathbf{L}(\mathbf{z}_{\mathrm{n}+\frac{1}{2}})\nabla\mathcal{E}(\mathbf{z}_{\mathrm{n}+\frac{1}{2}}) + \Delta t\mathbf{M}(\mathbf{z}_{\mathrm{n}+\frac{1}{2}})\nabla\mathcal{S}(\mathbf{z}_{\mathrm{n}+\frac{1}{2}}) \,. \tag{6.102}$$

Here, $\mathbf{z}_{\mathrm{n}}$ stands for the discrete vector of state variables at time t_n, and

$$\mathbf{z}_{\mathrm{n}+\frac{1}{2}} = \frac{1}{2}\left(\mathbf{z}_{\mathrm{n}} + \mathbf{z}_{\mathrm{n}+1}\right) \,.$$

Provided that the state variables $\mathbf{z}_{\mathrm{n}}$ are given, the state variables $\mathbf{z}_{\mathrm{n}+1}$ can be determined by solving (6.102).

6.4.1. Partially structure-preserving schemes

Next, we check whether, or under what conditions, structure-preserving properties hold in the discrete setting. In this connection, we shall see that the specific choice of the thermodynamic state variable $\tau \in \{\theta, \eta, u\}$ plays a crucial role. It can be easily verified that non-interaction conditions (6.90) and (6.92) are still satisfied in the sense that

$$\mathbf{M}(\mathbf{z}_{\mathrm{n}+\frac{1}{2}})\nabla\mathcal{E}(\mathbf{z}_{\mathrm{n}+\frac{1}{2}}) = \mathbf{0} \\ \mathbf{L}(\mathbf{z}_{\mathrm{n}+\frac{1}{2}})\nabla\mathcal{S}(\mathbf{z}_{\mathrm{n}+\frac{1}{2}}) = \mathbf{0} \tag{6.103}$$

To see whether the fundamental balance laws are correctly reproduced in the discrete formulation, we proceed along the lines of the time-continuous formulation in Section 6.3.2. In particular, concerning the balance of energy, similar to (6.89), we consider

$$\begin{aligned}\nabla\mathcal{E}(\mathbf{z}_{n+\frac{1}{2}})\cdot(\mathbf{z}_{n+1}-\mathbf{z}_n) &= \Delta t\nabla\mathcal{E}(\mathbf{z}_{n+\frac{1}{2}})\cdot\mathbf{L}(\mathbf{z}_{n+\frac{1}{2}})\nabla\mathcal{E}(\mathbf{z}_{n+\frac{1}{2}})\\ &+\Delta t\nabla\mathcal{E}(\mathbf{z}_{n+\frac{1}{2}})\cdot\mathbf{M}(\mathbf{z}_{n+\frac{1}{2}})\nabla\mathcal{S}(\mathbf{z}_{n+\frac{1}{2}})\,,\end{aligned}$$

where (6.102) has been used. In analogy to the time-continuous case the right-hand side of the above equation vanishes due to the skew-symmetry of $\mathbf{L}(\mathbf{z}_{n+\frac{1}{2}})$ and non-interaction condition $(6.103)_1$. Thus

$$\nabla\mathcal{E}(\mathbf{z}_{n+\frac{1}{2}})\cdot(\mathbf{z}_{n+1}-\mathbf{z}_n)=0\,.$$

On the other side,

$$\nabla\mathcal{E}(\mathbf{z}_{n+\frac{1}{2}})\cdot(\mathbf{z}_{n+1}-\mathbf{z}_n)\neq\mathcal{E}(\mathbf{z}_{n+1})-\mathcal{E}(\mathbf{z}_n) \tag{6.104}$$

in general. This inequality complies with the well-known fact that the mid-point rule is not capable to conserve nonlinear first integrals in general. However, there exists the exceptional case related to the choice $\tau = u$, for which (6.104) turns into an equality. This is due to the fact that for $\tau = u$ the total energy takes the form (6.96) and thus $\hat{\mathcal{E}}(\mathbf{z})$ is merely quadratic. Since the mid-point rule preserves quadratic first integrals (see [6, Sec. 4.4.2]), the choice $\tau = u$ yields a structure-preserving scheme which is capable to conserve total energy.

Concerning the evolution of total entropy, guided by (6.91), we consider

$$\begin{aligned}\nabla\mathcal{S}(\mathbf{z}_{n+\frac{1}{2}})\cdot(\mathbf{z}_{n+1}-\mathbf{z}_n) &= \Delta t\nabla\mathcal{S}(\mathbf{z}_{n+\frac{1}{2}})\cdot\mathbf{L}(\mathbf{z}_{n+\frac{1}{2}})\nabla\mathcal{E}(\mathbf{z}_{n+\frac{1}{2}})\\ &+\Delta t\nabla\mathcal{S}(\mathbf{z}_{n+\frac{1}{2}})\cdot\mathbf{M}(\mathbf{z}_{n+\frac{1}{2}})\nabla\mathcal{S}(\mathbf{z}_{n+\frac{1}{2}})\,,\end{aligned}$$

where again (6.102) has been used. Employing non-interaction condition $(6.103)_2$, we obtain

$$\nabla\mathcal{S}(\mathbf{z}_{n+\frac{1}{2}})\cdot(\mathbf{z}_{n+1}-\mathbf{z}_n)=\Delta t\nabla\mathcal{S}(\mathbf{z}_{n+\frac{1}{2}})\cdot\mathbf{M}(\mathbf{z}_{n+\frac{1}{2}})\nabla\mathcal{S}(\mathbf{z}_{n+\frac{1}{2}})\geq 0\,, \tag{6.105}$$

where the positive semi-definiteness of friction matrix $\mathbf{M}(\mathbf{z}_{n+\frac{1}{2}})$ has been taken as a basis. On the other hand, in analogy to (6.104), we have

$$\nabla\mathcal{S}(\mathbf{z}_{n+\frac{1}{2}})\cdot(\mathbf{z}_{n+1}-\mathbf{z}_n)\neq\mathcal{S}(\mathbf{z}_{n+1})-\mathcal{S}(\mathbf{z}_n)\,. \tag{6.106}$$

This implies that, despite the encouraging result (6.105), the mid-point scheme in general does not guarantee a non-decreasing entropy. However, there again is an exception related to the choice $\tau = \eta$. For this particular case, the total entropy takes the form (6.99) and thus $\tilde{\mathcal{S}}(\mathbf{z})$ is merely linear. Accordingly, the choice $\tau = \eta$ turns inequality (6.106) into an equality and the entropy-based mid-point scheme is therefore capable to correctly reproduce the second law of thermodynamics in the discrete setting.

We eventually verify that all mid-point-based schemes under consideration are capable to conserve angular momentum. The incremental change of angular momentum (6.94) can be written in the form

$$\mathbf{J}^{\mathrm{h}}_{\mathrm{n}+1} - \mathbf{J}^{\mathrm{h}}_{\mathrm{n}} = \mathbf{q}_{a_{\mathrm{n}+\frac{1}{2}}} \times \left(\mathbf{p}^{a}_{\mathrm{n}+1} - \mathbf{p}^{a}_{\mathrm{n}+1}\right) - \mathbf{p}^{a}_{\mathrm{n}+\frac{1}{2}} \times \left(\mathbf{q}_{a_{\mathrm{n}+1}} - \mathbf{q}_{a_{\mathrm{n}}}\right) . \tag{6.107}$$

Mid-point scheme (6.102) gives rise to

$$\begin{aligned} \mathbf{q}_{a_{\mathrm{n}+1}} - \mathbf{q}_{a_{\mathrm{n}}} &= \Delta t M_{ab} \mathbf{p}^{b}_{\mathrm{n}+\frac{1}{2}} , \\ \mathbf{p}^{a}_{\mathrm{n}+1} - \mathbf{p}^{a}_{\mathrm{n}+1} &= \Delta t \left(-\partial_{\mathbf{q}_a} \mathcal{E}(\mathbf{z}_{\mathrm{n}+\frac{1}{2}}) + \mathbf{l}^{a\cdot}_{\cdot b}(\mathbf{z}_{\mathrm{n}+\frac{1}{2}})\, \partial_{\tau_b} \mathcal{E}(\mathbf{z}_{\mathrm{n}+\frac{1}{2}}) \right) . \end{aligned} \tag{6.108}$$

Inserting from (6.108) into (6.107) yields

$$\mathbf{J}^{\mathrm{h}}_{\mathrm{n}+1} - \mathbf{J}^{\mathrm{h}}_{\mathrm{n}} = \Delta t \mathbf{q}_{a_{\mathrm{n}+\frac{1}{2}}} \times \left(-\partial_{\mathbf{q}_a} \mathcal{E}(\mathbf{z}_{\mathrm{n}+\frac{1}{2}}) + \mathbf{l}^{a\cdot}_{\cdot b}(\mathbf{z}_{\mathrm{n}+\frac{1}{2}})\, \partial_{\tau_b} \mathcal{E}(\mathbf{z}_{\mathrm{n}+\frac{1}{2}}) \right) - \Delta t M_{ab} \mathbf{p}^{a}_{\mathrm{n}+\frac{1}{2}} \times \mathbf{p}^{b}_{\mathrm{n}+\frac{1}{2}} . \tag{6.109}$$

Symmetry conditions (6.93) imply

$$\begin{aligned} \mathbf{0} &= \mathbf{q}_{a_{\mathrm{n}+\frac{1}{2}}} \times \partial_{\mathbf{q}_a} \mathcal{E}(\mathbf{z}_{\mathrm{n}+\frac{1}{2}}) , \\ \mathbf{0} &= \mathbf{q}_{a_{\mathrm{n}+\frac{1}{2}}} \times \mathbf{l}^{a\cdot}_{\cdot b}(\mathbf{z}_{\mathrm{n}+\frac{1}{2}}) . \end{aligned} \tag{6.110}$$

Inserting from (6.110) into (6.109) and taking into account the symmetry of M_{ab} together with the skew-symmetry of the vector product leads to the result $\mathbf{J}^{\mathrm{h}}_{\mathrm{n}+1} = \mathbf{J}^{\mathrm{h}}_{\mathrm{n}}$.

To summarize, depending on the choice for the thermodynamic state variable we get three alternative mid-point schemes which are partially structure-preserving. Correspondingly, the resulting schemes are denoted by $(\mathrm{EM})_u$ (energy-momentum scheme related to $\tau = u$), $(\mathrm{ME})_\eta$ (momentum-entropy scheme associated with $\tau = \eta$), and M_θ (momentum scheme related to $\tau = \theta$).

6.5. Numerical investigations

In this section, the alternative mid-point type schemes newly developed in the present Chapter are applied to representative numerical examples dealing with finite-strain thermo-viscoelastodynamics. Depending on the specific choice for the thermodynamic state variable $\tau \in \{\theta, \eta, u\}$, the following methods are applied:

$\tau = u$	$(\mathrm{EM})_u$	Energy-Momentum consistent scheme
$\tau = \eta$	$(\mathrm{ME})_\eta$	Momentum-Entropy consistent scheme
$\tau = \theta$	$(\mathrm{M})_\theta$	Momentum consistent scheme

Following the previous Chapters we shall focus on momentum maps associated with symmetries of the mechanical system at hand, and the balance laws associated with the two fundamental laws of thermodynamics. In this connection we also consider the functional

$$\mathcal{L} = \mathcal{E} - \theta_0 \mathcal{S} \,. \tag{6.111}$$

According to [146], for certain types of environments, $\mathcal{L}$ is a natural Lyapunov function and thus qualifies as estimate for the numerical stability of the schemes under consideration.

In each time step, the schemes emanating from (6.102) generate a system of nonlinear algebraic equations for the determination of the state variables $\mathbf{z}_{n+1}$. To this end, we apply a Multilevel-Newton algorithm, see [152] and references therein for more details.

6.5.1. Material model

In order to particularize the Helmholtz free energy density (6.28) used in the numerical examples we start from a temperature-based description. In particular, we consider

$$\begin{aligned} \overline{\psi}(\mathbf{C},\theta,\mathbf{C}_p^{-1}) &= \overline{\psi}^{\mathrm{eq}}(\mathbf{C},\theta) + \overline{\psi}^{\mathrm{neq}}(\mathbf{C},\theta,\mathbf{C}_p^{-1}) \,, \\ \overline{\psi}^{\mathrm{eq}}(\mathbf{C},\theta) &= \psi_1(\mathbf{C}) + \psi_2(\theta) - (\theta-\theta_0)\psi_3(J) \,, \\ \overline{\psi}^{\mathrm{neq}}(\mathbf{C},\theta,\mathbf{C}_p^{-1}) &= \psi_{1,\mathrm{visc}}(\mathbf{C},\mathbf{C}_p^{-1}) \,, \end{aligned} \tag{6.112}$$

where

$$\begin{aligned} \psi_1(\mathbf{C}) &= \frac{\mu}{2}\left(\mathbf{C}:\mathbf{I} - 3 - 2\log J - \frac{2}{3}(J-1)^2\right) + W_{\mathrm{vol}}(J) \,, \\ \psi_{1,\mathrm{visc}}(\mathbf{C},\mathbf{C}_p^{-1}) &= \frac{\mu_e}{2}\left(\mathbf{C}:\mathbf{C}_p^{-1} - 3 - 2\log J_e - \frac{2}{3}(J_e-1)^2\right) + W_{\mathrm{vol,visc}}(J_e) \,, \\ \psi_2(\theta) &= c\left(\theta - \theta_0 - \theta\log(\theta/\theta_0)\right) \,, \\ \psi_3(J) &= 3\beta W_{\mathrm{vol}}(J) \,, \\ W_{\mathrm{vol}}(J) &= \frac{\lambda + \frac{2}{3}\mu}{4}\left((\log J)^2 + (J-1)^2\right) \,, \\ W_{\mathrm{vol,visc}}(J_e) &= \frac{\lambda_e + \frac{2}{3}\mu_e}{4}\left((\log J_e)^2 + (J_e-1)^2\right) \,. \end{aligned} \tag{6.113}$$

Here, $J = \sqrt{\det \mathbf{C}}$ is the determinant of the deformation gradient and $J_e = \sqrt{\det(\mathbf{C}\mathbf{C}_p^{-1})}$. Further μ, μ_e, λ and λ_e are prescribed parameters, $c > 0$ is the specific heat capacity at constant deformation, β is the coefficient of thermal expansion, and θ_0 is the reference temperature. We refer to [1] and the references therein for a detailed investigation of the thermoelastic part of the specific Helmholtz free energy density (6.112).

Further, for the viscoelastic part we refer to [74] and the references therein. For simplicity we assume incompressible material behavior. Quasi incompressible material models for finite-strain thermo-viscoelasticy are considered in, e.g., [142, 153]. It is now a straightforward exercise to calculate further quantities emanating from (6.112), depending on the specific choice for the thermodynamic variable $\tau \in \{u, \theta, \eta\}$ (see also [1] for additional details). In particular, the *temperature-based formulation* yields

$$\begin{aligned}
\overline{\eta}(\mathbf{C}, \theta, \mathbf{C}_p^{-1}) &= c \log(\theta/\theta_0) + \psi_3(J) \,, \\
\overline{u}(\mathbf{C}, \theta, \mathbf{C}_p^{-1}) &= \psi_1(\mathbf{C}) + \psi_{1,\text{visc}}(\mathbf{C}, \mathbf{C}_p^{-1}) + c(\theta - \theta_0) + \theta_0 \psi_3(J) \,.
\end{aligned}$$

The formulation based on the *entropy density* gives

$$\begin{aligned}
\widetilde{\eta}(\mathbf{C}, \eta, \mathbf{C}_p^{-1}) &= \eta \,, \\
\widetilde{u}(\mathbf{C}, \eta, \mathbf{C}_p^{-1}) &= \psi_1(\mathbf{C}) + \psi_{1,\text{visc}}(\mathbf{C}, \mathbf{C}_p^{-1}) + c\left(\theta_0 e^{\frac{\eta - \psi_3(J)}{c}} - \theta_0\right) + \theta_0 \psi_3(J) \,.
\end{aligned}$$

Moreover, the formulation based on the *internal energy density* leads to

$$\begin{aligned}
\widehat{\eta}(\mathbf{C}, u, \mathbf{C}_p^{-1}) &= c \log\left(1 + \frac{u - \psi_1(\mathbf{C}) - \psi_{1,\text{visc}}(\mathbf{C}, \mathbf{C}_p^{-1})}{c\theta_0} - \frac{\psi_3(J)}{c}\right) + \psi_3(J) \,, \\
\widehat{u}(\mathbf{C}, u, \mathbf{C}_p^{-1}) &= u \,.
\end{aligned}$$

Concerning the constitutive equation for the material heat flux vector, we assume thermally isotropic material, with material conductivity tensor given by

$$\mathbf{K} = kJ\mathbf{C}^{-1} \,. \tag{6.114}$$

Here, k is a prescribed coefficient of thermal conductivity and, as before, $J = \sqrt{\det(\mathbf{C})}$. Finally, the constitutive equation for the isotropic fourth-order material inelastic flow tensor is given by (see [69] or [148] and references therein for the spatial representation of the isotropic fourth-order inelastic flow tensor)

$$\boldsymbol{\mathcal{N}} = \frac{1}{2\nu_D}\left(\left[\mathbf{C}^{-1} \odot \mathbf{I}\right]\mathbf{C} - \frac{1}{3}\mathbf{I} \otimes \mathbf{I}\right) + \frac{1}{9\nu_V}\mathbf{I} \otimes \mathbf{I} \,, \tag{6.115}$$

where $\nu_D > 0$ and $\nu_V > 0$ represent the deviatoric and volumetric viscosities, respectively, and where $(\mathbf{A} \odot \mathbf{B})^{A}{}_{B}{}^{CD} = \frac{1}{2}\left[(\mathbf{A})^{AC}(\mathbf{B})_{B}{}^{D} + (\mathbf{A})^{AD}(\mathbf{B})_{B}{}^{C}\right]$.

6.5.2. Flying L-shaped block

The first numerical example deals with the L-shaped block depicted in Fig. 6.4.

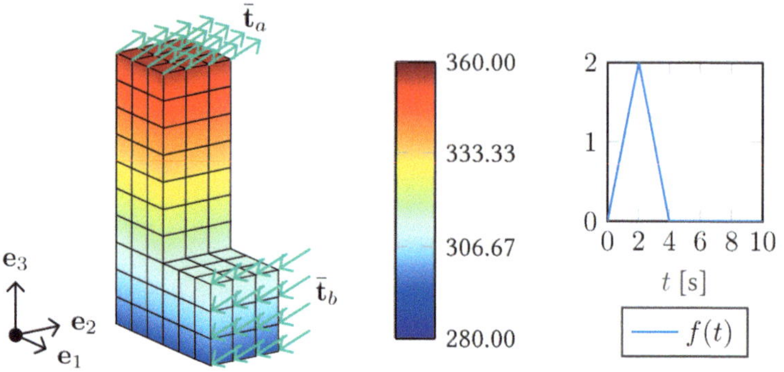

Figure 6.4.: L-shaped block: Discretised block with initial temperature distribution and mechanical boundary conditions (left), load function over time (right)

The spatial discretization of the block relies on 117 tri-linear finite elements leading to 224 nodes. The initial temperature field is varying linearly over the height (x_3 direction) of the block. In particular, at $x_3 = 0$, the initial temperature is prescribed as θ_a, while at $x_3 = h$, the temperature is prescribed as θ_b. The whole block is assumed to be thermally insulated ($\overline{q} = 0$ on $\partial_q \mathcal{B} = \partial \mathcal{B}$). Starting at rest, Piola traction vectors $\bar{\mathbf{t}}_a$ and $\bar{\mathbf{t}}_b$ are acting on two parts of the boundary surface of the block (Fig. 6.4). The external loads are applied in the form of a hat function over time. In particular, the traction vectors are given by

$$\bar{\mathbf{t}}_a = -\bar{\mathbf{t}}_b = f(t) \begin{pmatrix} 256/9 \\ 512/9 \\ 768/9 \end{pmatrix} \text{Pa}, \quad f(t) = \begin{cases} t & \text{for} \quad 0\text{s} \leq t \leq 2\text{s} \,, \\ 4-t & \text{for} \quad 2\text{s} \leq t \leq 4\text{s} \,, \\ 0 & \text{for} \quad t > 4\text{s} \,. \end{cases} \tag{6.116}$$

Table 6.1 provides a summary of the data used in the simulations. During the loading phase ($t \leq 4$s) the time step size for all simulations is $\Delta t = 0.05$, such that all systems start from the same energy-and entropy level directly after vanishing external loads. No Dirichlet boundary conditions are applied. Since in the initial loading phase the external forces are equilibrated, the total linear momentum of the block is a conserved quantity. In addition to that, after the loading phase ($t \geq 4$s) no external torque is acting on the block.

Lamé parameters	μ	997.5	Pa
	λ	4544	Pa
	μ_e	49.875	Pa
	λ_e	272.2	Pa
Specific heat capacity	c	100	$\mathrm{JK^{-1}m^{-3}}$
Expansion coefficient	β	$2.233 \cdot 10^{-4}$	$\mathrm{K^{-1}}$
Thermal conductivity	k	10	$\mathrm{WK^{-1}m^{-1}}$
Viscosities	ν_D	500	$\mathrm{Jsm^{-3}}$
	ν_V	100	$\mathrm{Jsm^{-3}}$
Ref. temperature	θ_0	293.15	K
Mass density	ρ	100	$\mathrm{kgm^{-3}}$
Initial temperature	θ_a	290	K
	θ_b	360	K
Geometry	h	10	m
	b	3	m
Newton tolerance globoal	ε_g	10^{-8}	-
Newton tolerance local	ε_l	10^{-9}	-
Simulation duration	T	300	s
Time step	Δt	0.05, 0.5, 0.6	s

Geometry

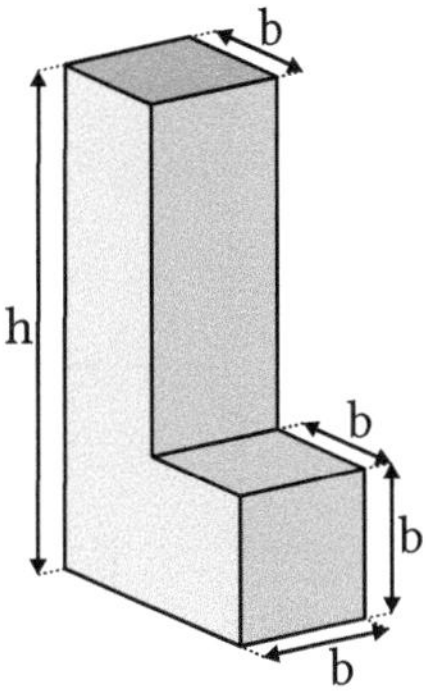

Table 6.1.: L-shaped block: Data used in the simulations

Consequently, the total angular momentum is conserved as well. All of the integrators under consideration are capable to exactly conserve both momentum maps (up to numerical round-off), independent of the chosen time step. This can be observed from Fig. 6.5, where representative numerical results of the $(\mathrm{EM})_u$ integrator are shown. After the loading phase the total energy must be a conserved quantity. As expected, the $(\mathrm{EM})_u$ scheme does exactly reproduce this conservation law (up to numerical round-off), see Fig. 6.6. In contrary, the schemes $(\mathrm{M})_\theta$ and $(\mathrm{ME})_\eta$ are not capable of conserving the total energy for larger time step sizes. Depending on the time step size, both schemes tend to increase the energy which can be observed from Fig. 6.7. Typically, such energy blow-ups eventually lead to a failure of the iterative (Newton-Raphson) solution procedure. In the diagrams the break down of the simulation is indicated by vertical lines.

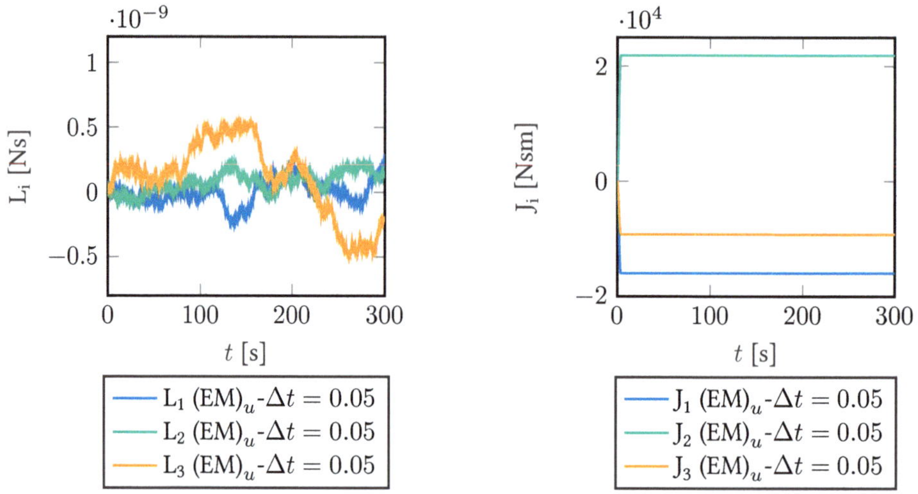

Figure 6.5.: L-shaped block: Algorithmic conservation of linear momentum $(\mathrm{EM})_u$ scheme (left), Total discrete angular momentum $(\mathrm{EM})_u$ scheme (right)

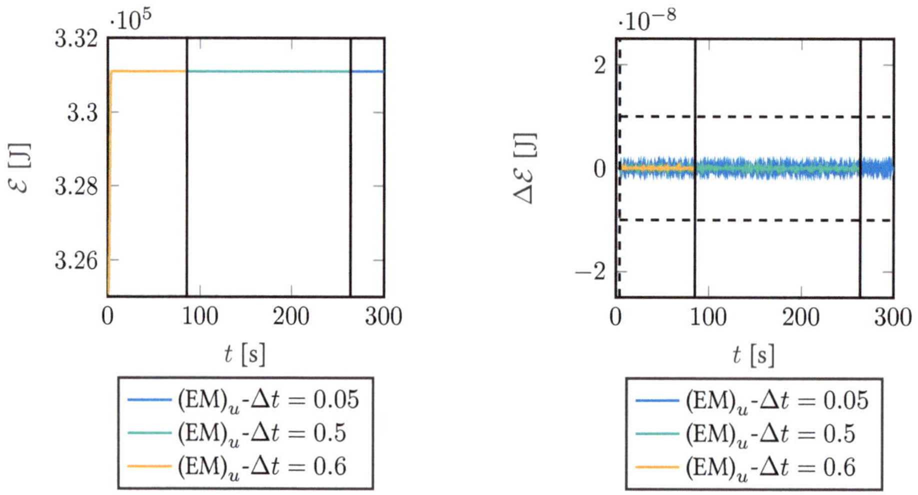

Figure 6.6.: L-shaped block: Total energy $(\mathrm{EM})_u$ scheme (left), Incremental change of total energy $(\mathrm{EM})_u$ scheme (right)

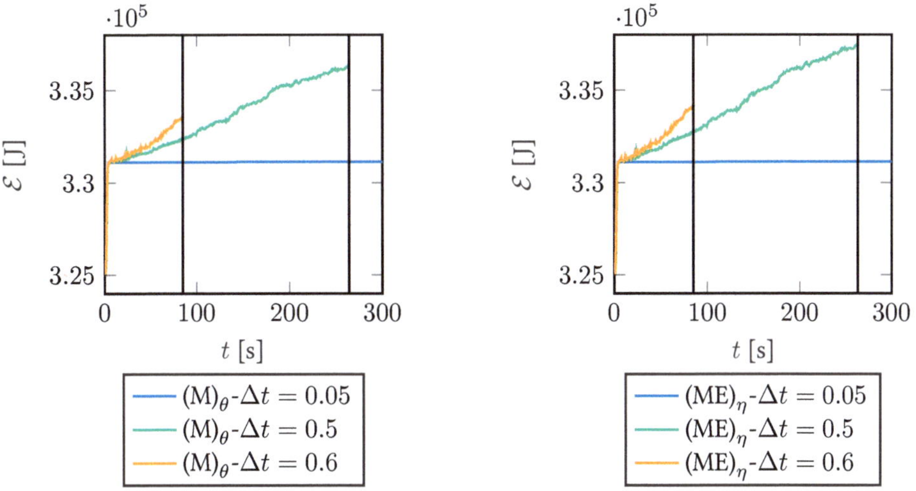

Figure 6.7.: L-shaped block: Total energy (M)$_\theta$ scheme (left), Total energy (ME)$_\eta$ scheme (right)

Regardless of the capability of the (EM)$_u$ scheme to conserve the total energy, the simulation still breaks down at about the same point in time as the break down of (M)$_\theta$ and (ME)$_\eta$ occurs. The numerical instability of the (EM)$_u$ is accompanied by a nonphysical decrease of the total entropy as can be observed from Fig. 6.8. In fact, the total entropy ought to be a non-decreasing function of time. In contrast to (EM)$_u$ and (M)$_\theta$, (ME)$_\eta$ is capable to correctly adhere to the second law of thermodynamics, independent of the time step (Fig. 6.9).

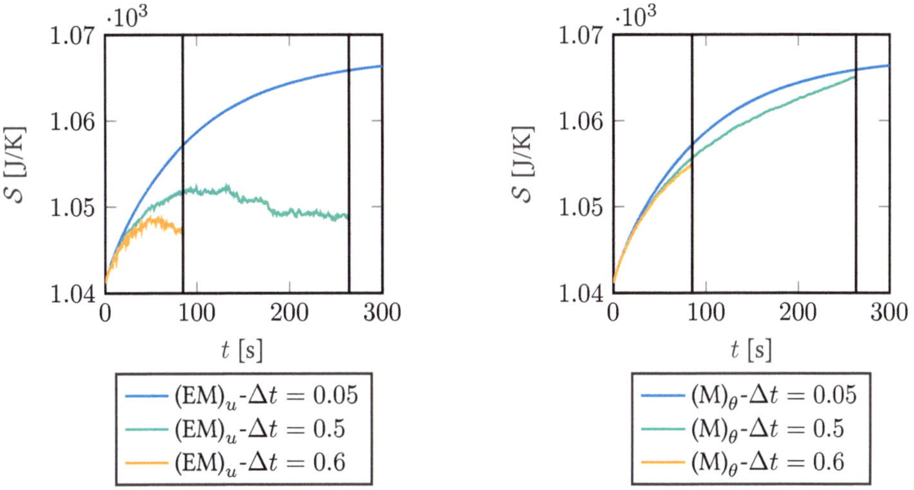

Figure 6.8.: L-shaped block: Total entropy (EM)$_u$ scheme (left), Total entropy (M)$_\theta$ scheme (right)

Indeed, in each time step the total entropy does increase, as can be observed from Fig. 6.9.

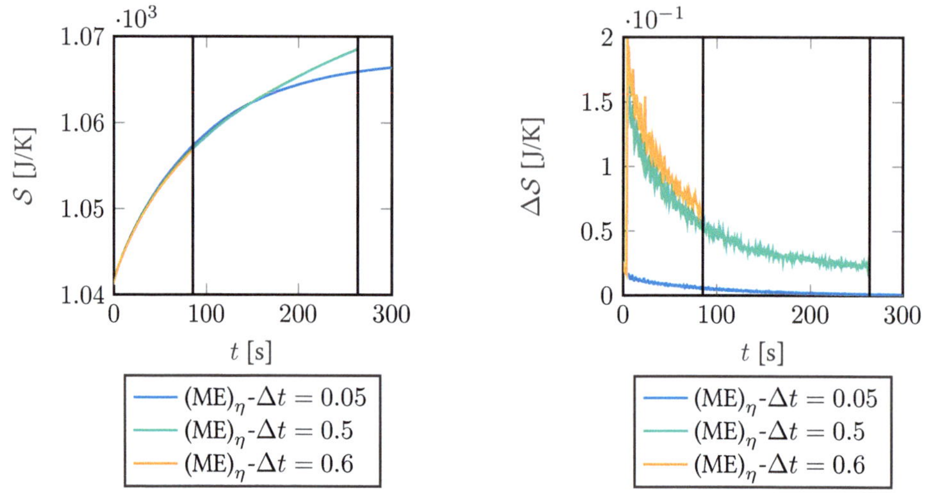

Figure 6.9.: L-shaped block: total entropy $(\mathrm{ME})_\eta$ scheme (left), Incremental change of total entropy $(\mathrm{ME})_\eta$ scheme (right)

For sufficient small time step sizes, the incremental change of total entropy is governed by

$$\mathcal{S}^{\mathrm{h}}_{\mathrm{n+1}} - \mathcal{S}^{\mathrm{h}}_{\mathrm{n}} = \underbrace{\Delta t \int_{\mathcal{B}} \mathcal{D}^{\mathrm{h}}_{\mathrm{cond}_{\mathrm{n}+\frac{1}{2}}} \,\mathrm{d}V}_{\text{Contribution due to } \mathcal{D}_{\mathrm{cond}}} + \underbrace{\Delta t \int_{\mathcal{B}} \mathcal{D}^{\mathrm{h}}_{\mathrm{inel}_{\mathrm{n}+\frac{1}{2}}} \,\mathrm{d}V}_{\text{Contribution due to } \mathcal{D}_{\mathrm{inel}}} \,, \tag{6.117}$$

for all mid-point based schemes. The two contributions to the discrete evolution of the incremental change of total entropy are visualized in Fig. 6.10 and Fig. 6.11.

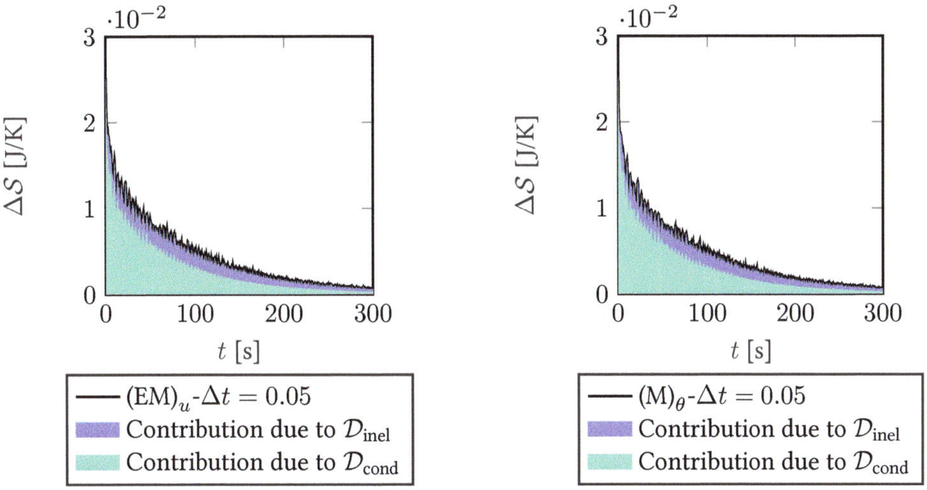

Figure 6.10.: L-shaped block: Contributions to incremental change of entropy $(\mathrm{EM})_u$ scheme (left), Contributions to incremental change of entropy $(\mathrm{M})_\theta$ scheme (right)

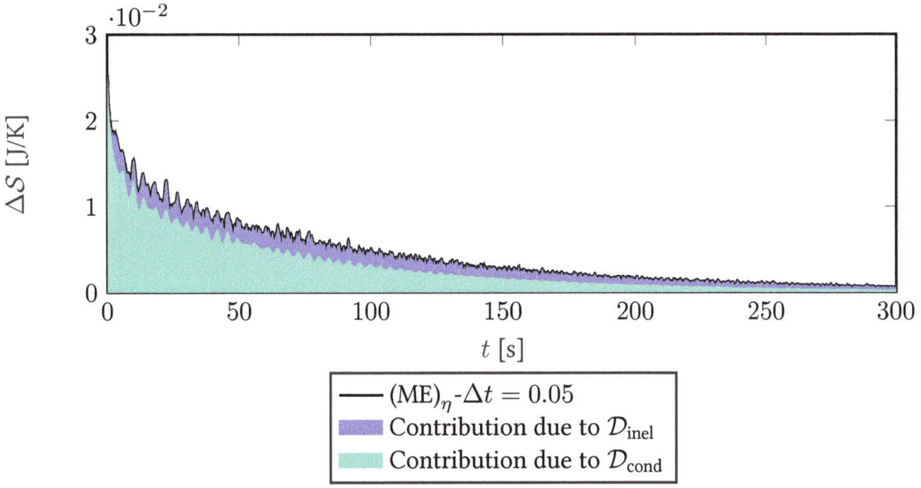

Figure 6.11.: L-shaped block: Contributions to incremental change of entropy $(\mathrm{ME})_\eta$ scheme

After the loading phase, the environment of the present example can be characterized as thermally perfect in the sense of [146]. Thus $\mathcal{L}$ defined in (6.111) plays the role of a Lyapunov function that has to decrease with time. However, the partially structure-preserving schemes $(\mathrm{EM})_u$, $(\mathrm{ME})_\eta$ and $(\mathrm{M})_\theta$ do not correctly reproduce this behavior, as can be seen from Figs. 6.12 and 6.13. That is, depending on the time step and the duration of the simulation, all of the schemes inevitably exhibit numerical instabilities characterized by increasing values of $\mathcal{L}$.

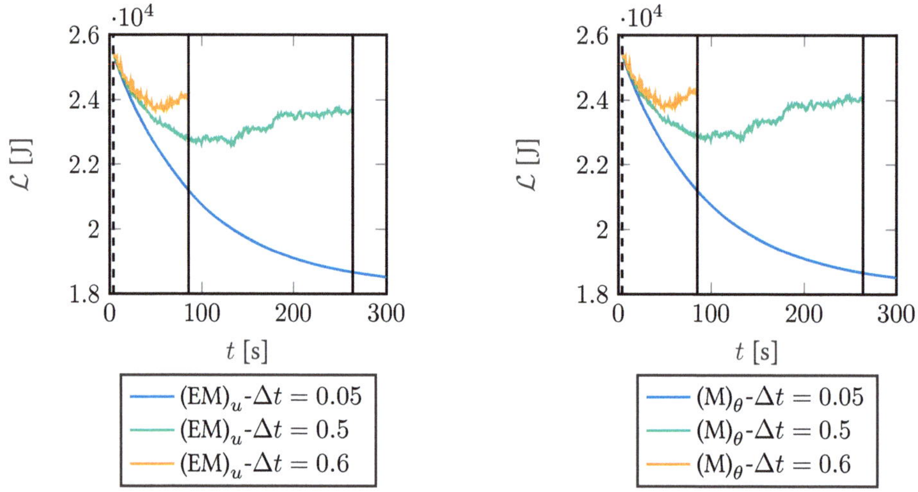

Figure 6.12.: L-shaped block: Lyapunov function (EM)$_u$ scheme (left), Lyapunov function (M)$_\theta$ scheme (right)

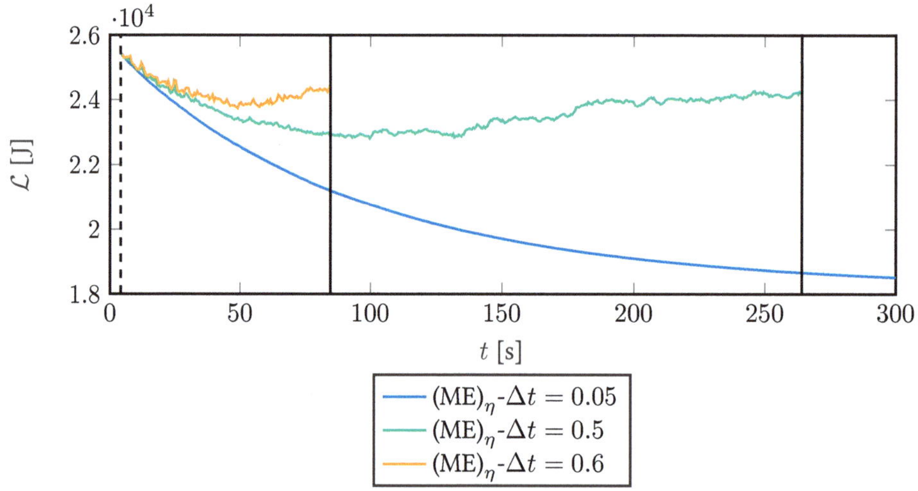

Figure 6.13.: L-shaped block: Lyapunov function (ME)$_\eta$ scheme

Eventually, the motion of the L-shaped block is illustrated in Fig. 6.14 with snapshots at successive points in time. In addition to that, the distribution of the temperature over the block is shown.

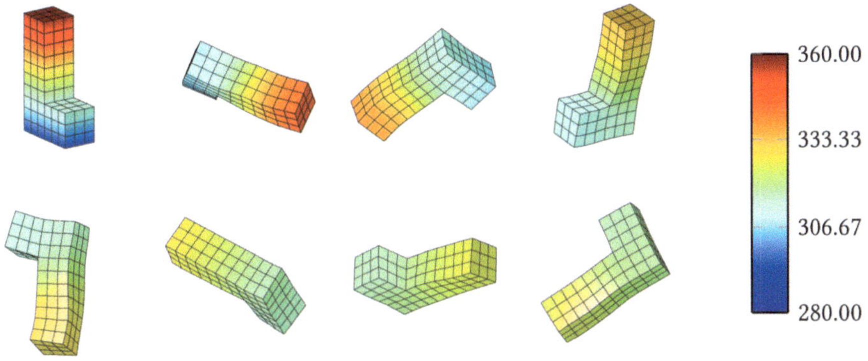

Figure 6.14.: L-shaped block: Snapshots of the motion along with the temperature distribution over the block at $t \in \{0, 40, 80, 120, 160, 200, 240, 280\}$s, obtained with the $(\mathrm{EM})_u$ scheme and time step $\Delta t = 0.05$s

6.5.3. Rotating disc

The second example deals with a rotating disc subjected to prescribed heat flow over part of the boundary surface (Fig. 6.15). The spatial discretization of the disc is based on 200 tri-linear finite elements leading to a total of 360 nodes.

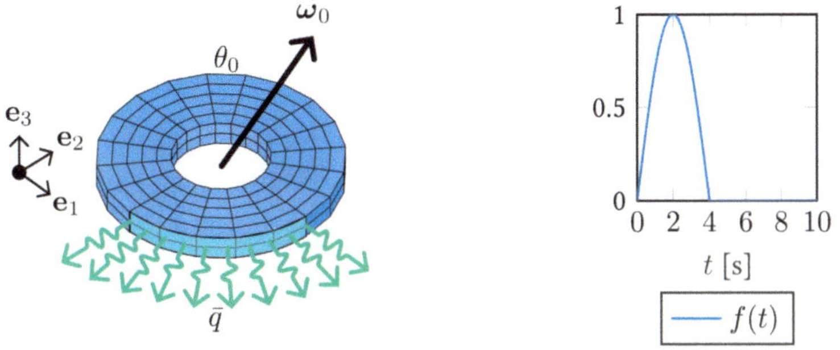

Figure 6.15.: Rotating disc: Initial configuration and thermal boundary conditions (left), function $f(t)$ for the prescribed heat flow over part of the boundary surface (right)

The initial velocity distribution over the disc results from a prescribed angular velocity $\boldsymbol{\omega}_0 \in \mathbb{R}^3$ and is given by

$$\mathbf{v}_0(\mathbf{X}) = \boldsymbol{\omega}_0 \times \mathbf{X}, \qquad \boldsymbol{\omega}_0 = \begin{pmatrix} 1 \\ 1 \\ 1 \end{pmatrix} \frac{1}{\mathrm{s}} .$$

The initial temperature of the disc is homogeneously distributed and equal to the reference temperature θ_0. In an initial period of time, $t \in [0, 4]$s, heat flow is prescribed over one quarter of the lateral boundary surface (Fig. 6.15). In particular, the heat flow into the disc is described by

$$\overline{q} = -\frac{2000W}{\pi \mathrm{m}^2} f(t), \qquad f(t) = \begin{cases} \sin(\frac{\pi}{4\mathrm{s}} t) & \text{for} \quad 0 \leq t \leq 4\mathrm{s}\,, \\ 0 & \text{for} \quad t > 4\mathrm{s}\,. \end{cases}$$

A plot of function $f(t)$ can be found in Fig. 6.15. The rest of the boundary surface of the disc is assumed to be thermally insulated ($\overline{q} = 0$). Note that the prescribed heat flow vanishes after $t = 4$s. For $t > 4$s, the complete boundary surface of the disc is assumed to be thermally insulated ($\overline{q} = 0$ on $\partial_q \mathcal{B} = \partial \mathcal{B}$). Then the environment of the disc is thermally perfect in the sense of [146]. A summary of the data used in the simulations of the rotating disc can be found in Table 6.2. During the loading phase ($t \leq 4$s) the time step size for all simulations is $\Delta t = 0.04$s, such that all systems start from the same energy-and entropy level directly after vanishing external loads.

Material parameters	λ	3000	Pa	Geometry
	μ	750	Pa	
	μ_e	120	Pa	
	λ_e	37.5	Pa	
Specific heat capacity	c	150	$\mathrm{JK}^{-1}\mathrm{m}^{-3}$	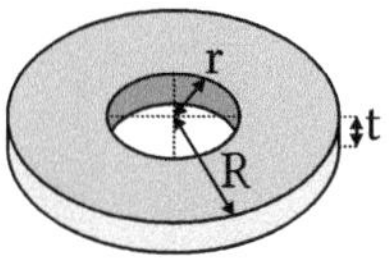
Expansion coefficient	β	$1 \cdot 10^{-4}$	K^{-1}	
Thermal conductivity	k	20	$\mathrm{WK}^{-1}\mathrm{m}^{-1}$	
Viscosities	ν_D	50	Jsm^{-3}	
	ν_V	10	Jsm^{-3}	
Ref. temperature	θ_0	293.15	K	
Mass density	ρ	8.93	kgm^{-3}	
Radius	r	0.8	m	
	R	2	m	
Thickness	t	0.4	m	
Newton tolerance global	ε_g	10^{-8}	-	
Newton tolerance local	ε_l	10^{-9}	-	
Simulation time	T	30	s	
Time step	Δt	0.04, 0.08, 0.1	s	

Table 6.2.: Rotating disc: Data used in the simulations

Due to the fact that neither external loads act on the disc, nor displacement boundary conditions are imposed, the mechanical system at hand has translational and rotational symmetry. Consequently, the corresponding momentum maps are first integrals of the motion. All integrators under consideration are capable to conserve the respective momentum map. Representative numerical results are shown in Fig. 6.16.

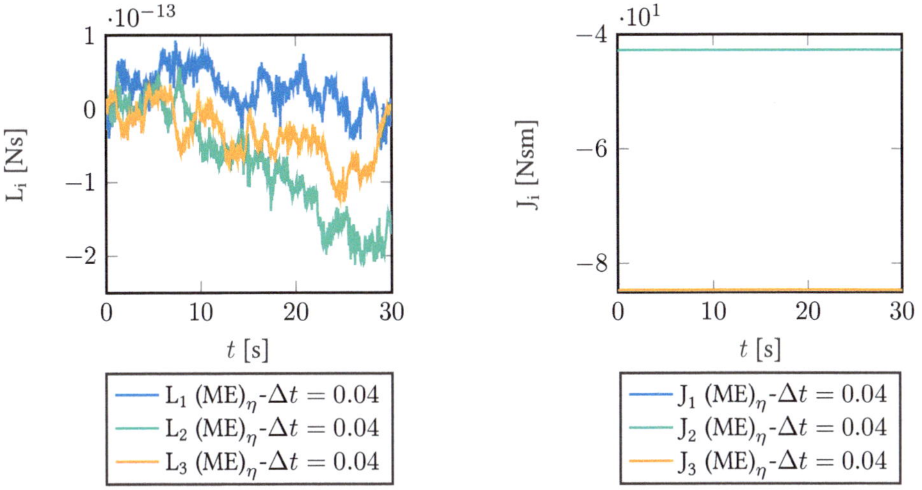

Figure 6.16.: Rotating disc: Algorithmic conservation of linear momentum $(\text{EM})_\eta$ scheme (left), Total discrete angular momentum $(\text{EM})_\eta$ scheme (right)

Since heat flow into the system is prescribed in the initial time period $[0, 4]$s, the total energy is expected to increase. For $t > 4$s, the system is closed and the total energy should stay constant. Again the $(\text{EM})_u$ scheme is capable to correctly reproduce the first law (Fig. 6.17).

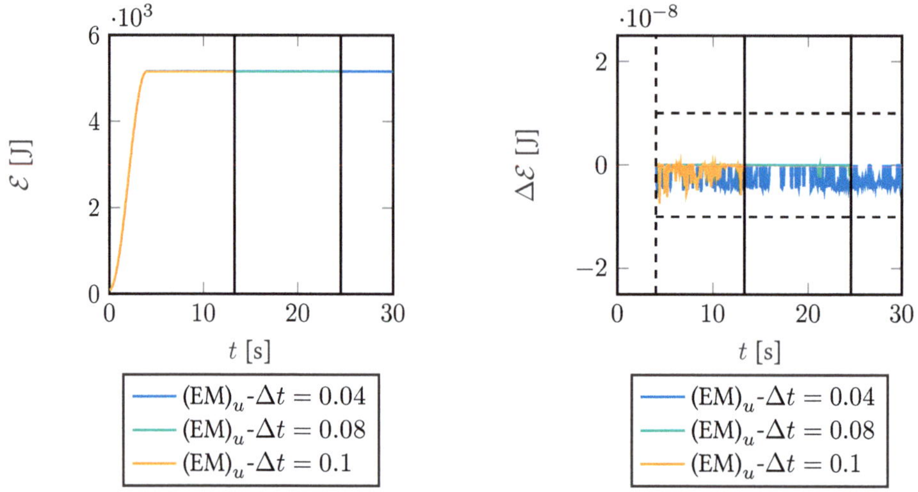

Figure 6.17.: Rotating disc: Total energy $(\mathrm{EM})_u$ scheme (left), Incremental change of total energy $(\mathrm{EM})_u$ scheme (right)

However, despite the algorithmic energy conservation (for $t > 4\mathrm{s}$), the $(\mathrm{EM})_u$ scheme is not devoid of numerical instabilities, depending on the time step. The corresponding point in time of the break down of the simulation is indicated with a vertical line in the diagrams. At about the same points in time, $(\mathrm{ME})_\eta$ and $(\mathrm{M})_\theta$ break down as well (Fig. 6.18). For these schemes the break down is accompanied by a sudden increase of the total energy leading to the divergence of the Newton-Raphson iterations. For the considered duration of the simulation ($t \in [0, 30]\mathrm{s}$), a time step of $\Delta t = 0.04\mathrm{s}$ is small enough to retain numerical stability of the three partially structure-preserving schemes at hand.

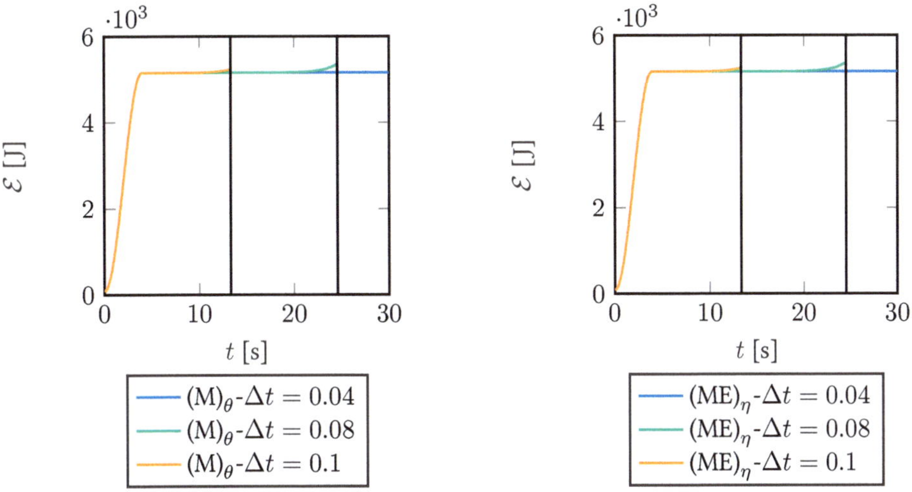

Figure 6.18.: Rotating disc: Total energy (M)$_\theta$ scheme (left), Total energy (ME)$_\eta$ scheme (right)

Due to the prescribed heat flow into the disc, the total entropy of the disc is expected to increase in the initial time period $[0, 4]$s. For $t > 4$s, the system is closed and the total entropy should be a non-decreasing function of time. The (EM)$_u$ scheme does not correctly reproduce the second law as can be seen from Fig. 6.19. Accordingly, the divergence of the iterative solution procedure is accompanied by a nonphysical decrease of the total entropy. The (M)$_\theta$ closely adheres to the second law as can be observed from Fig. 6.19.

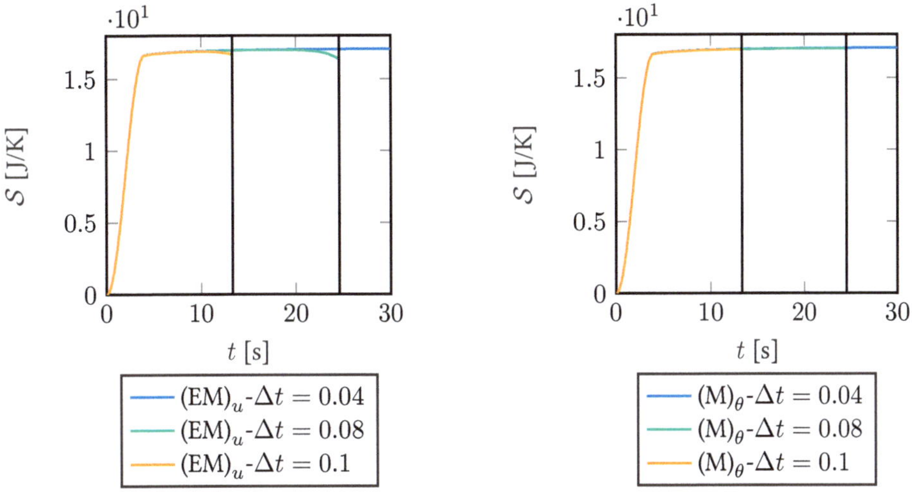

Figure 6.19.: Rotating disc: Total entropy (EM)$_u$ scheme (left), Total entropy (M)$_\theta$ scheme (right)

As expected, the $(\mathrm{ME})_\eta$ scheme is capable to exactly satisfy the second law, independent of the time step. This can be observed from Fig. 6.20. In particular, Fig. 6.20 (right) confirms that the change per time step of the total entropy is always positive.

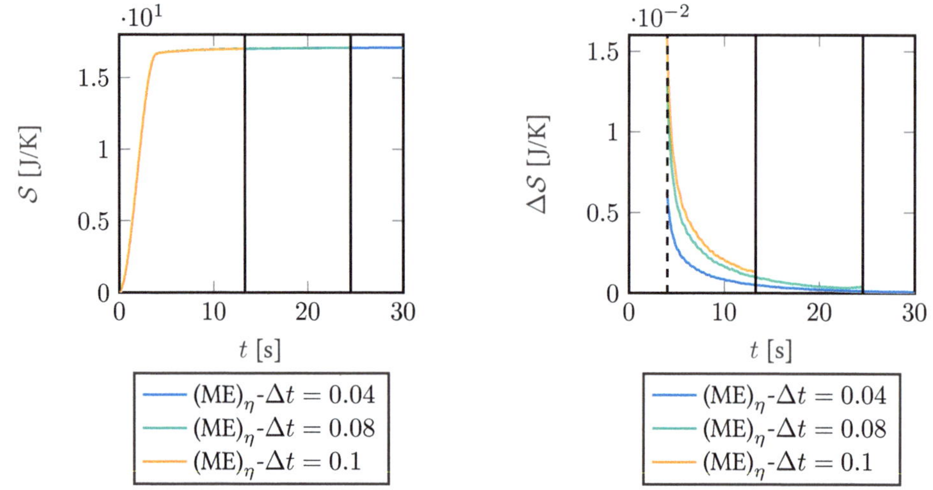

Figure 6.20.: Rotating disc: total entropy $(\mathrm{ME})_\eta$ scheme (left), Incremental change of total entropy $(\mathrm{ME})_\eta$ scheme (right)

In addition, the two contributions to the discrete evolution of the incremental change of total entropy are visualized in Fig. 6.21 and Fig. 6.22 for the mid-point based schemes. Most of the contribution is due to conduction of heat, only about 5.9% is due to inelastic deformations in the given example.

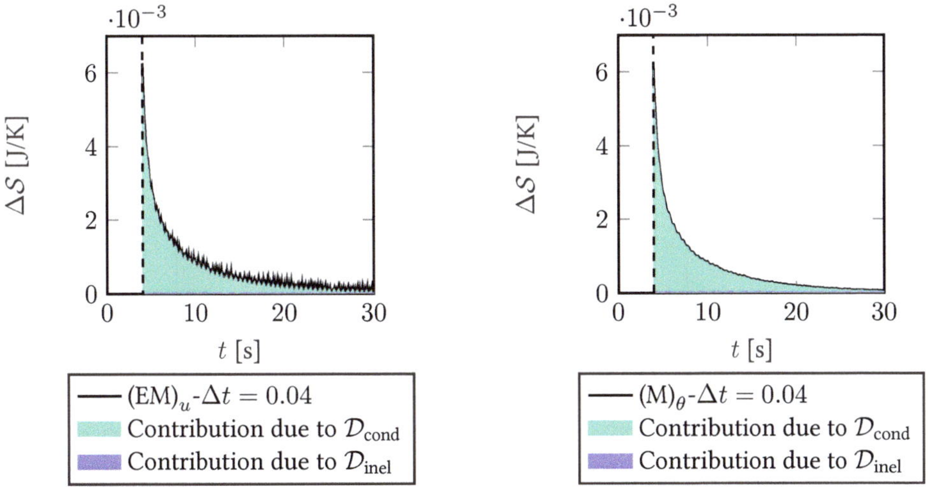

Figure 6.21.: Rotating disc: Contributions to incremental change of entropy $(\mathrm{EM})_u$ scheme (left), Contributions to incremental change of entropy $(\mathrm{M})_\theta$ scheme (right)

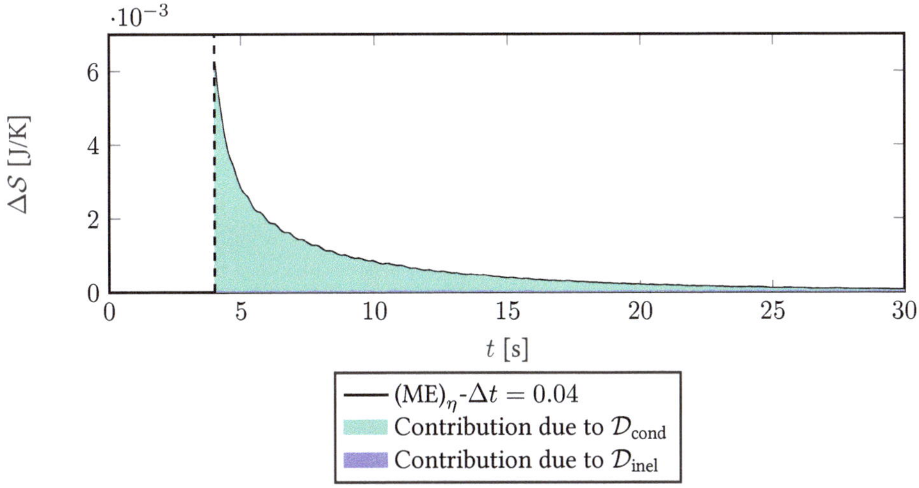

Figure 6.22.: Rotating disc: Contributions to incremental change of entropy $(\mathrm{ME})_\eta$ scheme

To shed further light on the numerical stability of the present schemes, we consider the Lyapunov function $\mathcal{L}$ defined in (6.111). For $t > 4\mathrm{s}$ the system is closed and the function $\mathcal{L}$ should decrease with time. As expected, the partially structure-preserving schemes $(\mathrm{EM})_u$, $(\mathrm{ME})_\eta$ and $(\mathrm{M})_\theta$ can not guarantee to correctly reproduce this behavior, depending on the size of the time step and the duration of the simulation (Figs. 6.23 and 6.24). In particular, it can be seen that the smallest time step, $\Delta t = 0.04\mathrm{s}$, yields a stable numerical simulation, at least in the considered time interval $[0, 30]$s. However, for larger time steps $\Delta t = 0.08$s and $\Delta t = 0.1$s, the numerical instability of each scheme becomes visible through the increase of $\mathcal{L}$.

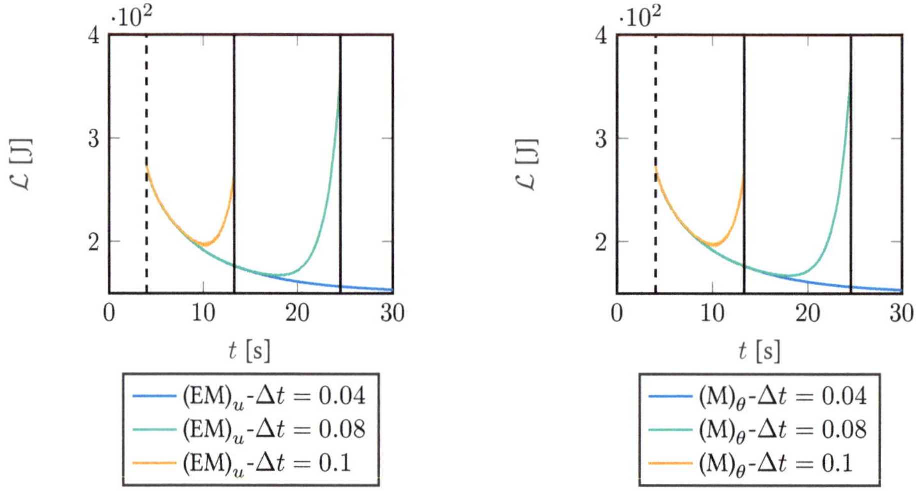

Figure 6.23.: Rotating disc: Lyapunov function (EM)$_u$ scheme (left), Lyapunov function (M)$_\theta$ scheme

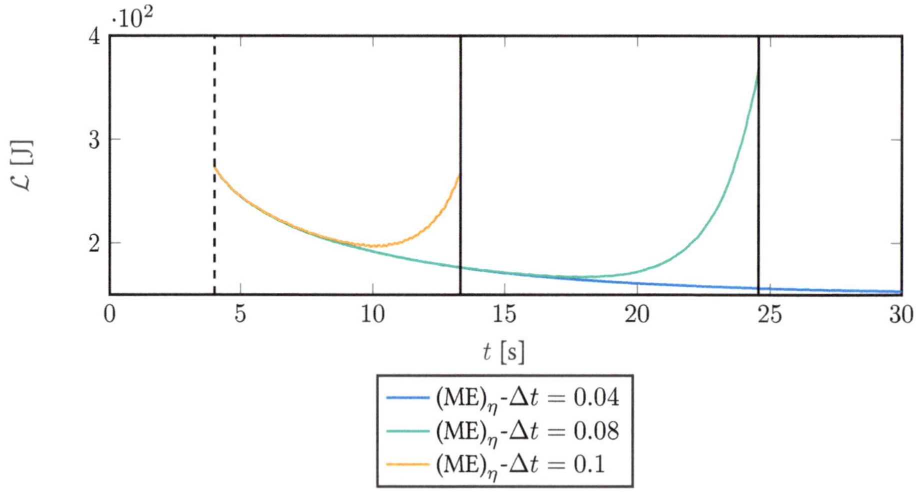

Figure 6.24.: Rotating disc: Lyapunov function (ME)$_\eta$ scheme

Eventually, the motion of the disc is illustrated in Fig. 6.25 with snapshots at successive points in time. In addition to that, the distribution of the temperature over the disc is shown.

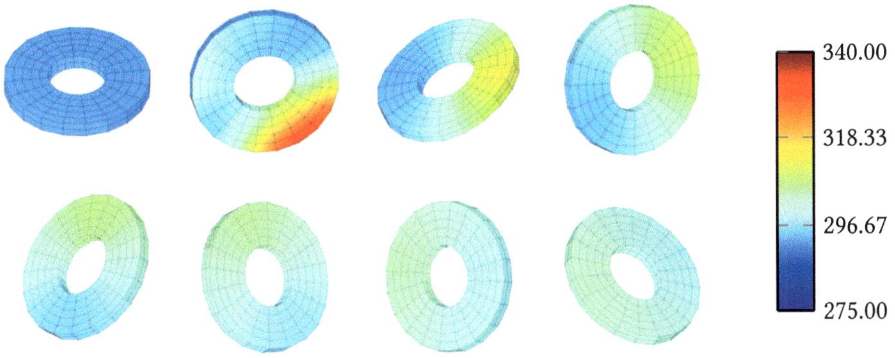

Figure 6.25.: Rotating disc: Snapshots of the motion along with the temperature distribution over the block at $t \in \{0, 4, 8, 12, 16, 20, 24, 28\}$s, obtained with the (ME)$_\eta$ scheme and time step $\Delta t = 0.04$s

7. Summary and outlook

7.1. Summary[1]

This thesis has addressed the formulation, analysis, implementation, and verification of second-order accurate structure-preserving schemes by utilizing the GENERIC structure.

First, we have analyzed in Chapter 3 energy and momentum conserving schemes in the context of molecular dynamics. Conserving schemes have been employed for more than four decades for the accurate and robust approximation of Hamiltonian problems in mechanics. However, in the field of molecular dynamics, where time integrators are routinely employed and are key to their usefulness, the use of conserving schemes has barely been explored. These second-order implicit methods are an interesting alternative to other commonly used integrators, and we have proven that, in addition to exhibiting exact energy conservation, they are more robust than the midpoint rule, the canonical second-order implicit method. We focused on the design of energy-momentum (EM) schemes for molecular dynamics in the view of three issues that are characteristic and unique to these problems: (i) the numerical treatment of dynamics in periodic domains, (ii) the discretization of three-body potentials, and (iii) the study of interatomic functional potentials. None of these three topics had been previously studied in the context of conserving schemes, to the authors' knowledge. However, the three of them are key for their implementation and clearly have an impact on their performance. Some of the most practical results of this Chapter are new expressions for EM approximations in fairly general problems in molecular dynamics. These approximations account for the three key issues mentioned before and can be shown to preserve linear momentum and energy exactly, while exhibiting a remarkable robustness.

In addition, in Chapter 4, we developed a GENERIC-based variational formulation of the IBVP for large-strain thermo-elastodynamics. The weak form of the problem directly emanates from the GENERIC evolution equation for open systems. This formulation makes possible the free choice of the thermodynamic state variable among the three options: (i) the absolute temperature θ, (ii) the entropy density η, and (iii) the internal energy density u. The GENERIC-based weak form is particularly well suited for the design of structure-preserving numerical schemes. This observation already holds true for the discretization in time employing of the standard mid-point rule. Depending on the specific

[1] The summary is based on [1–4]

choice of the thermodynamic state variable, specific balance laws are correctly reproduced in the discrete setting, independent of the size of the time-step. In particular, by using the internal energy density as the thermodynamic state variable, the resulting scheme is capable of consistently reproducing the balance of energy and thus has been termed energy-momentum (or $(\mathrm{EM})_u$) scheme. Similarly, choosing the entropy density as state variable leads to an entropy consistent method, termed momentum-entropy (or $(\mathrm{ME})_\eta$) scheme. However, none of the three newly developed mid-point-based schemes does preserve all of the key balance laws under discretization. In particular, it was shown that numerical instabilities can occur if the time-step size is too large for a prescribed duration of the simulation. This leads to the conclusion that all major balance laws should be maintained in the discrete setting in order to enhance the numerical stability.

Consequently, novel fully structure-preserving numerical methods for finite-strain thermoelasticity with heat conduction were developed in Chapter 5, starting from the newly developed GENERIC-based weak form. The proposed EME schemes enable the free choice of the thermodynamic state in the same manner. Each choice of the thermodynamic variable $\tau \in \{\eta, \theta, u\}$ leads to a corresponding $(\mathrm{EME})_\tau$ scheme. We have shown that two modifications of the mid-point-based approach are crucial for obtaining numerically stable EME schemes: (i) the introduction of specific projections in the wake of the space discretization and, (ii) the replacement of the standard derivatives of the internal energy density and the entropy density, respectively, by appropriate discrete derivatives. The in Chapter 2 introduced notion *GENERIC-consistent space discretization* is crucial with regard to the discretization in space. This notion asserts that a consistent space discretization should fit into the GENERIC framework for discrete systems. In particular, a GENERIC consistent space discretization automatically inherits the balance laws for energy and entropy, respectively, from the underlying continuous formulation. We have seen that the discretization in space, relying on standard Lagrangian shape functions, necessitates specific projections to reach a GENERIC-consistent space discretization. Even though the structure-preserving properties of the EME schemes are independent of the specific choice for $\tau \in \{\eta, \theta, u\}$, the absolute temperature can be regarded as the preferred choice for the constitutive description. In addition to that, temperature Dirichlet boundary conditions can be applied in the standard manner by using the temperature-based formulation. It was shown that the EME schemes lead to a significant improvement in the numerical stability when compared to mid-point type schemes.

Finally, in Chapter 6 the GENERIC-based approach was extended to more involved coupled thermomechanical problems, which also account for viscous (inelastic) deformations. Starting from a GENERIC-based formulation of large-strain thermo-viscoelasticity, we have developed alternative structure-preserving schemes based on the implicit mid-point rule. Depending on the choice of the thermodynamic variable, the plain mid-point rule yields partially structure-preserving schemes just as in the case for thermoelasticity. Also, in analogy to this case as well, these schemes cannot prevent numerical instabilities, as was shown in the numerical examples.

Upon summarizing the findings of the newly proposed GENERIC-based schemes, the following features are eminent:

- The underlying GENERIC formulation leads to characteristic expressions such as those for the first and second Piola-Kirchhoff stress tensors, the Mandel stress tensor, and the absolute temperature.
- The transformation properties of the GENERIC description facilitate the use of alternative thermodynamic variables such as the absolute temperature, the internal energy density, and the entropy density used in the present work.
- The newly proposed material form of the inelastic dissipative bracket makes it possible to include viscoelastic effects into the GENERIC formulation, which are often used nonlinear evolution laws for the internal variables associated with inelastic deformations.
- A GENERIC-consistent space discretization is essential when transitioning from infinitesimal dimensional to finite dimensional systems. This means that the evolution equations for the state variables of the semi-discrete system fit into the GENERIC framework for discrete systems. This is an essential prerequisite for the development of fully structure-preserving EME schemes.

7.2. Outlook

Based on the work contained within this thesis, some of the following lines of research are currently in progress or seem worthy of investing:

- The GENERIC-based semi-discrete weak form for the thermo-viscoelastic problem is well suited for the design of structure-preserving schemes because the GENERIC structure is preserved after the discretization in space. Thus, starting from the GENERIC-consistent space discretization it is verly likely that EME numerical schemes with enhanced numerical stability and robustness will emerge when applying the discrete gradient operator in analogy to Chapters 4 and 5.
- Volumetric constraints of incompressible rubber-like materials (e.g., polycarbonate) should be accounted for within the GENERIC framework, because viscous dissipative features for such materials are usually associated with isochoric deformations, whereas the thermal coupling is naturally associated with volumetric deformations. The resulting GENERIC-based EME methods would greatly extend the scope of structure-preserving methods to thermodynamical systems with constraints.
- The extension to non-smooth dissipative systems seems very promising, as effects like plasticity or damage are clearly irreversible and, therefore, the numerical method itself should align with this property. Incorporating an EME consistent time integration scheme would guarantee that the dissipated heat during such

irreversible transformations is not transferred to the mechanical field.[2] A first attempt in this direction can be found in [124] for a thermo-elastoplastic spring, but the transcription to infinitesimal dimensional systems is, to the author's knowledge, yet to be made.

- Despite their good properties, structure-preserving numerical methods have not yet been implemented in common software packages; one reason for this is the very elaborate implementation of the classical discrete gradient operator. A promising attempt to circumvent this problem is the usage of an alternative discrete gradient operator introduced in [154] (see [155] for several contributions in this direction). This alternative version utilizes a polyconvex structure for the Helmholtz free energy density function which, paired with the tensor-cross product introduced in [156], leads to a very simple and elegant structure for the discrete gradient operator and could consequently makes structure-preserving numerical methods more appealing for commercial software packages.
- Interestingly, the class of "GENERIC-integrators" has recently been coined [157] and can be viewed as an extension of symplectic integrators for Hamiltonian systems to the dissipative regime. They may link the present work to the class of variational integrators.
- Finally, another interesting avenue for future study is the development of mixed finite elements within the GENERIC framework. The combination of GENERIC-based EME methods with mixed finite elements gives rise to even more stability and accuracy, as mixed finite elements are known for their improved performance compared to classical finite elements like the ones used in this thesis. Finally, a further improvement of the spatial resolution could be obtained by incorporating the framework of isogeometrical analysis [158].

[2] Solely energy-consistent schemes correctly balance the amount of energy contained within the system but are generally not concerned about the direction of energy transfer and thus do not generally satisfy the second law of thermodyamics (e.g., see results of $(\mathrm{EM})_u$ methods in Sections 4.5 and 6.5).

A. Appendix to Chapter 4[1]

A.1. Linear thermoelasticity

We verify that the Helmholtz free energy density (4.36) yields an expression for the linearized stress tensor which is consistent with the theory of linear thermoelasticity. According to Section 4.2.4, the temperature-based expression for the first Piola-Kirchhoff stress tensor reads

$$\overline{\mathbf{P}} = \partial_{\nabla\varphi}\overline{u} - \theta\partial_{\nabla\varphi}\overline{\eta}$$

Taking into account the functions $\overline{\eta}(\nabla\varphi, \theta)$, and $\overline{u}(\nabla\varphi, \theta)$ summarized in Section 4.2.3, the first Piola-Kirchhoff stress tensor can be recast in the form

$$\overline{\mathbf{P}}(\mathbf{F}, \theta) = D\psi_1(\mathbf{F}) - (\theta - \theta_0)\psi_3'(J)\mathrm{cof}(\mathbf{F}) \tag{A.1}$$

It can be easily verified that the reference configuration is stress-free in the sense that $\overline{\mathbf{P}}(\mathbf{I}, \theta_0) = \mathbf{0}$. To perform the linearization of the stress tensor (A.1) about the stress-free reference configuration, consider one-parameter families of the deformation gradient and the temperature of the form

$$\begin{aligned} \mathbf{F}^\varepsilon &= \mathbf{I} + \varepsilon\nabla\mathbf{u} \\ \theta^\varepsilon &= \theta_0 + \varepsilon(\theta - \theta_0) \end{aligned}$$

Here, the displacement field is denoted by $\mathbf{u}$. Note that $\overline{\mathbf{P}}(\mathbf{F}^0, \theta^0) = \mathbf{0}$. The linearized stress tensor can now be calculated via

$$\boldsymbol{\sigma} := \left.\frac{d}{d\varepsilon}\overline{\mathbf{P}}(\mathbf{F}^\varepsilon, \theta^\varepsilon)\right|_{\varepsilon=0}$$

Using (A.1) along with the relationships

$$\begin{aligned} \left.\frac{d}{d\varepsilon}\det(\mathbf{F}^\varepsilon)\right|_{\varepsilon=0} &= \mathrm{tr}(\nabla\mathbf{u}) \\ \left.\frac{d}{d\varepsilon}\mathrm{cof}(\mathbf{F}^\varepsilon)\right|_{\varepsilon=0} &= \mathrm{tr}(\nabla\mathbf{u})\mathbf{I} - \nabla\mathbf{u}^{\mathrm{T}} \end{aligned}$$

a straightforward calculation yields

$$\boldsymbol{\sigma} = \mu\left(\nabla\mathbf{u} + \nabla\mathbf{u}^{\mathrm{T}}\right) + \lambda\mathrm{tr}(\nabla\mathbf{u})\mathbf{I} - \beta(\theta - \theta_0)(3\lambda + 2\mu)\mathbf{I}$$

[1] This Appendix is based on [1]

Introducing the infinitesimal strain tensor $\varepsilon = \frac{1}{2}(\nabla\mathbf{u} + \nabla\mathbf{u}^{\mathrm{T}})$, the linearized stress tensor can be rewritten as

$$\boldsymbol{\sigma} = 2\mu\varepsilon + \lambda\mathrm{tr}(\varepsilon)\mathbf{I} - 3K\beta(\theta - \theta_0)\mathbf{I} \tag{A.2}$$

Expression (A.2) for the stress tensor coincides with the generalized Duhamel-Neumann form of Hookes's law for isotropic behavior in the framework of linear thermoelasticity (cf. [159, Sec. 6.2]). Accordingly, λ and μ can be identified as the Lamé constants, $K = \lambda + \frac{2}{3}\mu$ is the bulk modulus, and β is the linear coefficient of thermal expansion.

A.2. Lyapunov function

Following [146] we consider the functional

$$\mathcal{L}' = \mathcal{E}' - \theta_0\mathcal{S}' \tag{A.3}$$

where θ_0 is the reference temperature. Note that the total energy, $\mathcal{E}'$, and the total entropy, $\mathcal{S}'$, have been defined in (4.39) and (4.41), respectively. The time derivative of (A.3) is given by

$$\frac{\mathrm{d}\mathcal{L}'}{\mathrm{d}t} = \frac{\mathrm{d}\mathcal{E}'}{\mathrm{d}t} - \theta_0\frac{\mathrm{d}\mathcal{S}'}{\mathrm{d}t}$$

Taking into account the results of Section 4.3.2, the last equation can be recast in the form

$$\begin{aligned}
\frac{\mathrm{d}\mathcal{L}'}{\mathrm{d}t} &= -\left(\{\mathcal{E}',\mathcal{E}'\}_{\mathrm{boun}} + [\mathcal{E}',\mathcal{S}']_{\mathrm{boun}}\right) - \theta_0\left([\mathcal{S}',\mathcal{S}'] - [\mathcal{S}',\mathcal{S}']_{\mathrm{boun}}\right) \\
&= \int_{\partial\mathcal{B}} \rho^{-1}\mathbf{p}\cdot\mathbf{PN}\,\mathrm{d}A - \int_{\partial\mathcal{B}} \mathbf{N}\cdot\mathbf{Q}\,\mathrm{d}A - \theta_0\left([\mathcal{S}',\mathcal{S}'] - \int_{\partial\mathcal{B}}\frac{1}{\theta}\mathbf{N}\cdot\mathbf{Q}\,\mathrm{d}A\right) \\
&= \mathcal{P}_{\mathrm{ext}} - \theta_0\left([\mathcal{S}',\mathcal{S}'] + \mathcal{G}_e\right)
\end{aligned}$$

Here, $\mathcal{P}_{\mathrm{ext}} = \int_{\partial\mathcal{B}} \rho^{-1}\mathbf{p}\cdot\bar{\mathbf{t}}\,\mathrm{d}A$ is the power expended on $\mathcal{B}$ by external Piola tractions $\bar{\mathbf{t}} = \mathbf{PN}$ acting on $\partial\mathcal{B}$. Moreover,

$$\mathcal{G}_e = \int_{\partial\mathcal{B}} \frac{\theta - \theta_0}{\theta_0\theta}\mathbf{N}\cdot\mathbf{Q}\,\mathrm{d}A \tag{A.4}$$

is the entropy produced at the interface between the body and the environment. The environment is called *thermally perfect* in the sense of [146], if $\mathcal{G}_e = 0$. In the numerical examples we consider the case that $\mathcal{G}_e = 0$ and $\mathcal{P}_{\mathrm{ext}} = 0$, so that

$$\frac{\mathrm{d}\mathcal{L}'}{\mathrm{d}t} = -\theta_0[\mathcal{S}',\mathcal{S}'] \leq 0$$

Then $\mathcal{L}'$ is non-increasing over time and thus a natural Lyapunov function.

A.3. Convergence criteria

To terminate the iterative solution procedure we use the norm of the residual vector as convergence criterion for both the $(\mathrm{ME})_\eta$ scheme and the $(\mathrm{M})_\theta$ scheme. Since the $(\mathrm{EM})_u$ scheme is capable to satisfy the balance of energy, we define an alternative termination criterion based on the discrete balance of energy:

$$|(\mathcal{E}^{\mathrm{h}}_{\mathrm{n}+1} - \mathcal{E}^{\mathrm{h}}_{\mathrm{n}}) - \Delta t \left(\mathcal{P}^{\mathrm{h,ext}}_{\mathrm{n}+\frac{1}{2}} + \mathcal{Q}^{\mathrm{h,int}}_{\mathrm{n}+\frac{1}{2}}\right)| \leq \varepsilon \tag{A.5}$$

Here, ε is the prescribed tolerance, $\mathcal{E}^{\mathrm{h}}_{\mathrm{n}}$ is the discrete version of the total energy (4.39) at time t_{n}, $\mathcal{P}^{\mathrm{h,ext}} = \int_{\partial\mathcal{B}} \rho^{-1} \mathbf{p}^{\mathrm{h}}_{\mathrm{n}+\frac{1}{2}} \cdot \overline{\mathbf{t}}^{\mathrm{h}}_{\mathrm{n}+\frac{1}{2}} \, \mathrm{d}A$ is the power expended on the body by external nodal forces due to external tractions, and $\mathcal{Q}^{\mathrm{h,int}} = -\int_{\partial\mathcal{B}} \overline{q}^{\mathrm{h}}_{\mathrm{n}+\frac{1}{2}} \, \mathrm{d}A$ is the heat flux across the boundary.

B. Appendix to Chapter 5[1]

B.1. Notes on the implementation

We provide a short outline of the implementation of the present EME schemes. To this end, the residual vector associated with node a can be written in the form

$$\mathbf{R}^{a^{\mathrm{T}}} = \left(\mathbf{R}_{\varphi}^{a^{\mathrm{T}}},\ \ \mathbf{R}_{\mathrm{p}}^{a^{\mathrm{T}}},\ \ R_{\tau}^{a},\ \ R_{u}^{a},\ \ R_{\eta}^{a}\right) \tag{B.1}$$

The nodal contributions to (B.1) emanate from weak form (5.65) by inserting the finite element approximations for the test functions leading to the expressions

$$\begin{aligned}
\mathbf{R}_{\varphi}^{a} =& \int_{\mathcal{B}} N^{a}\left(\frac{\boldsymbol{\varphi}_{\mathrm{n}+1}^{\mathrm{h}}-\boldsymbol{\varphi}_{\mathrm{n}}^{\mathrm{h}}}{\Delta t}-\mathbf{v}_{\mathrm{n}+\frac{1}{2}}^{\mathrm{h}}\right)\mathrm{d}V \\
\mathbf{R}_{\mathrm{p}}^{a} =& \int_{\mathcal{B}}\left(N^{a}\left(\rho\frac{\mathbf{v}_{\mathrm{n}+1}^{\mathrm{h}}-\mathbf{v}_{\mathrm{n}}^{\mathrm{h}}}{\Delta t}-\mathbf{b}\right)+\mathbf{F}_{\mathrm{n}+\frac{1}{2}}^{\mathrm{h}}\mathbf{S}_{\mathrm{algo}}^{\mathrm{h}}\nabla N^{a}\right)\mathrm{d}V-\int_{\partial_{\sigma}\mathcal{B}} N^{a}\overline{\mathbf{t}}\,\mathrm{d}A \\
R_{\tau}^{a} =& \int_{\mathcal{B}} N^{a}\left(\frac{\tau_{\mathrm{n}+1}^{\mathrm{h}}-\tau_{\mathrm{n}}^{\mathrm{h}}}{\Delta t}+\nabla\mathbf{v}_{\mathrm{n}+\frac{1}{2}}^{\mathrm{h}}:\left(\frac{2}{\Pi_{h}(\mathsf{D}_{\tau}\eta^{\mathrm{h}})}\mathbf{F}_{\mathrm{n}+\frac{1}{2}}^{\mathrm{h}}\mathsf{D}_{\mathbf{C}}\eta^{\mathrm{h}}\right)\right)\mathrm{d}V \\
&-\int_{\mathcal{B}}\nabla\left(\frac{N^{a}}{\Pi_{h}(\mathsf{D}_{\tau}u^{\mathrm{h}})}\right)\cdot\mathbf{Q}_{\mathrm{algo}}^{\mathrm{h}}\,\mathrm{d}V+\int_{\partial_{q}\mathcal{B}}\frac{N^{a}}{\Pi_{h}(\mathsf{D}_{\tau}u^{\mathrm{h}})}\overline{q}\,\mathrm{d}A \\
R_{u}^{a} =& H^{ab}(\mathsf{D}_{\tau}u)_{b}-\int_{\mathcal{B}} N^{a}\mathsf{D}_{\tau}u^{\mathrm{h}}\,\mathrm{d}V \\
R_{\eta}^{a} =& H^{ab}(\mathsf{D}_{\tau}\eta)_{b}-\int_{\mathcal{B}} N^{a}\mathsf{D}_{\tau}\eta^{\mathrm{h}}\,\mathrm{d}V
\end{aligned} \tag{B.2}$$

where, according to (5.66),

$$\begin{aligned}
\mathbf{S}_{\mathrm{algo}}^{\mathrm{h}} &= 2\left(\mathsf{D}_{\mathbf{C}}u^{\mathrm{h}}-\Theta_{\mathrm{algo}}^{\mathrm{h}}\mathsf{D}_{\mathbf{C}}\eta^{\mathrm{h}}\right) \\
\mathbf{Q}_{\mathrm{algo}}^{\mathrm{h}} &= (\Theta_{\mathrm{algo}}^{\mathrm{h}})^{2}\mathbf{K}_{\mathrm{algo}}^{\mathrm{h}}\nabla\left(\frac{1}{\Theta_{\mathrm{algo}}^{\mathrm{h}}}\right) = -\mathbf{K}_{\mathrm{algo}}^{\mathrm{h}}\nabla\Theta_{\mathrm{algo}}^{\mathrm{h}} \\
\Theta_{\mathrm{algo}}^{\mathrm{h}} &= \frac{\Pi_{h}(\mathsf{D}_{\tau}u^{\mathrm{h}})}{\Pi_{h}(\mathsf{D}_{\tau}\eta^{\mathrm{h}})} = \frac{N^{a}(\mathsf{D}_{\tau}u)_{a}}{N^{b}(\mathsf{D}_{\tau}\eta)_{b}}
\end{aligned}$$

[1] This Appendix is based on [2]

Note that eventually (B.2) leads to a nonlinear algebraic system of equations, $\mathbf{R}^a = \mathbf{0}$, $a = 1, \dots, \mathrm{N}$, for the determination of the nodal state variables $\mathbf{q}_{a_{n+1}}$, $\mathbf{v}_{a_{n+1}}$, and $\tau_{a_{n+1}}$ along with the nodal quantities $(\mathsf{D}_\tau u)_a$, $(\mathsf{D}_\tau \eta)_a$ related to the projections $\Pi_h(\mathsf{D}_\tau u^{\mathrm{h}})$ and $\Pi_h(\mathsf{D}_\tau \eta^{\mathrm{h}})$, cf. (5.72). As usual, we apply Newton's method for that purpose. The consistent linearization of $\mathbf{R}^a$ can be performed in a straightforward way. In this connection, the projection formulas $\Pi_h(\mathsf{D}_\tau u^{\mathrm{h}})$ and $\Pi_h(\mathsf{D}_\tau \eta^{\mathrm{h}})$ do not pose any additional difficulties. For example, the linearization of the algorithmic temperature assumes the form

$$\Delta\Theta^{\mathrm{h}}_{\mathrm{algo}} = \Theta^{\mathrm{h}}_{\mathrm{algo}} N^a \left(\frac{\Delta(\mathsf{D}_\tau u)_a}{\Pi_h(\mathsf{D}_\tau u^{\mathrm{h}})} - \frac{\Delta(\mathsf{D}_\tau \eta)_a}{\Pi_h(\mathsf{D}_\tau \eta^{\mathrm{h}})} \right)$$

Moreover, the material gradient of the algorithmic temperature is given by

$$\nabla\Theta^{\mathrm{h}}_{\mathrm{algo}} = \Theta^{\mathrm{h}}_{\mathrm{algo}} \nabla N^a \left(\frac{(\mathsf{D}_\tau u)_a}{\Pi_h(\mathsf{D}_\tau u^{\mathrm{h}})} - \frac{(\mathsf{D}_\tau \eta)_a}{\Pi_h(\mathsf{D}_\tau \eta^{\mathrm{h}})} \right)$$

leading to the corresponding linearized form

$$\begin{aligned} \Delta(\nabla\Theta^{\mathrm{h}}_{\mathrm{algo}}) = {} & \Theta^{\mathrm{h}}_{\mathrm{algo}} \nabla N^a \left(\delta^b_a - \frac{(\mathsf{D}_\tau \eta)_a}{\Pi_h(\mathsf{D}_\tau \eta^{\mathrm{h}})} N^b \right) \frac{\Delta(\mathsf{D}_\tau u)_b}{\Pi_h(\mathsf{D}_\tau u^{\mathrm{h}})} \\ & - \Theta^{\mathrm{h}}_{\mathrm{algo}} \nabla N^a \left(\delta^b_a - 2\frac{(\mathsf{D}_\tau \eta)_a}{\Pi_h(\mathsf{D}_\tau \eta^{\mathrm{h}})} N^b + \frac{(\mathsf{D}_\tau u)_a}{\Pi_h(\mathsf{D}_\tau u^{\mathrm{h}})} N^b \right) \frac{\Delta(\mathsf{D}_\tau \eta)_b}{\Pi_h(\mathsf{D}_\tau \eta^{\mathrm{h}})} \end{aligned}$$

where δ^b_a is the Kronecker delta. Concerning the termination criterion for the Newton iterations we apply the nonstandard form

$$\left| \mathcal{E}^{\mathrm{h}}_{\mathrm{n+1}} - \mathcal{E}^{\mathrm{h}}_{\mathrm{n}} + \Delta t \left(\int_{\partial_q \mathcal{B}} \bar{q}_{\mathrm{n}+\frac{1}{2}} \,\mathrm{d}A - \int_{\partial_\sigma \mathcal{B}} \mathbf{v}^{\mathrm{h}}_{\mathrm{n}+\frac{1}{2}} \cdot \bar{\mathbf{t}}_{\mathrm{n}+\frac{1}{2}} \,\mathrm{d}A \right) \right| \leq \varepsilon \tag{B.3}$$

which is made possible due to the energy consistency of the present $(\mathrm{EME})_\tau$ schemes. In (B.3), ε denotes the prescribed numerical tolerance for the Newton iterations. The left-hand side of (B.3) can be directly linked to the discrete balance of energy (5.80). Note that here, the standard contribution of the external tractions has been taken into account.

C. Appendix to Chapter 6[1]

C.1. Rotational symmetry

Due to invariance under superposed rigid rotations the present problem satisfies symmetry properties (6.93). To see this, we consider a rotation tensor $\mathbf{Q}^{\varepsilon} = \exp(\varepsilon\widehat{\boldsymbol{\xi}})$, expressed through the Rodrigues formula [24]. Here, ε is a scalar parameter and $\widehat{\boldsymbol{\xi}}$ is a skew-symmetric tensor with associated vector $\boldsymbol{\xi} \in \mathbb{R}^3$, such that $\widehat{\boldsymbol{\xi}}\mathbf{a} = \boldsymbol{\xi} \times \mathbf{a}$ for all $\mathbf{a} \in \mathbb{R}^3$. Now, a superposed rigid motion gives rise to rotated nodal position vectors

$$\mathbf{q}_a^{\varepsilon} = \mathbf{Q}^{\varepsilon}\mathbf{q}_a \,. \tag{C.1}$$

Note that $\mathbf{q}_a^0 = \mathbf{q}_a$ and $\frac{d}{d\varepsilon}\mathbf{Q}^{\varepsilon}\big|_{\varepsilon=0} = \widehat{\boldsymbol{\xi}}$. Nodal pattern (C.1) of the rigidly rotated discrete system gives rise to the corresponding expression for the discrete deformation gradient (cf. $(6.53)_1$)

$$\mathbf{F}^{\mathrm{h},\varepsilon} = \mathbf{q}_a^{\varepsilon} \otimes \nabla N^a = \mathbf{Q}^{\varepsilon}\mathbf{F}^{\mathrm{h}} \,. \tag{C.2}$$

Furthermore, the corresponding right Cauchy-Green deformation tensor follows from (6.3) and is given by

$$\mathbf{C}^{\mathrm{h},\varepsilon} = (\mathbf{F}^{\mathrm{h},\varepsilon^{\mathrm{T}}})\mathbf{F}^{\mathrm{h},\varepsilon} = \mathbf{C}^{\mathrm{h}} \,, \tag{C.3}$$

where property $(\mathbf{Q}^{\varepsilon})^{\mathrm{T}} = (\mathbf{Q}^{\varepsilon})^{-1}$ of the rotation tensor has been taken into account. A frame-indifferent formulation implies that the internal energy density takes the form $u(\mathbf{F}, \tau, \mathbf{C}_p^{-1}) = u(\mathbf{C}, \tau, \mathbf{C}_p^{-1})$. Thus, taking the derivative of $u(\mathbf{F}^{\mathrm{h},\varepsilon}, \tau, \mathbf{C}_p^{-1}) = u(\mathbf{C}^h, \tau, \mathbf{C}_p^{-1})$ with respect to parameter ε and subsequently setting $\varepsilon = 0$ yields

$$\begin{aligned} 0 &= \frac{d}{d\varepsilon}\bigg|_{\varepsilon=0} u\left(\mathbf{F}_g^{\mathrm{h},\varepsilon}, \tau_g^{\mathrm{h}}, (\mathbf{C}_p^{-1})_g\right) \\ &= \partial_{\mathbf{F}} u_g : \frac{d}{d\varepsilon}\bigg|_{\varepsilon=0} \mathbf{q}_a^{\varepsilon} \otimes \nabla N_g^a \\ &= \partial_{\mathbf{F}} u_g : \left(\widehat{\boldsymbol{\xi}}\mathbf{q}_a \otimes \nabla N_g^a\right) \\ &= (\boldsymbol{\xi} \times \mathbf{q}_a) \cdot \partial_{\mathbf{F}} u_g \nabla N_g^a \\ &= \boldsymbol{\xi} \cdot \left(\mathbf{q}_a \times \partial_{\mathbf{F}} u_g \nabla N_g^a\right) \,, \end{aligned} \tag{C.4}$$

[1] This Appendix is based on [4]

where u_g is given by $(6.52)_1$. Due to the arbitrariness of $\boldsymbol{\xi} \in \mathbb{R}^3$ (and vanishing body forces), the last equation coincides with symmetry condition $(6.93)_1$, that is

$$\mathbf{q}_a \times \partial_{\mathbf{q}_a} \mathcal{E} = \mathbf{0} \,.$$

Note that this condition holds as well for the specific choice $\tau = u$, due to $(6.97)_1$.

Similarly, symmetry condition $(6.93)_2$ results from the frame-indifference of the entropy density function $\eta(\mathbf{F}, \tau, \mathbf{C}_p^{-1}) = \eta(\mathbf{C}, \tau, \mathbf{C}_p^{-1})$, or from $(6.100)_1$ in the case $\tau = \eta$.

Declaration of Authorship

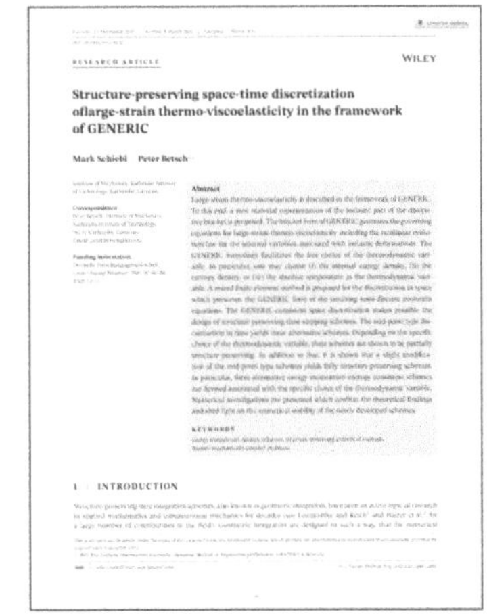
RESEARCH ARTICLE

WILEY

Structure-preserving space-time discretization oflarge-strain thermo-viscoelasticity in the framework of GENERIC

Mark Schiebl | Peter Betsch

Abstract

KEYWORDS

1 | INTRODUCTION

Citation:
Schiebl, M. & Betsch, P. Structure-preserving space-time discretization of large-strain thermo-viscoelasticity in the framework of GENERIC. *International Journal for Numerical Methods in Engineering,* 1–41 (2021)
Declaration:
Mark Schiebl (MS) worked on the theory together with Peter Betsch (PB). MS was responsible for the numerical implementation. Moreover, MS performed the numerical verification of the newly developed formalism. Furthermore, MS and PB co-worked on the manuscript. Finally, the article was reviewed by all authors (MS, PB).

ORIGINAL PAPER

Energy-momentum conserving integration schemes for molecular dynamics

Mark Schiebl · Ignacio Romero

Abstract

Keywords

1 Introduction

Citation:
Schiebl, M. & Romero, I. Energy-momentum conserving integration schemes for molecular dynamics. *Computational Mechanics* **67,** 915–935 (2021)
Declaration:
Mark Schiebl (MS) worked on the theory together with Ignacio Romero (IR). MS was responsible for the numerical implementation. Moreover, MS performed the numerical verification of the newly developed formalism. Furthermore, MS and IR co-worked on the manuscript. Finally, the article was reviewed by all authors (MS, IR).

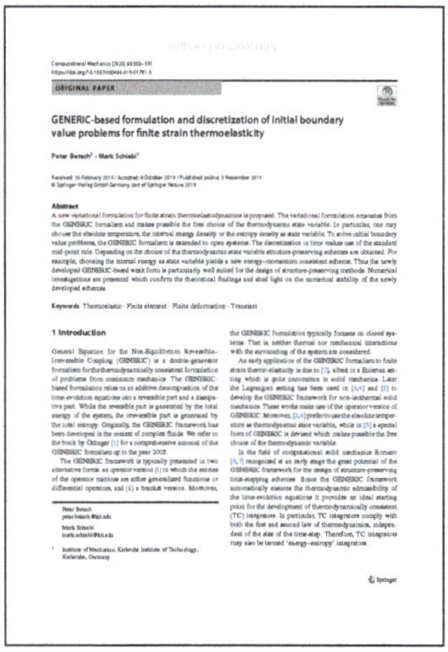
ORIGINAL PAPER

GENERIC-based formulation and discretization of initial boundary value problems for finite strain thermoelasticity

Peter Betsch · Mark Schiebl

Citation:
Betsch, P. & Schiebl, M. GENERIC-based formulation and discretization of initial boundary value problems for finite strain thermoelasticity. *Computational Mechanics* **65,** 503–531 (2020)

Declaration:
Mark Schiebl (MS) worked on the theory together with Peter Betsch (PB). MS was responsible for the numerical implementation. Moreover, MS performed the numerical verification of the newly developed formalism. Furthermore, MS and PB co-worked on the manuscript. Finally, the article was reviewed by all authors (MS, PB).

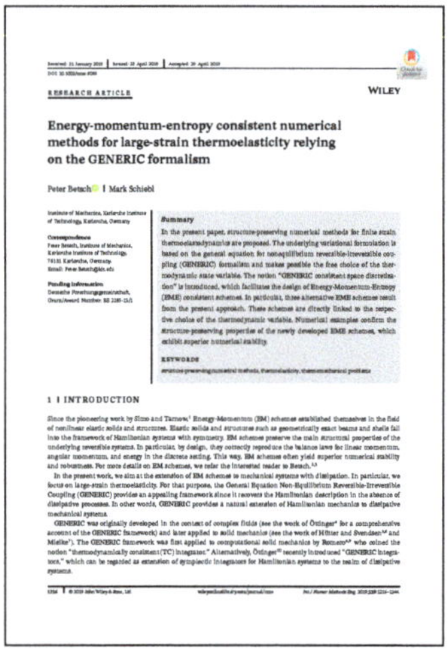
RESEARCH ARTICLE

WILEY

Energy-momentum-entropy consistent numerical methods for large-strain thermoelasticity relying on the GENERIC formalism

Peter Betsch | Mark Schiebl

1 | INTRODUCTION

Citation:
Betsch, P. & Schiebl, M. Energy-momentum-entropy consistent numerical methods for large-strain thermoelasticity relying on the GENERIC formalism. *International Journal for Numerical Methods in Engineering* **119,** 1216–1244 (2019)

Declaration:
Mark Schiebl (MS) worked on the theory together with Peter Betsch (PB). MS was responsible for the numerical implementation. Moreover, MS performed the numerical verification of the newly developed formalism. Furthermore, MS and PB co-worked on the manuscript. Finally, the article was reviewed by all authors (MS, PB).

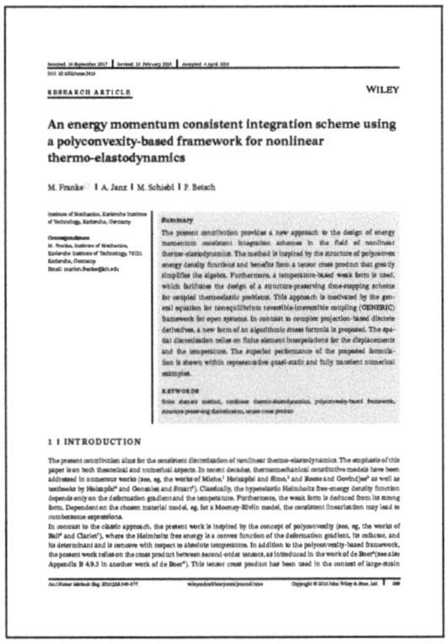

RESEARCH ARTICLE

WILEY

An energy momentum consistent integration scheme using a polyconvexity-based framework for nonlinear thermo-elastodynamics

M. Franke | A. Janz | M. Schiebl | P. Betsch

Summary

The present contribution provides a new approach to the design of energy momentum consistent integration schemes in the field of nonlinear thermo-elastodynamics. The method is inspired by the structure of polyconvex energy density functions and benefits from a tensor cross product that greatly simplifies the algebra. Furthermore, a temperature-based weak form is used, which facilitates the design of a structure-preserving time-stepping scheme for coupled thermoelastic problems. This approach is motivated by the general equation for nonequilibrium reversible-irreversible coupling (GENERIC) framework for open systems. In contrast to complex projection-based discrete derivatives, a new form of an algorithmic stress formula is proposed. The spatial discretization relies on finite element interpolations for the displacements and the temperature. The superior performance of the proposed formulation is shown within representative quasi-static and fully transient numerical examples.

1 | INTRODUCTION

The present contribution aims for the consistent discretization of nonlinear thermo-elastodynamics. The emphasis of this paper is on both theoretical and numerical aspects. In recent decades, thermomechanical constitutive models have been addressed in numerous works (see, eg, the works of Miehe,[1] Holzapfel and Simo,[2] and Reese and Govindjee[3] as well as textbooks by Holzapfel[4] and Gonzalez and Stuart[5]). Classically, the hyperelastic Helmholtz free-energy density function depends only on the deformation gradient and the temperature. Furthermore, the weak form is deduced from its strong form. Dependent on the chosen material model, eg, for a Mooney-Rivlin model, the consistent linearization may lead to cumbersome expressions.

In contrast to the classic approach, the present work is inspired by the concept of polyconvexity (see, eg, the works of Ball[6] and Ciarlet[7]), where the Helmholtz free energy is a convex function of the deformation gradient, its cofactor, and its determinant and is concave with respect to absolute temperature. In addition to the polyconvexity-based framework, the present work relies on the cross product between second-order tensors, as introduced in the work of de Boer[8] (see also Appendix B 4.9.3 in another work of de Boer[9]). This tensor cross product has been used in the context of large-strain

Citation:

Franke, M., Janz, A., Schiebl, M. & Betsch, P. An energy momentum consistent integration scheme using a polyconvexity-based framework for nonlinear thermo-elastodynamics. *International Journal for Numerical Methods in Engineering* **115,** 549–577 (2018)

Declaration:

Mark Schiebl (MS) worked on the theory together with Marlon Franke (MF) and Alexander Janz (AJ). MS, MF and AJ were responsible for the numerical implementation. Moreover, MS, MF and AJ performed the numerical verification of the newly developed formalism. Furthermore, MS, MF and AJ co-worked on the manuscript supported by Peter Betsch (PB). Finally, the article was reviewed by all authors (MS, MF, AJ, PB).

Publications and talks

Journal articles

- Schiebl, M. & Betsch, P. Structure-preserving space-time discretization of large-strain thermo-viscoelasticity in the framework of GENERIC. *International Journal for Numerical Methods in Engineering,* 1–41 (2021)
- Schiebl, M. & Romero, I. Energy-momentum conserving integration schemes for molecular dynamics. *Computational Mechanics* **67,** 915–935 (2021)
- Betsch, P. & Schiebl, M. GENERIC-based formulation and discretization of initial boundary value problems for finite strain thermoelasticity. *Computational Mechanics* **65,** 503–531 (2020)
- Betsch, P. & Schiebl, M. Energy-momentum-entropy consistent numerical methods for large-strain thermoelasticity relying on the GENERIC formalism. *International Journal for Numerical Methods in Engineering* **119,** 1216–1244 (2019)
- Franke, M., Janz, A., Schiebl, M. & Betsch, P. An energy momentum consistent integration scheme using a polyconvexity-based framework for nonlinear thermo-elastodynamics. *International Journal for Numerical Methods in Engineering* **115,** 549–577 (2018)

Conference proceedings

- Schiebl, M. & Betsch, P. *Energy-Momentum-Entropy consistent numerical methods for thermomechanically solids based on the GENERIC formalism* in *Proceedings of the 6th European Conference on Computational Mechanics, Glasgow, UK* (2018)
- Betsch, P. & Schiebl, M. *Variational Formulations for large strain thermo-elastodynamics based on the GENERIC formalism* in *Proceedings of the 6th European Conference on Computational Mechanics, Glasgow, UK* (2018)
- Schiebl, M. & Betsch, P. A general framework for the thermodynamically consistent time integration of open nonlinear thermoelastic systems. *Proceedings in Applied Mathematics and Mechanics* **18** (2018)

- Ströhle, T., Betsch, P. & Schiebl, M. *Galerkin-based mixed time finite elements for structural dynamics* in *PAMM - Proceedings in Applied Mathematics and Mechanics* **18** (2018)
- Schiebl, M., Betsch, P. & Hesch, C. *A temperature-based GENERIC approach for the thermodynamically consistent integration of thermoelastic solids* in *PAMM - Proceedings in Applied Mathematics and Mechanics* **17** (2017)
- Schiebl, M., Betsch, P. & Hesch, C. *Thermodynamically consistent integration of coupled thermoelastic systems* in *PAMM - Proceedings in Applied Mathematics and Mechanics* **16** (2016)

Conference talks

- Schiebl, M., Romero, I.: Structure-preserving integration of molecular dynamics. 14th World Congress in Computational Mechanics (WCCM), Virtual Congress, January 11-15, 2021
- Schiebl, M., Betsch, P.: GENERIC based thermodynamically consistent discretisation in space and time for open thermomechanical systems. VIIIth International Conference on Computational Methods for Coupled Problems in Science and Engineering, COUPLED PROBLEMS 2019, Sitges (Barcelona), Spain, June 3-5, 2019
- Schiebl, M., Betsch, P.: GENERIC consistent discretisation in space and Energy-Momentum-Entropy consistent schemes for open thermomechanical systems. 90th Annual Meeting of the International Association of Applied Mathematics and Mechanics (GAMM), Vienna, Austria, Februrary 18 - 22, 2019
- Schiebl, M., Betsch, P.: GENERIC-based numerical methods for the thermodynamically consistent simulation of coupled thermomechanical solids. 13th World Congress in Computational Mechanics (WCCM), New York City, NY, USA, July 22-27, 2018
- Schiebl, M., Betsch, P.: Energy-momentum-entropy consistent numerical methods for thermomechanically solids based on the GENERIC formalism. 6th European Conference on Computational Mechanics (ECCM), Glasgow, UK, June 11-15, 2018
- Schiebl, M., Betsch, P.: Comparison of general thermodynamically consistent time integration methods for open nonlinear thermoelastic systems. 89th Annual Meeting of the International Association of Applied Mathematics and Mechanics (GAMM), Munich, Germany, March 19 - 23, 2018
- Schiebl, M., Betsch, P.: A temperature-based approach for the thermodynamically consistent integration of thermoelastic solids using the GENERIC framework for open systems. 3rd International Conference on Computational Models for Solids and Fluids - ECCOMAS MSF 2017, Ljubljana, Slovenia, September 20 - 22, 2017

- Schiebl, M., Betsch, P., Hesch, C.: Thermodynamically consistent integration of thermoelastic solids using a temperature-based GENERIC framework. VIIth International Conference on Computational Methods for Coupled Problems in Science and Engineering, COUPLED PROBLEMS 2017, Rhodes Island, Greece, June 12-14, 2017
- Schiebl, M., Betsch, P., Hesch, C.: A temperature-based GENERIC approach for the thermodynamically consistent integration of thermoelastic solids. 88th Annual Meeting of the International Association of Applied Mathematics and Mechanics (GAMM), Weimar, Germany, March 06 - 10, 2017
- Schiebl, M., Betsch, P., Hesch, C.: Thermodynamically consistent integration of coupled thermoelastic systems. 87th Annual Meeting of International Association of Applied Mathematics and Mechanics (GAMM), Braunschweig, Germany, March 07 - 11, 2016

Bibliography

[1] Betsch, P. & Schiebl, M. GENERIC-based formulation and discretization of initial boundary value problems for finite strain thermoelasticity. *Computational Mechanics* **65,** 503–531 (2020).

[2] Betsch, P. & Schiebl, M. Energy-momentum-entropy consistent numerical methods for large-strain thermoelasticity relying on the GENERIC formalism. *International Journal for Numerical Methods in Engineering* **119,** 1216–1244 (2019).

[3] Schiebl, M. & Romero, I. Energy-momentum conserving integration schemes for molecular dynamics. *Computational Mechanics* **67,** 915–935 (2021).

[4] Schiebl, M. & Betsch, P. Structure-preserving space-time discretization of large-strain thermo-viscoelasticity in the framework of GENERIC. *International Journal for Numerical Methods in Engineering,* 1–41 (2021).

[5] Sanz-Serna, J. M. & Calvo, M. P. *Numerical Hamiltonian Problems* (Dover Publications Inc., 2018).

[6] Leimkuhler, B. & Reich, S. *Simulating Hamiltonian Dynamics* (Cambridge University Press, 2004).

[7] Hairer, E., Lubich, C. & Wanner, G. *Geometric Numerical Integration* (Springer Science & Business Media, 2006).

[8] De Vogelaere, R. Methods of integration which preserve the contact transformation property of the Hamilton equations. *Technical report (University of Notre Dame. Department of Mathematics)* (1956).

[9] Ruth, R. D. A canonical integration technique. *IEEE Transactions on Nuclear Science* **30,** 2669–2671 (1983).

[10] Candy, J. & Rozmus, W. A symplectic integration algorithm for separable Hamiltonian functions. *Journal of Computational Physics* **92,** 230–256 (1991).

[11] McLachlan, R. I. & Atela, P. The accuracy of symplectic integrators. *Nonlinearity* **5,** 541–562 (1992).

[12] Feng, K. & Qin, M.-z. *The symplectic methods for the computation of hamiltonian equations* in *Numerical Methods for Partial Differential Equations* (eds Zhu, Y.-I. & Guo, B.-y.) (Springer Berlin Heidelberg, Berlin, Heidelberg, 1987), 1–37.

[13] Channell, P. J. & Scovel, C. Symplectic integration of Hamiltonian systems. *Nonlinearity* **3,** 231–259 (1990).

[14] Feng, K. Difference schemes for Hamiltonian formalism and symplectic geometry. *Journal of Computational Mathematics* **4,** 279–289 (1986).

[15] Sanz-Serna, J. M. Runge-Kutta schemes for Hamiltonian systems. *Bit* **28,** 877–883 (1988).

[16] Lasagni, F. M. Canonical Runge-Kutta methods. *ZAMP Zeitschrift für angewandte Mathematik und Physik / Journal of Applied Mathematics and Physics* **39,** 952–953 (1988).

[17] Zhong, G. & Marsden, J. E. Lie-Poisson Hamilton-Jacobi theory and Lie-Poisson integrators. *Physics Letters A* **133,** 134–139 (1988).

[18] Maeda, S. Lagrangian formulation of discrete systems and concept of difference space. *Math. Japon* **27,** 345–356 (1982).

[19] Veselov, A. P. Integrable discrete-time systems and difference operators. *Functional Analysis and Its Applications* **22,** 83–93 (1988).

[20] Moser, J. & Veselov, A. P. Discrete versions of some classical integrable systems and factorization of matrix polynomials. *Communications in Mathematical Physics* **139,** 217–243 (1991).

[21] Marsden, J. E. & West, M. Discrete mechanics and variational integrators. *Acta Numerica 2001* **10,** 357–514 (2001).

[22] Marsden, J. E., Patrick, G. W. & Shkoller, S. Multisymplectic geometry, variational integrators, and nonlinear PDEs. *Communications in Mathematical Physics* **199,** 351–395 (1998).

[23] Lew, A. J., Marsden, J. E., Ortiz, M. & West, M. Variational time integrators. *International Journal for Numerical Methods in Engineering* **60,** 153–212 (2004).

[24] Marsden, J. E. & Ratiu, T. S. *Introduction to mechanics and symmetry* (Springer New York, 1999).

[25] Marsden, J. E., Pekarsky, S. & Shkoller, S. Symmetry reduction of discrete Lagrangian mechanics on Lie groups. *Journal of Geometry and Physics* **36,** 140–151 (2000).

[26] Mata Almonacid, P. Explicit symplectic momentum-conserving time-stepping scheme for the dynamics of geometrically exact rods. *Finite Elements in Analysis and Design* **96,** 11–22 (2015).

[27] Demoures, F., Gay-Balmaz, F., Kobilarov, M. & Ratiu, T. S. Multisymplectic Lie group variational integrator for a geometrically exact beam in R3. *Communications in Nonlinear Science and Numerical Simulation* **19,** 3492–3512 (2014).

[28] Grinspun, E., Hirani, A. N., Desbrun, M. & Schröder, P. Discrete shells. *Proceedings of the 2003 ACM SIGGRAPH/Eurographics Symposium on Computer Animation, SCA 2003,* 62–68 (2003).

[29] Wolff, S. & Bucher, C. Asynchronous collision integrators: Explicit treatment of unilateral contact with friction and nodal restraints. *International Journal for Numerical Methods in Engineering* **95,** 562–586 (2013).

[30] Betsch, P., Hesch, C., Sänger, N. & Uhlar, S. Variational Integrators and Energy-Momentum Schemes for Flexible Multibody Dynamics. *Journal of Computational and Nonlinear Dynamics* **5,** 1–11 (2010).

[31] Jiménez, F., Kobilarov, M. & Martín De Diego, D. Discrete variational optimal control. *Journal of Nonlinear Science* **23,** 393–426 (2013).

[32] Wang, L., Hong, J., Scherer, R. & Bai, F. Dynamics and variational integrators of stochastic Hamiltonian systems. *International Journal of Numerical Analysis and Modeling* **6,** 586–602 (2009).

[33] Leyendecker, S., Marsden, J. E. & Ortiz, M. Variational integrators for constrained dynamical systems. *ZAMM Zeitschrift fur Angewandte Mathematik und Mechanik* **88,** 677–708 (2008).

[34] Gawlik, E. S., Mullen, P., Pavlov, D., Marsden, J. E. & Desbrun, M. Geometric, variational discretization of continuum theories. *Physica D: Nonlinear Phenomena* **240,** 1724–1760 (2011).

[35] Lew, A. J. & Mata, P. in *Structure-preserving Integrators in Nonlinear Structural Dynamics and Flexible Multibody Dynamics* (ed Betsch, P.) 201–291 (Springer, 2016).

[36] Maugin, G. A. & Kalpakides, V. K. A Hamiltonian formulation for elasticity and thermoelasticity. *Journal of Physics A: Mathematical and General* **35,** 10775–10788 (2002).

[37] Mata, P. & Lew, A. J. Variational integrators for the dynamics of thermo-elastic solids with finite speed thermal waves. *Journal of Computational Physics* **257,** 1423–1443 (2014).

[38] Kern, D., Bär, S. & Groß, M. Variational Integrators for Thermomechanical Coupled Dynamic Systems with Heat Conduction. *PAMM - Proceedings in Applied Mathematics and Mechanics* **14,** 47–48 (2014).

[39] Kern, D., Romero, I., Martín, S. C. & García Orden, J. C. Performance Assessment of Variational Integrators for Thermomechanical Problems. *Journal of Theoretical and Applied Mechanics (Bulgaria)* **48,** 3–23 (2018).

[40] LaBudde, R. A. & Greenspan, D. Energy and momentum conserving methods of arbitrary order for the numerical integration of equations of motion - I. Motion of a Single of particles. *Numerische Mathematik* **25,** 323–346 (1976).

[41] LaBudde, R. A. & Greenspan, D. Energy and momentum conserving methods of arbitrary order for the numerical integration of equations of motion - II. Motion of a System of Particles. *Numerische Mathematik* **26,** 1–16 (1976).

[42] Simo, J. C. & Wong, K. K. Unconditionally stable algorithms for rigid body dynamics that exactly preserve energy and momentum. *International Journal for Numerical Methods in Engineering* **31,** 19–52 (1991).

[43] Simo, J. C., Tarnow, N. & Wong, K. K. Exact energy-momentum conserving algorithms and symplectic schemes for nonlinear dynamics. *Computer Methods in Applied Mechanics and Engineering* **100,** 63–116 (1992).

[44] Greenspan, D. Conservative Numerical Methods for $\ddot{x} = f(x)$. *Journal of Computational Physics* **56,** 28–41 (1984).

[45] Simo, J. C. & Tarnow, N. The discrete energy-momentum method. Conserving algorithms for nonlinear elastodynamics. *Journal of Applied Mathematics and Physics* **43,** 757–792 (1992).

[46] Gonzalez, O. *Design and analysis of conserving integrators for nonlinear hamiltonian systems with symmetry.* PhD thesis (Stanford University Stanford, CA, 1996).

[47] Gonzalez, O. Exact energy and momentum conserving algorithms for general models in nonlinear elasticity. *Computer Methods in Applied Mechanics and Engineering* **190,** 1763–1783 (2000).

[48] Laursen, T. A. & Meng, X. N. A new solution procedures for application of energy-conserving algorithms to general constitutive models in nonlinear elastodynamics. *Computer Methods in Applied Mechanics and Engineering* **190,** 6309–6322 (2001).

[49] Simo, J. C. & Tarnow, N. A new energy and momentum conserving algorithm for the non-linear dynamics of shells. *International Journal for Numerical Methods in Engineering* **37,** 2527–2549 (1994).

[50] Zhong, H. G. & Crisfield, M. A. An energy-conserving co-rotational procedure for the dynamics of shell structures. *Engineering Computations* **15,** 552–576 (1998).

[51] Campello, E. d. M. B., Pimenta, P. M. & Wriggers, P. An exact conserving algorithm for nonlinear dynamics with rotational DOFs and general hyperelasticity. Part 2: Shells. *Computational Mechanics* **48,** 195–211 (2011).

[52] Simo, J. C., Tarnow, N. & Doblare, M. Non-linear dynamics of three-dimensional rods: Exact energy and momentum conserving algorithms. *International Journal for Numerical Methods in Engineering* **38,** 1431–1473 (1995).

[53] Romero, I. & Armero, F. An objective finite element approximation of the kinematics of geometrically exact rods and its use in the formulation of an energy-momentum conserving scheme in dynamics. *International Journal for Numerical Methods in Engineering* **54,** 1683–1716 (2002).

[54] Pimenta, P. M., Campello, E. d. M. B. & Wriggers, P. An exact conserving algorithm for nonlinear dynamics with rotational DOFs and general hyperelasticity. Part 1: Rods. *Computational Mechanics* **42,** 715–732 (2008).

[55] García Orden, J. C. & Goicolea, J. M. Conserving Properties in Constrained Dynamics of Flexible Multibody Systems. *Multibody System Dynamics* **4,** 225–244 (2000).

[56] Betsch, P. & Uhlar, S. Energy-momentum conserving integration of multibody dynamics. *Multibody System Dynamics* **17,** 243–289 (2007).

[57] McLachlan, R. I., Quispel, G. R. W. & Robidoux, N. Geometric integration using discrete gradients. *Philosophical Transactions of the Royal Society A: Mathematical, Physical and Engineering Sciences* **357,** 1021–1045 (1999).

[58] Celledoni, E., Grimm, V., McLachlan, R. I., McLaren, D. I., O'Neale, D., Owren, B. & Quispel, G. R. W. Preserving energy resp. dissipation in numerical PDEs using the "Average Vector Field" method. *Journal of Computational Physics* **231,** 6770–6789 (2012).

[59] Betsch, P. *Structure-preserving Integrators in Nonlinear Structural Dynamics and Flexible Multibody Dynamics* (Springer, 2016).

[60] Newmark, N. M. A method of computation for structural dynamics. *Journal of the engineering mechanics division* **85,** 67–94 (1959).

[61] Hilber, H. M., Hughes, T. J. R. & Taylor, R. L. Improved numerical dissipation for time integration algorithms in structural dynamics. *Earthquake Engineering & Structural Dynamics* **5,** 283–292 (1977).

[62] Wilson, E. L. *A computer program for the dynamic stress analysis of underground structures.* Technical report (California Univ Berkeley Structural Engineering Lab, 1968).

[63] Kuhl, D. & Crisfield, M. A. Energy-conserving and decaying algorithms in non-linear structural dynamics. *International Journal for Numerical Methods in Engineering* **45,** 569–599 (1999).

[64] Bauchau, O. A. & Theron, N. J. Energy decaying scheme for nonlinear elastic multi-body systems. *Computers and Structures* **59,** 317–330 (1996).

[65] Kuhl, D. & Ramm, E. Generalized Energy-Momentum Method for non-linear adaptive shell dynamics. *Computer Methods in Applied Mechanics and Engineering* **178,** 343–366 (1999).

[66] Armero, F. & Romero, I. On the formulation of high-frequency dissipative time-stepping algorithms for nonlinear dynamics. Part II: Second-order methods. *Computer Methods in Applied Mechanics and Engineering* **190,** 6783–6824 (2001).

[67] Armero, F. & Romero, I. On the formulation of high-frequency dissipative time-stepping algorithms for nonlinear dynamics. Part I: Low-order methods for two model problems and nonlinear elastodynamics. *Computer Methods in Applied Mechanics and Engineering* **190,** 2603–2649 (2001).

[68] Mehrmann, V. & Morandin, R. *Structure-preserving discretization for port-Hamiltonian descriptor systems* in *Proceedings of the IEEE Conference on Decision and Control* (Institute of Electrical and Electronics Engineers Inc., 2019), 6863–6868.

[69] Groß, M. & Betsch, P. Energy-momentum consistent finite element discretization of dynamic finite viscoelasticity. *International Journal for Numerical Methods in Engineering* **81,** 1341–1386 (2010).

[70] Conde Martín, S., García Orden, J. C. & Romero, I. Energy-consistent time integration for nonlinear viscoelasticity. *Computational Mechanics* **54,** 473–488 (2014).

[71] Armero, F. & Zambrana-Rojas, C. Volume-preserving energy-momentum schemes for isochoric multiplicative plasticity. *Computer Methods in Applied Mechanics and Engineering* **196,** 4130–4159 (2007).

[72] Groß, M. & Betsch, P. Galerkin-based energy-momentum consistent time-stepping algorithms for classical nonlinear thermo-elastodynamics. *Mathematics and Computers in Simulation* **82,** 718–770 (2011).

[73] Franke, M., Janz, A., Schiebl, M. & Betsch, P. An energy momentum consistent integration scheme using a polyconvexity-based framework for nonlinear thermoelastodynamics. *International Journal for Numerical Methods in Engineering* **115,** 549–577 (2018).

[74] Groß, M. *Higher-order accurate and energy-momentum consistent discretisation of dynamic finite deformation thermo-viscoelasticity.* Habilitation thesis (Universität Siegen, 2009).

[75] García Orden, J. C. & Romero, I. Energy-Entropy-Momentum integration of discrete thermo-visco-elastic dynamics. *European Journal of Mechanics, A/Solids* **32,** 76–87 (2012).

[76] Groß, M., Bartelt, M. & Betsch, P. Structure-preserving time integration of non-isothermal finite viscoelastic continua related to variational formulations of continuum dynamics. *Computational Mechanics* **62,** 123–150 (2018).

[77] Erler, N. & Groß, M. Energy-momentum conserving higher-order time integration of nonlinear dynamics of finite elastic fiber-reinforced continua. *Computational Mechanics* **55,** 921–942 (2015).

[78] Groß, M., Dietzsch, J. & Bartelt, M. Variational-based higher-order accurate energy–momentum schemes for thermo-viscoelastic fiber-reinforced continua. *Computer Methods in Applied Mechanics and Engineering* **336,** 353–418 (2018).

[79] Groß, M. & Dietzsch, J. Variational-based locking-free energy–momentum schemes of higher-order for thermo-viscoelastic fiber-reinforced continua. *Computer Methods in Applied Mechanics and Engineering* **343,** 631–671 (2019).

[80] Groß, M., Dietzsch, J. & Röbiger, C. Non-isothermal energy–momentum time integrations with drilling degrees of freedom of composites with viscoelastic fiber bundles and curvature–twist stiffness. *Computer Methods in Applied Mechanics and Engineering* **365,** 1–52 (2020).

[81] Badlyan, A. M., Maschke, B. M., Beattie, C. & Mehrmann, V. Open physical systems: from GENERIC to port-Hamiltonian systems. *arXiv preprint arXiv:1804.04064* **1,** 1–8 (2018).

[82] Grmela, M. & Öttinger, H. C. Dynamics and thermodynamics of complex fluids. I. Development of a general formalism. *Physical Review E* **56,** 6620–6632 (1997).

[83] Öttinger, H. C. & Grmela, M. Dynamics and thermodynamics of complex fluids. II. Illustrations of a general formalism. *Physical Review E - Statistical, nonlinear, biological and soft matter physics* **56,** 6633–6655 (1997).

[84] Öttinger, H. C. & Beris, A. N. Thermodynamically consistent reptation model without independent alignment. *Journal of Chemical Physics* **110,** 6593–6596 (1999).

[85] Grmela, M., Bousmina, M. & Palierne, J.-F. On the rheology of immiscible blends. *Rheologica Acta* **40,** 560–569 (2001).

[86] Ramazani S. A., A., Ait-Kadi, A. & Grmela, M. Rheology of fiber suspensions in viscoelastic media: Experiments and model predictions. *Journal of Rheology* **45,** 945–962 (2001).

[87] Hütter, M. Thermodynamically consistent incorporation of the Schneider rate equations into two-phase models. *Physical Review E - Statistical Physics, Plasmas, Fluids, and Related Interdisciplinary Topics* **64,** 1–11 (2001).

[88] Ilg, P. & Öttinger, H. C. Nonequilibrium relativistic thermodynamics in bulk viscous cosmology. *Physical Review D - Particles, Fields, Gravitation and Cosmology* **61,** 1–10 (2000).

[89] Español, P., Serrano, M. & Öttinger, H. C. Thermodynamically admissible form for discrete hydrodynamics. *Physical Review Letters* **83,** 4542–4545 (1999).

[90] Liu, Q. & De Kee, D. Modeling of diffusion through polymeric membranes. *Rheologica Acta* **44,** 287–294 (2005).

[91] Gu, J. F., Grmela, M. & Bousmina, M. Mesohydrodynamics of membrane suspensions. *Journal of Non-Newtonian Fluid Mechanics* **165,** 75–83 (2010).

[92] Hernandez, Q., Badias, A., Gonzalez, D., Chinesta, F. & Cueto, E. Structure-preserving neural networks. *arXiv preprint arXiv:2004.04653,* 1–19 (2020).

[93] Öttinger, H. C. The geometry and thermodynamics of dissipative quantum systems. *Epl* **94,** 1–6 (2011).

[94] Öttinger, H. C. *Beyond Equilibrium Thermodynamics* (John Wiley & Sons, 2005).

[95] Öttinger, H. C. GENERIC: Review of Successful Applications and a Challenge for the Future. *arXiv preprint arXiv:1810.08470,* 73–78 (2018).

[96] Hütter, M. & Tervoort, T. A. Finite anisotropic elasticity and material frame indifference from a nonequilibrium thermodynamics perspective. *Journal of Non-Newtonian Fluid Mechanics* **152,** 45–52 (2008).

[97] Hütter, M. & Tervoort, T. A. Continuum damage mechanics: Combining thermodynamics with a thoughtful characterization of the microstructure. *Acta Mechanica* **201,** 297–312 (2008).

[98] Hütter, M. & Svendsen, B. On the Formulation of Continuum Thermodynamic Models for Solids as General Equations for Non-equilibrium Reversible-Irreversible Coupling. *Journal of Elasticity* **104,** 357–368 (2011).

[99] Hütter, M. & Svendsen, B. Thermodynamic model formulation for viscoplastic solids as general equations for non-equilibrium reversible-irreversible coupling. *Continuum Mechanics and Thermodynamics* **24,** 211–227 (2012).

[100] Mielke, A. Formulation of thermoelastic dissipative material behavior using GENERIC. *Continuum Mechanics and Thermodynamics* **23,** 233–256 (2011).

[101] Romero, I. Thermodynamically consistent time-stepping algorithms for non-linear thermomechanical systems. *International Journal for Numerical Methods in Engineering* **79,** 706–732 (2009).

[102] Romero, I. Algorithms for coupled problems that preserve symmetries and the laws of thermodynamics. Part I: Monolithic integrators and their application to finite strain thermoelasticity. *Computer Methods in Applied Mechanics and Engineering* **199,** 1841–1858 (2010).

[103] Ilg, P. Thermodynamically consistent coarse graining the non-equilibrium dynamics of unentangled polymer melts. *Journal of Non-Newtonian Fluid Mechanics* **165,** 973–979 (2010).

[104] Ilg, P. & Kröger, M. Molecularly derived constitutive equation for low-molecular polymer melts from thermodynamically guided simulation. *Journal of Rheology* **55,** 69–93 (2011).

[105] Romero, I. A characterization of conserved quantities in non-equilibrium thermodynamics. *Entropy* **15,** 5580–5596 (2013).

[106] Krüger, M., Groß, M. & Betsch, P. An energy-entropy-consistent time stepping scheme for nonlinear thermo-viscoelastic continua. *ZAMM - Journal of Applied Mathematics and Mechanics / Zeitschrift für Angewandte Mathematik und Mechanik* **96,** 141–178 (2016).

[107] Toxvaerd, S. Energy conservation in molecular dynamics. *Journal of Computational Physics* **52,** 214–216 (1983).

[108] Rapaport, D. C. *The Art of Molecular Dynamics Simulation* (Cambridge University Press, 2004).

[109] Skeel, R. D., Hardy, D. J. & Phillips, J. C. Correcting mesh-based force calculations to conserve both energy and momentum in molecular dynamics simulations. *Journal of Computational Physics* **225,** 1–5 (2007).

[110] Allen, M. P. & Tildesley, D. J. *Computer Simulation of Liquids* (Clarendon Press, 1989).

[111] Tuckermann, M. E. *Statistical Mechanics: Theory and Molecular Simulation* (Oxford University Press, 2010).

[112] Salueña, C. & Avalos, J. B. Molecular dynamics algorithm enforcing energy conservation for microcanonical simulations. *Physical Review E - Statistical, nonlinear, biological and soft matter physics* **89,** 1–10 (2014).

[113] Griebel, M., Knapek, S. & Zumbusch, G. *Numerical Simulation in Molecular Dynamics* (eds Barth, T. J., Griebel, M., E.Keyes, D., M.Nieminen, R., Roose, D. & Schlick, T.) (Springer, Berlin, 2007).

[114] Brooks, B. R., Bruccoleri, R. E., Olafson, B. D., States, D. J., Swaminathan, S. & Karplus, M. CHARMM: A program for macromolecular energy, minimization, and dynamics calculations. *Journal of Computational Chemistry* **4,** 187–217 (1983).

[115] Stillinger, F. H. & Weber, T. A. Computer simulation of local order in condensed phases of silicon. *Physical Review B: Condensed Matter and Materials Physics* **31,** 5262–5271 (1985).

[116] Daw, M. S. & Baskes, M. I. Embedded-atom method: Derivation and application to impurities, surfaces and other defects in metals. *Physical Review B: Condensed Matter and Materials Physics* **29,** 6443–6453 (1984).

[117] Daw, M. S. Model of metallic cohesion: The embedded-atom method. *Physical Review B: Condensed Matter and Materials Physics* **39,** 7441–7452 (1989).

[118] Baskes, M. I. Modified embedded-atom potentials for cubic materials and impurities. *Physical Review B: Condensed Matter and Materials Physics* **46,** 2727–2742 (1992).

[119] Daw, M. S., Foiles, S. M. & Baskes, M. I. The embedded-atom method: a review of theory and applications. *Materials Science Reports* **9,** 251–310 (1993).

[120] Tadmor, E. B. & Miller, R. E. *Modeling materials: continuum, atomistc and multiscale techniques* (Cambridge University Press, 2011).

[121] Lennard-Jones, J. E. On the determination of molecular fields. II. From the equation of state of gas. *Proceedings of the Royal Society A: Mathematical, Physical and Engineering Sciences* **106,** 463–477 (1924).

[122] Conde Martín, S., Betsch, P. & García Orden, J. C. A temperature-based thermodynamically consistent integration scheme for discrete thermo-elastodynamics. *Communications in Nonlinear Science and Numerical Simulation* **32,** 63–80 (2016).

[123] Conde Martín, S. & García Orden, J. C. On Energy–Entropy–Momentum integration methods for discrete thermo-visco-elastodynamics. *Computers and Structures* **181,** 3–20 (2017).

[124] Portillo, D., García Orden, J. C. & Romero, I. Energy–entropy–momentum integration schemes for general discrete non-smooth dissipative problems in thermomechanics. *International Journal for Numerical Methods in Engineering* **112,** 776–802 (2017).

[125] Gonzalez, O. Time Integration and Discrete Hamiltonian Systems. *Journal of Nonlinear Science* **6,** 449–467 (1996).

[126] Goldstein, H., Poole, C. & Safko, J. *Classical Mechanics* (Addison Wesley (Amsterdam), 2002).

[127] Kuzkin, V. A. On angular momentum balance for particle systems with periodic boundary conditions. *ZAMM - Journal of Applied Mathematics and Mechanics / Zeitschrift für Angewandte Mathematik und Mechanik* **95,** 1290–1295 (2015).

[128] Haile, J. M., Johnston, I., Mallinckrodt, A. J. & McKay, S. Molecular dynamics simulation: elementary methods. *Computers in Physics* **7,** 625 (1993).

[129] Romero, I. An analysis of the stress formula for energy-momentum methods in nonlinear elastodynamics. *Computational Mechanics* **50,** 603–610 (2012).

[130] Kim, S. Issues on the choice of a proper time step in molecular dynamics. *Physics Procedia* **53,** 60–62 (2014).

[131] Plimpton, S. Fast Parallel Algorithms for Short-Range Molecular Dynamics. *Journal of Computational Physics* **117,** 1–19 (1995).

[132] Kreher, D. L. & Stinson, D. R. *Combinatorial Algorithms: Generation, Enumeration, and Search* (Taylor & Francis, 1998).

[133] Baskes, M. I. Many-Body Effects in fcc Metals: A Lennard-Jones Embedded-Atom Potential. *Physical Review Letters* **83,** 2588–2591 (1999).

[134] Srinivasan, S. G. & Baskes, M. I. On the Lennard-Jones EAM potential. *Proceedings of the Royal Society A: Mathematical, Physical and Engineering Sciences* **460,** 1649–1672 (2004).

[135] Beris, A. N. & Edwards, B. J. *Thermodynamics of flowing systems: with internal microstructure* (Oxford University Press, 1994).

[136] Gonzalez, O. & Stuart, A. M. *A First Course in Continuum Mechanics* (Cambridge University Press, 2008).

[137] Edwards, B. J. An Analysis of Single and Double Generator Thermodynamic Formalisms for the Macroscopic Description of Complex Fluids. *Journal of Non-Equilibrium Thermodynamics* **23,** 301–333 (1998).

[138] Öttinger, H. C. Nonequilibrium thermodynamics for open systems. *Physical Review E - Statistical, nonlinear, biological and soft matter physics* **73,** 1–10 (2006).

[139] Marsden, J. E. & Hughes, T. J. R. *Mathematical Foundations of Elasticity* (Courier Corporation, 1994).

[140] Miehe, C. Entropic thermoelasticity at finite strains. Aspects of the formulation and numerical implementation. *Computer Methods in Applied Mechanics and Engineering* **120,** 243–269 (1995).

[141] Balzani, D., Gandhi, A., Tanaka, M. & Schröder, J. Numerical calculation of thermo-mechanical problems at large strains based on complex step derivative approximation of tangent stiffness matrices. *Computational Mechanics* **55,** 861–871 (2015).

[142] Netz, T. & Hartmann, S. A monolithic finite element approach using high-order schemes in time and space applied to finite strain thermo-viscoelasticity. *Computers and Mathematics with Applications* **70,** 1457–1480 (2015).

[143] Holzapfel, G. A. & Simo, J. C. Entropy elasticity of isotropic rubber-like solids at finite strains. *Computer Methods in Applied Mechanics and Engineering* **132,** 17–44 (1996).

[144] Hesch, C. & Betsch, P. Energy-momentum consistent algorithms for dynamic thermomechanical problems—Application to mortar domain decomposition problems. *International Journal for Numerical Methods in Engineering* **86,** 1277–1302 (2011).

[145] Hughes, T. J. R. *The finite element method: linear static and dynamic finite element analysis* (Courier Corporation, 2012).

[146] Gurtin, M. E. Thermodynamics and stability. *Archive for Rational Mechanics and Analysis* **59,** 63–96 (1975).

[147] Holzapfel, G. A. *Nonlinear solid mechanics: a continuum approach for engineering* (John Wiley & Sons, Inc., 2000).

[148] Reese, S. & Govindjee, S. A theory of finite viscoelasticity and numerical aspects. *International Journal of Solids and Structures* **35,** 3455–3482 (1998).

[149] Sidoroff, F. Nonlinear viscoelastic model with intermediate configuration. *Journal de Mecanique* **13,** 679–713 (1974).

[150] Reese, S. & Govindjee, S. Theoretical and Numerical Aspects in the Thermo-Viscoelastic Material Behaviour of Rubber-Like Polymers. *Mechanics of Time-Dependent Materials* **1,** 357–396 (1998).

[151] Budday, S., Ovaert, T. C., Holzapfel, G. A., Steinmann, P. & Kuhl, E. *Fifty Shades of Brain: A Review on the Mechanical Testing and Modeling of Brain Tissue* **4,** (Springer Netherlands, 2019).

[152] Hartmann, S. A remark on the application of the Newton-Raphson method in non-linear finite element analysis. *Computational Mechanics* **36,** 100–116 (2005).

[153] Schröder, J., Lion, A. & Johlitz, M. in *State of the Art and Future Trends in Material Modeling* (eds Altenbach, H. & Öchsner, A.) 325–348 (Springer International Publishing, Cham, 2019).

[154] Betsch, P., Janz, A. & Hesch, C. A mixed variational framework for the design of energy–momentum schemes inspired by the structure of polyconvex stored energy functions. *Computer Methods in Applied Mechanics and Engineering* **335,** 660–696 (2018).

[155] Janz, A. *Structure-preserving space-time discretization in a mixed framework for multi-field problems in large strain elasticity.* PhD thesis (Karlsruhe Institut für Technologie, 2019).

[156] De Boer, R. *Vektor- und Tensorrechnung für Ingenieure* (Springer, 1982).

[157] Öttinger, H. C. GENERIC Integrators: Structure Preserving Time Integration for Thermodynamic Systems. *Journal of Non-Equilibrium Thermodynamics* **43,** 89–100 (2018).

[158] Cottrell, J. A., Hughes, T. J. R. & Bazilevs, Y. *Isogeometric analysis: toward integration of CAD and FEA* (John Wiley & Sons, 2009).

[159] Malvern, L. E. *Introduction to the Mechanics of a Continuous Medium* (Prentice Hall, 1969).

Schriftenreihe des Instituts für Mechanik
ISSN 2363-4936

Band 1 Marlon Franke
Discretisation techniques for large deformation computational contact elastodynamics.
ISBN 978-3-7315-0278-4

Band 2 Philipp Hempel
Constitutive modeling of amorphous thermoplastic polymers with special emphasis on manufacturing processes.
ISBN 978-3-7315-0550-1

Band 3 Martin Helbig
Mehrskalenmodellierung von Schädigung in gummimodifizierten thermoplastischen Kunststoffen.
ISBN 978-3-7315-0571-6

Band 4 Maik Dittmann
Isogeometric analysis and hierarchical refinement for multi-field contact problems.
ISBN 978-3-7315-0616-4

Band 5 Yinping Yang
Numerical methods for the inverse dynamics simulation of underactuated mechanical systems.
ISBN 978-3-7315-0626-3

Band 6 Alexander Janz
Structure-preserving space-time discretization in a mixed framework for multi-field problems in large strain elasticity.
ISBN 978-3-7315-0926-4

Band 7 Mark Georg Schiebl
Thermodynamically consistent space-time discretization of non-isothermal mechanical systems in the framework of GENERIC.
ISBN 978-3-7315-1117-5

Die Bände sind unter www.ksp.kit.edu als PDF frei verfügbar oder als Druckausgabe bestellbar.